# STRESS WAVES
# IN ANELASTIC SOLIDS

INTERNATIONAL UNION OF THEORETICAL
AND APPLIED MECHANICS

# STRESS WAVES
# IN ANELASTIC SOLIDS

SYMPOSIUM
HELD AT BROWN UNIVERSITY, PROVIDENCE, R. I.
APRIL 3–5, 1963

EDITORS

## HERBERT KOLSKY     WILLIAM PRAGER

BROWN UNIVERSITY     IBM-FORSCHUNGSLABORATORIUM
PROVIDENCE, R. I., U.S.A.     ZÜRICH, SCHWEIZ

WITH 145 FIGURES

SPRINGER-VERLAG
BERLIN/GÖTTINGEN/HEIDELBERG
1964

ISBN-13: 978-3-642-88290-6      e-ISBN-13: 978-3-642-88288-3
DOI: 10.1007/978-3-642-88288-3

© by·Springer Verlag, Berlin / Göttingen / Heidelberg 1964
Softcover reprint of the hardcover 1st edition 1964

Library of Congress Catalog Card Number: 64-19 866

Titel-Nr. 1224

# Preface

Although the subject of wave propagation in solids has a long history, the classical theory of elastic waves having been developed in the nineteenth century by STOKES, POISSON, RAYLEIGH and KELVIN, the last two decades have seen a remarkable revival of interest in this subject among both theoreticians and experimenters. There are a number of reasons for this; first, experimental methods for the generation and detection of high frequency mechanical waves have become available only with the advent of electronic techniques and of high speed photographic recording apparatus. Secondly, the appearance of new materials, such as plastics, the mechanical properties of which depend very markedly on the rate at which stresses are applied to them, has led to studies of their response to high frequency mechanical waves, with a view to correlating their microscopic structure with their mechanical behavior. Finally, engineers have become more and more concerned with the response of conventional engineering materials, such as metals, to large impulsive forces applied for very short periods of time. This interest arises both in military developments and in problems of impact and of shock absorption in engineering structures. A proper understanding of all these problems requires a knowledge of the nature of stress wave propagation in elastic and anelastic solids alike, and, while in recent years considerable progress has been made on various aspects of elastic wave propagation, it is in the study of stress waves in anelastic solids that many of the most radical advances have been achieved, and where the most novel and far-reaching results may be expected in the future.

A number of distinct types of wave propagation in anelastic solids have been investigated and, although the phenomena observed in practical situations do not always conform to the idealized models which are postulated before theoretical predictions can be made, the theoretical work has received experimental confirmation in a number of the problems, and the experiments have, in turn, shown effects which have led to further theoretical advances.

Three types of anelastic wave have received particular attention; viz, plastic waves, viscoelastic waves, and shock waves. Plastic waves can be propagated in a material, such as a metal, which exhibits the phenomenon of yielding when stressed beyond its proportional limit. The theory of the propagation of such waves was first considered by DONNELL in 1930, and was subsequently developed during World War II

by a number of workers, including VON KÁRMÁN in this country, G. I.
TAYLOR in England, and RAHKMATULIN in the USSR. The theory,
as originally conceived, was based on a non-linear stress-strain relationship which was independent of the rate at which stresses were applied
to the metal. Subsequent experimental work has shown that the time-rate
dependence of the stress-strain relation has a considerable influence on
the nature of the wave propagation. Although MALVERN has made an
important first step in this direction, a theoretical approach which takes
such time dependence into account leads to rather involved mathematical
analysis. Further theoretical advances now await reliable experimental
data on the time-rate dependence of the stress-strain behavior of real
materials, and a number of experimentalists in many countries are at
present carrying out such investigations with various metals.

The mechanical properties of viscoelastic solids, such as plastics and
rubber, have been studied extensively in a number of countries during
recent years, and the subject of rheology is, to a large extent, devoted
to the description of such viscoelastic behavior. One offshoot of these
studies has been a consideration of the propagation of stress waves
through such materials. The problems involved are of theoretical interest, in that one is here dealing with media which are dispersive with
respect to both velocity and attenuation. The study of the propagation,
reflection and refraction of stress waves under these conditions leads to
a number of problems which are not only of mathematical and physical
interest, but also of practical importance in their bearing on the use of
high polymers as shock absorbers, and the response of complete viscoelastic structures to rapid mechanical loading.

For materials which are linearly viscoelastic, the basic theory has
been developed, and many one-dimensional problems have been solved.
Much of the present effort is being directed towards three-dimensional
problems, in which propagation of stress waves depends on the dilatational
properties of the viscoelastic solid, as well as on its response to shear
deformations.

The third type of anelastic waves which have been studied are termed
shock waves, and these arise when very large stresses are applied to solids
and lateral movement is prevented. Such conditions are normally encountered in "explosive" loading, or during the impact of high speed projectiles. The shock waves arise because the effective bulk modulus of the
material increases with increasing pressure. The importance of these shock
waves lies, on the theoretical side, in obtaining the equation of state of
solids at pressures which cannot be achieved in any other way, and, on
the practical side, in obvious military and mining applications.

It may be seen that the field of stress wave propagation in anelastic
solids is one in which advances are being made in many diverse directions

and where many important problems are in the process of being solved. It is a subject in which mathematicians, engineers, physicists and geophysicists are involved. The International Symposium on Stress Waves in Anelastic Solids was planned to bring workers from these fields together to survey and discuss recent work and probable future developments.

The Symposium was one of a series of international symposia in mechanics organized by the International Union on Theoretical and Applied Mechanics. It was supported financially by IUTAM, the National Science Foundation (Washington D. C., USA), and Brown University.

The International Scientific Committee responsible for the planning of the Symposium consisted of

HERBERT KOLSKY, Brown University, Providence, R. I., USA (Chairman),

N. CRISTESCU, Mathematical Institute of the Academy of the People's Republic, Bucarest, Rumania,

H. GEOFFREY HOPKINS, Royal Armament Research and Development Establishment, Fort Halstead, Sevenoaks, Kent, England,

ROBERT MAZET, Université de Paris et Office National d'Etudes et de Recherches Aéronautiques, Châtillon-sous-Bagneux, France,

CHAIM L. PEKERIS, The Weizmann Institute of Science, Rehovoth, Israel,

NIKITA V. ZVOLINSKY, Institute of Earth Physics, Moscow, USSR.

The local arrangements were in the hands of committee consisting of Professors H. KOLSKY (Chairman), E. H. LEE, E. T. ONAT, W. PRAGER, P. S. SYMONDS, and R. TRUELL of Brown University.

In line with the general IUTAM policy for these symposia, the number of persons invited as speakers or observers was kept small.

The program was as follows:

## Wednesday, April 3

*Session 1* 9:00 a. m.

Chairman: R. J. EICHELBERGER, Aberdeen Proving Ground, U.S.A.

N. DAVIDS and H. CALVIT
Pennsylvania State University, University Park, Pa., U.S.A.
"Some dynamical applications of shock wave propagation in solids"

G. E. DUVALL
Stanford Research Institute, Menlo Park, California, U.S.A.
"Propagation of plane shock waves in a stress-relaxing medium"

J. JACQUESSON
Université de Poitiers, Poitiers, France
"Précurseurs élastoplastiques d'ondes de choc en milieu solide isotrope"

W. NOWACKI and S. KALISKI
Polish Academy of Sciences, Warsaw, Poland
"The propagation of magnetoelastic disturbances in viscoelastic bodies"

*Session 2* 1:30 p. m.

Chairman: R. D. MINDLIN, Columbia University, New York, U.S.A.

J. D. CAMBPELL and D. B. TAYLOR
Oxford University Engineering Laboratory, Oxford, England
"On the numerical solution of a wave-propagation problem in the theory of dislocation motion"

R. MAZET
Office National d'Etudes et de Recherches Aéronautiques, Châtillon-sous-Bagneux, France
"Vibrations d'une structure lamellaire sous fluage localisé"

G. N. SAVIN
Institute of Mechanics, Kiev, U.S.S.R.
"Dynamics of an anelastic string of variable length"[1]

H. ZORSKI
Polish Academy of Science, Warsaw, Poland
"Anelastic waves in thin plates"

Cocktail Party — 6:00 p. m. University Club
Banquet — 7:00 p. m. University Club

Thursday, April 4

*Session 3* 9:00 a. m.

Chairman: H. P. TARDIF, CARDE, Quebec, Canada

G. BIANCHI
Politecnico di Milano, Italy
"Some experimental and theoretical studies on the propagation of longitudinal plastic waves in a strain-rate-dependent material"

N. CRISTESCU
Institut de Mathématique, Bucarest, Rumania
"Some problems of the mechanics of extensible strings"[1]

H. G. HOPKINS
Royal Armament Research and Development Establishment, Fort Halstead, Kent, England
"Mechanical waves and strain-rate effects in metals"

S. C. HUNTER and I. A. JOHNSON
Royal Armament Research and Development Establishment, Fort Halstead, Kent, England
"The propagation of small amplitude elastic-plastic waves in pre-stressed cylindrical bars"

*Session 4* 1:30 p. m.

Chairman: J. N. GOODIER, Stanford University, Stanford, California, U.S.A.

J. F. BELL
The Johns Hopkins University, Baltimore, Md., U.S.A.
"The initiation of finite amplitude waves in annealed metals"

E. DAVID, R. SCHALL and H. SCHARDIN
Franco-German Research Institute, St. Louis, France and Ernst Mach Institute, Freiburg, Germany
"Visualization of wave propagation in impulse-loaded bars"

---

[1] Presented by title.

JOHN S. RINEHART
  Colorado School of Mines, Golden, Colorado, U.S.A.
  "Transient stress wave boundary interactions"
N. ZVOLINSKY and G. V. RYKOV
  Institute of Geophysics, Moscow, U.S.S.R.
  "Reflection of a plastic wave at an obstacle[1]"

Informal Reception
Brown University Computing Laboratory 5:30—6:30 p. m.

Friday, April 5

Session 5　9:00 a. m.
Chairman: K. W. HILLIER, Imperial Chemical Industries, Ltd., England

C.-C. CHAO and J. D. ACHENBACH
  Stanford University, Stanford, California, U.S.A.
  "A simple viscoelastic analogy for stress waves"
E. H. LEE and D. T. LIU
  Stanford University, Stanford, California, U.S.A.
  "An example of the influence of yield on high pressure wave propagation"
J. MIKLOWITZ
  California Institute of Technology, Pasadena, California, U.S.A.
  "Pulse propagation in a viscoelastic solid with geometric dispersion"
J. MANDEL
  Ecole Polytechnique, Paris, France
  "Propagation des surfaces de discontinuité dans un milieu élastoplastique"

Session 6　1:30 p. m.
Chairman: W. T. KOITER, Technical University, Delft, Netherlands

W. GOLDSMITH and C. T. AUSTIN
  University of California, Berkeley, California, U.S.A.
  "Some dynamic characteristics of rocks"
S. S. GRIGORIAN
  Institute of Mechanics, Moscow University, Moscow, USSR
  "On some simplifications in the description of the motion of a soft soil"[1]
W. HEIERLI
  Eidgenössische Technische Hochschule, Zürich, Switzerland
  "One-dimensional inelastic wave propagation in soils: experimental and theoretical investigations"

At the banquet Professor W. T. KOITER, Treasurer of IUTAM, thanked the National Science Foundation and Brown University for their co-sponsorship of the Symposium and expressed appreciation of the contributions of the speakers and the work of the International Scientific Committee and the Committee on Arrangements.

---

[1] Presented by title.

The list of the participants in the Symposium follows:

ABLOW, CLARENCE M., Stanford Res. Inst., Menlo Park, Calif.
ACHENBACH, JAN D., Northwestern University, Evanston, Ill.
ALLISON, FLOYD E., Aberdeen Proving Ground, Md.
BACKMAN, MARVIN E., 58 B Ringgold, China Lake, Calif.
BELL, JAMES F., The Johns Hopkins University, Baltimore, Md.
BIANCHI, GIOVANNI, Politecnico di Milano, Milano, Italy
BROOKS, P., CARDE, Quebec, P. Q., Canada
BYCROFT, GEORGE N., Stanford Res. Inst., Menlo Park, Calif.
CALVIT, HARRY H., Pennsylvania State University, Univ. Park, Pa.
CAMPBELL, JOHN D., Oxford University, Oxford, England
CHAO, CHI-CHANG, Stanford University, Stanford, Calif.
CHAREST, JACQUES, Colorado School of Mines, Golden, Colorado
CHOU, PEI CHI, Drexel Institute of Technology, Philadelphia, Pa.
CHU, BOA-TEH, Brown University, Providence, R. I.
DAVID, ERWIN, Postfach des ISL, 7858 Weil am Rhein, Germany
DAVIDS, NORMAN, Pennsylvania State University, Univ. Park, Pa.
DRUCKER, DANIEL C., Brown University, Providence, R. I.
DUFFY, JACQUES, Brown University, Providence, R. I.
DUVALL, GEORGE E., Poulter Labs., Stanford Res. Inst., Menlo Park, Calif.
EICHELBERGER, ROBERT J., Aberdeen Proving Ground, Md.
GAUS, MICHAEL P., NSF, Washington 25, D. C.
GOLDSMITH, WERNER, University of California, Berkeley, Calif.
GOODIER, JAMES N., 506 Mayfield Ave., Stanford, Calif.
HANDELMAN, GEORGE H., Rensselaer Polytechnic Inst., Troy, N. Y.
HEIERLI, WERNER, Eichhalde 19, Zürich, Switzerland
HILLIER, KENNETH, W., ICI Fibres Div., Harrogate, Yorks, England
HOPKINS, HARRY G., ARDE, B. Div., Fort Halstead, Sevenoaks, Kent, England
HUNTER, STEPHEN C., ARDE, Fort Halstead, Sevenoaks, Kent, England
JACQUESSON, JEAN, ENSMA, Poitiers, France
JONES, J. P., Aerospace Corp., El Segundo, Calif.
KOITER, WERNER T., Mekelweg 2, Delft, Netherlands
KOLSKY, H., Brown University, Providence, R. I.
KRAFFT, JOSEPH M., US Naval Research Laboratory, Washington 25, D. C.
LEE, ERASTUS H., Stanford University, Stanford, Calif.
LEE, TUNG-MING, P. O. Box 282, Hanover, N. H.
LIEBOWITZ, HAROLD, ONR-Washington, D. C.
LINDHOLM, ULRIC E., Southwest Research Inst., 8500 Culebra Rd., San Antonio, Texas
LIU, DAVID T., Lockheed Missiles and Space Co., Palo Alto, Calif.
LONG, RALPH H., jr., 3405 Pennsylvania St., W. Hyattsville, Md.
MALVERN, LAWRENCE E., Michigan State University, East Lansing, Mich.
MANDEL, JEAN, Avenue du Colonel Bonnet No. 16, Paris, France
MARTIN, JOHN B., Brown University, Providence, R. I.
MAZET, ROBERT, ONERA, Châtillon-sous-Bagneux (Seine), France
MIKLOWITZ, JULIUS, California Inst. of Technology, Pasadena, Calif.
MINDLIN, RAYMOND D., RFD No. 2, Katonah, N. Y.
NOWACKI, W., Polish Academy of Sciences, Warsaw, Poland
ONAT, E. T., Brown University, Providence, R. I.
OWENS, ROBERT H., Math. Sci. Sec., NSF, Washington, D. C.
POMERANTZ, JACOB, Air Force Office of Scientific Research, OAR Washington, D. C.
PRAGER, WILLIAM, Brown University, Providence, R. I.

RINEHART, JOHN S., Colorado School of Mines, Golden, Colorado
RIPPERGER, E. A., The University of Texas, Austin, Texas
SHIELD, RICHARD T., Brown University, Providence, R. I.
SIMON, RALPH, Battelle Memorial Inst., Columbus 1, Ohio
STERNBERG, E., Brown University, Providence, R. I.
SYMONDS, P. S., Brown University, Providence, R. I.
TARDIFF, HENRI P., CARDE, Quebec, P. Q., Canada
TRUELL, ROHN, Brown University, Providence, R. I.
WHITTIER, J. S., Aerospace Corp., El Segundo, Calif.
WILLIAMS, ARTHUR O., Brown University, Providence, R. I.
ZORSKI, HENRYK, Polish Academy of Sciences, Warsaw, Poland

Providence, R. I., July 1963

**H. Kolsky · W. Prager**

# Contents

# Some Dynamical Applications
# of Shock Wave Propagation in Solids

By **Norman Davids** and **Harry H. Calvit**

The Pennsylvania State University, U.S.A.

## Abstract

A theory of crater formation by impact is compared with the process of shock propagation in solids, especially for waves with spherical symmetry. Previous approaches which led to a 2/5-th power law for penetration versus velocity also lead, on the basis of the similarity transformation known as the method of progressing waves, to results on the non-steady motion of metallic spheres initiated by explosive blasts in a spherical cavity. Although a polytropic equation of state is required to satisfy the constant energy condition, the quasi-polytropic state is investigated as far as possible. Integrations are carried out for four metals and $R-T$-diagrams, and pressure and radius relations are obtained. An expansion suitable for the initial stage of crater formation is described together with some numerical calculations, and the impulsive nature of the shock process is discussed, as well as the limitations on the method of progressing waves.

## 1. Introduction

The study of the anelastic behavior of materials leads in many directions. One of them is that in which the disturbing intensity or pressure is so rapid, or in which the time duration is so small that we are said to be in the "hydro-dynamic" range of behavior. By this is meant, that all shear resistance is supposed to become negligible, whether the material actually is known to melt or not. The natural mathematical description of this state would be expected to be similar to that of supersonic flow in gases. We shall discuss in this paper two related applications of this idea which will show some of the limitations of unmodified gas-dynamical approaches for solids.

Two prominent features of gas-dynamical descriptions are valid for solid materials as well: 1. existence of an equation of state of the material, and 2. the presence of shock waves, i.e. non-linear discontinuity fronts started by very intense localized disturbances. We may state in a general way that shock waves in solids behave similarly to those in gases for very high pressures, but depart from them for intermediate and low pressures (magnitudes to be given below).

The theory and observation of shocks in solids is far less advanced than in gases not only because of the difficulty of observation in opaque materials but also on account of the difficulty of obtaining plane waves, avoiding the complications of cylindrical or spherical geometry.

In this paper we shall describe some recent work in spherical blast waves. We shall also compare the approaches used with that applied in a recently published study of shock waves in solid craters [3]. In a recent thesis by Y. K. HUANG, some of the problems connected with an equation of state for metals are considered.

The use of equations of state for metals and the hydrodynamic theory has been the subject of contributions in the recent literature. DUVALL and ZWOLINSKI in [7] have developed equations of state (pressure-volume relations) and applied them to the description of shock wave phenomena in solids. DRUMMOND [8] developed a method for generating shock waves through the use of explosives. For copper, ALLEN, MORRISON, RAY, and ROGERS [9] developed the analysis of shock fronts, wedge angles and Mach lines for penetrating projectiles. GOLDSMITH [10] devotes a section of his book to the hydrodynamic theory of shock propagation in solids and summarizes equations of state data to date. RINEHART and WHITE [11] studied cratering in plaster of Paris, and MAURER and RINEHART [11] analyzed impact crater formation in rock. Also worthy of mention is the comprehensive article by HOPKINS [13], which contains an analysis of plastic deformation in expanding spherical cavities.

## 2. The Blast Problem and Cratering Problem Compared

If an explosive is detonated in a small cavity of a thick metal sphere, the sequence of effects is somewhat as follows:

(a) *Initial Stage.* Here the detonation wave of the exploding gas starts contact with the solid and then generates a shock wave in the solid. This stage might be considered as terminated when the density in the solid has dropped to its free space value, at the inner cavity.

(b) *Expansion Stage.* The compressed solid expands radially outward and actually forms the cavity. This stage is dominated mostly by inertia forces.

(c) *Final Stage.* Here the shock wave decays, permanent deformation of the cavity stops, and the material has undergone some permanent plastic strain.

For these effects to be observed properly, it is necessary for the sphere to be big enough not to rupture, so that the blast sees essentially an infinite sphere. Thus 7″ spheres with a 1″ cavity have been used for tests in an experimental program at the Ballistic Research Laboratory (BRL), Aberdeen Proving Ground. It should be noted at the outset that the phenomena that result from detonation of the explosive in the inner cavity, are complex because different effects predominate in different parts of the material. Thus, there is an innermost zone or spherical shell where very large radial displacements have occurred under temperatures and pressures far beyond the range of conventional mechanical behavior. The material is in some type of "fluid" state in this zone, and there is relatively little tendency for cracks to initiate there. Next, there is an intermediate zone of the sphere where displacements have dropped to elastic ranges and where brittleness seems to have returned, as evidenced by the many small tension and shear cracks which have formed there. A few of the stronger cracks which get started may penetrate into the adjacent regions and ultimately reach the boundaries. Finally, there is an outermost zone dominated by the effect of the external boundary of the sphere. Here reflection effects such as scabbing cracks are often observed.

The transitions between successive zones are not exact, but in some specimens, rather surprisingly enough, are fairly sharply delineated.

In this paper we shall limit our study to the innermost shock zone. Our direct aim is to provide a description of the shock process in the metal. More specifically a useful theory must furnish a time for the duration of the process, the size of the zone influenced by the shock front, values for the displacements, and thermodynamic variables of pressure, density, and temperature in the material. A useful tool is the $R-T$ diagram which shows the path made by a set of concentric spherical shells. Such a diagram is possible because of the single space coordinate.

The first stage lasts up to about 2.5 micro-seconds. The second turns out to be relatively long and can take up to about 100 micro-seconds or even longer. It must, of course, be understood that these phases need not be distinctly separated events in time, especially the terminating phase of the expansion.

The problem of the shock expansion of spherical cavities is closely related to that of crater formation by hypervelocity projectiles. The features that we have outlined above are essentially present in the crater problem as well. However, the region around an impact cavity is not radially symmetrical; thus the problem requires two space coordinates because of the tangential flow. There is a zone ahead of the projectile where the target material is probably being triaxially compressed. This zone is bounded in front by an advancing shock wave and gradually shades off at the sides into another zone where the material, in some liquid or plastic state, flows sideways along the crater. Ultimately, as the angle of deviation from the direction of impact is increased, the material undergoes permanent plastic deformation and hills up to form the lip of the crater. The transition between the two zones is not meant to be exact in such a schematic picture. The cratering problem thus leads to effects in the shock-compression zone somewhat similar to those that exist around the spherical cavity. This leads naturally to the consideration of a simplified radially-symmetric hemispherical crater. In effect, this extends the shock front all the way round to $\theta = 90°$, with spherical symmetry and radial particle motion. The validity of this model, and a comparison of values for crater diameter and volume, with some experimental data by CHARTERS and SUMMERS [6], is described by the author in [3].

Theories which include the sideways flow have been slow to develop, so that meanwhile, the hemispherical model is useful in a provisional capacity.

The crater problem has been receiving considerably more attention recently than the spherical blast problem, both experimentally and theoretically. In pointing out some basic similarities, we feel that the blast problem is worthy of more intensive study because it exhibits the fundamental processes present in cratering. It has the decided advantage, that, except for the presence of a plug, the arrangement for the blasts has spherical symmetry, and we may confidently assert that purely radial motion occurs, so that all the physical quantities of the problem depend on only one space coordinate. Slight departures, which occur in practice, because of the plug or because of asymmetrical detonation, are not important to the problem.

## 3. The Initial Stage

We lack direct data on the initial stage of crater formation. HEAD-INGTON and JAUNZEMIS (in a paper not yet published) have analyzed this stage upon the basis of conservation laws and the known equations of state of the materials. The completion of such a detailed study would provide the initial and boundary conditions for stage 2. However, for the "similarity" solutions we can be satisfied with simpler assumptions about this phase: 1. the impact is so rapid that flow has not yet started, 2. the impact is essentially an instantaneous explosion (negligible pressure rise-time, and 3. the process is adiabatic.

## 4. Basic Mathematical Equations of Shock Waves

In the blast wave problem we certainly have spherical symmetry, that is, a radial and time coordinate only. For the cratering problem, such a restriction, whatever the degree of validity, results in an analysis which is common with that of the blast wave. However, since edge effects are being neglected, the results will be limited to some angle much less than 90° to the normal. We may make the following assumptions about the medium:

1. Thermodynamic equilibrium holds, i.e. changes of state are *adiabatic*. By this we mean that entropy is constant along a "particle path", i.e. for a fixed element of the medium (but not throughout the medium).

2. The medium is a perfect fluid, i.e., any rigidity or shear effects are neglected.

3. The effects of entropy changes are negligible, i.e., the pressure is a function of the density alone (the medium is then said to be *barotropic*).

4. The total energy available for the motion is fixed.

The spherical blast, of course, introduces no net momentum into the system, whereas a crater impact imposes the momentum of the projectile. This offers the alternative conservation law statement for momentum instead of energy. The presence of momentum is one of the main differences between the blast and the crater problems.

The conservation laws for an element of material, expressed in Lagrangian form, i.e., along particle paths, are as follows:

$$d\varrho/dt = -\varrho(\partial u/\partial r) - 2u\varrho/r = -(\varrho/r^2)\,\partial(r^2u)/\partial r \quad \text{(mass)}, \qquad (4.1)$$

$$du/dt = -\partial p/\varrho\,\partial r \quad \text{(momentum)}, \qquad (4.2)$$

$$f(p,\varrho) = 0 \quad \text{or} \quad df/dt = 0 \quad \text{(equation of state)}. \qquad (4.3)$$

These equations, in Eulerian form, with subscripts denoting partial derivatives, become

$$\varrho_t + u\varrho_r + \varrho u_r + 2u\varrho/r = 0 \quad \text{(mass)}, \qquad (4.4)$$

$$u_t + uu_r + p_r/\varrho = 0 \quad \text{(momentum)}, \qquad (4.5)$$

$$f_t(p,\varrho) + uf_r(p,\varrho) = 0 \quad \text{(state)}. \qquad (4.6)$$

The third of these equations is not quite equivalent to an equation of state, since it only expresses the fact that the entropy is constant along

the path of an element, and does not imply its constancy throughout.
This is a difference from the case of plane waves; another difference from
the equations of one-dimensional flow is the additional term $2u\varrho/r$
occurring in (4.1) and (4.4), which stands essentially for the spherical
attenuation of the wave. This term, of course, is very important in the
problem.

The geometry of the disturbance is shown by an $R—T$ diagram
(Fig. 1). Here the solid lines represent the motion of the points of a

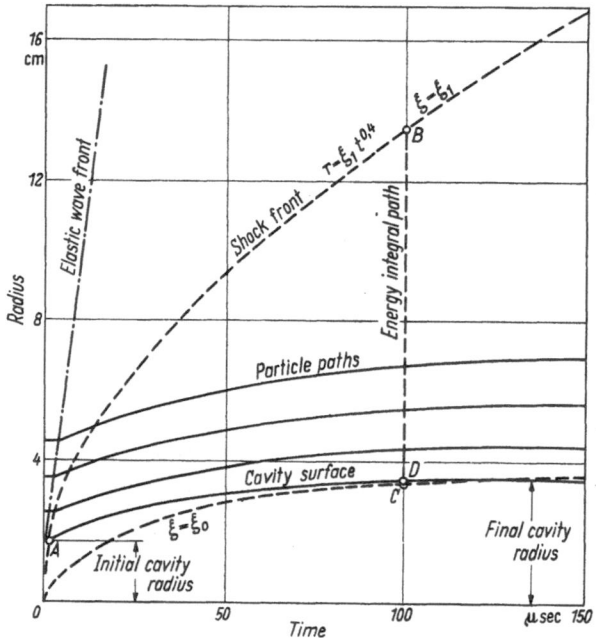

Fig. 1. Cavity expansion and particle trajectories.

spherical surface, referred to as a "particle". We note the shock front
which propagates through the material at the head of the disturbance.
The solution to the system of partial differential eqs. (4.4)—(4.6) applies
between this surface and the cavity boundary. Certain boundary conditions, to be discussed later, must be satisfied. However, the difficulty
of the problem is that here, unlike in the conventional boundary value
problems, the boundary curves are themselves unknown, and must be
found as part of the problem.

## 5. The Method of Progressing Waves

The idea of this and similar mathematical methods is to reduce the partial differential equations to ordinary ones. This is accomplished by assuming the specific form for the shock front curve and embedding it in a one-parameter family of curves. These curves are called "progressing waves". For general details of the method, see [1], p. 419—433. The method was developed by G. I. TAYLOR for detonation problems and used by R. G. NEWTON [4] and L. I. SEDOV [5] to analyze blast shock problems.

We shall assume in this part of the analysis that the medium has a so-called "quasi-polytropic" equation of state

$$p = A \varrho^{\gamma} - B, \qquad (5.1\,\text{a})$$

with $A$ a constant or possibly a function of the entropy and

$$B = A \varrho_0^{\gamma}. \qquad (5.1\,\text{b})$$

The progressing wave analysis proceeds essentially in a similar way to that for a polytropic gas if we replace $p$ by the *excess pressure*

$$\bar{p} = p + B. \qquad (5.1\,\text{c})$$

The progressing-wave solutions to the partial differential equations (4.4) to (4.6) are defined to be of the form

$$u = t^{\beta} \xi U(\xi) \qquad (5.2\,\text{a})$$

with $\xi = r t^{-\alpha}$,

$$\varrho = t^{\delta} D(\xi), \qquad (5.2\,\text{b})$$

$$\bar{p}/\varrho = t^{\epsilon} \xi^2 P(\xi), \qquad (5.2\,\text{c})$$

where $\alpha$, $\beta$, $\delta$, $\epsilon$ are parameters, and $U$, $D$, $P$ functions to be determined. By introducing the variable $\xi$ we have defined geometrically a family of surfaces $\xi = \text{const.}$ in the $r,t$-plane which will play an important role in the analysis. For a polytropic medium we shall see that the shock front belongs to this family of surfaces.

We now explore these solutions mathematically by substituting the expressions (5.1) and (5.2) into the equations of motion (4.4)—(4.6), from which it is found that the time $t$ can be explicitly separated. It then becomes possible (see [3] for details) to eliminate $t$ by the following conditions on the parameters:

$$\epsilon = 2\beta, \qquad \beta = \alpha - 1. \qquad (5.3)$$

With this we obtain, after simplification, a system of three ordinary differential equations for the functions $U(\xi)$, $D(\xi)$, $P(\xi)$ and two as-yet free parameters $\alpha$ and $\delta$. The substitutions (5.2) which may appear artificial, thus have some justification.

The next step is to solve for $\xi U'$ and $\xi P'$ and then to obtain by division, the ordinary differential equation

$$dP/dU = \xi P'/\xi U' = P[N(U) + PQ(U)]/[R(U) + PS(U)], \quad (5.4\,a)$$

where, after simplification,

$$
\begin{aligned}
N(U) &= \gamma U(3\alpha - 1 - 2U) + (3 - \alpha)U - 2\alpha, \\
Q(U) &= [2\beta - (\gamma - 1)\delta]/(U - \alpha) + 2\gamma, \\
R(U) &= U(U - \alpha)(1 - U), \\
S(U) &= \delta + 2\beta + 3U\gamma.
\end{aligned}
\qquad (5.4\,b)
$$

This is the basic differential equation for progressing waves. After the appropriate solution has been found for $P = P(U)$, the function $\xi = \xi(U)$ is found by an additional quadrature and the density function $D(\xi)$ from

$$\xi D'/D = -(\delta + \xi U' + 3U)/(U - \alpha). \qquad (5.5)$$

These progressing wave solutions, as we shall see, provide under certain conditions a sufficiently general mathematical description of an expanding cavity reasonably consistent with the given conditions of initiation of the process. There remains the problem of choosing the two parameters $\alpha$ and $\delta$. For this we have two alternative possibilities:

(a) the motion is *isentropic*,

(b) the motion is *adiabatic*, with constant total energy.

If (a) holds, then from Eqs. (5.1 c) and (5.2),

$$\bar{p}\varrho^{-\gamma} = (t^\epsilon \xi^2 P)(t^\delta D)^{(1-\gamma)} = A$$

or
$$t^{[\epsilon + ](1-\gamma)\delta]} \xi^2 P D^{(1-\gamma)} = A$$

requiring, for independence of time, that

$$\epsilon + (1 - \gamma)\delta = 0. \qquad (5.6)$$

This condition is not, in general, satisfied by a spherical wave since such a wave must attenuate with increasing radius to small stress levels. Instead, we have condition (b), that is, constant total energy. This is a reasonable one for the cavity expansion process, because of its short duration, provided certain secondary effects are neglected. Starting from the equation of state $p = A\varrho^{\gamma} - B$, the work done on the material, in becoming compressed from an initial volume $V_0$ to a volume $V$, is

$$E(V) = -\int_{V_0}^{V} p\, dV = -\varrho_0 V_0 \int_{\varrho_0}^{\varrho} (A\varrho^{\gamma} - B)(-d\varrho/\varrho^2)$$

$$= [B + \bar{p}/(\gamma - 1)] V - \gamma B V_0/(\gamma - 1),$$

where we have used the relation $\varrho V = \varrho_0 V_0$. In this form the energy $E = 0$ for $V = V_0$, but if we allow the material to possess the constant initial energy $\gamma B V_0/(\gamma - 1)$, then we can write the simple form

$$E(V) = [B + \bar{p}/(\gamma - 1)] V. \tag{5.7}$$

With $\xi = \xi_1$ representing the shock front at a time $t$, the total energy in the fluid shell (potential + kinetic) at time $t$ is given by

$$E(t) = \int_{r_0}^{r_1} [B + \bar{p}/(\gamma - 1)] 4\pi r^2 dr + \int_{r_0}^{r_1} \frac{1}{2} \varrho u^2 4\pi r^2 dr, \tag{5.8}$$

where $r_0 = \xi_0 t^{\alpha}$ is the inner radius of the shell (Fig. 1) and $r_1 = \xi_1 t^{\alpha}$ is the location of the shock front. Using the substitutions in (5.2), the energy expression becomes

$$E(t) = 4\pi \int_{r_0}^{r_1} (t^{\delta+\epsilon} P/(\gamma - 1) + \frac{1}{2} t^{\delta+2\beta} U^2) \xi^2 D(\xi) r^2 dr + 4\pi B \int_{r_0}^{r_1} r^2 dr.$$

Substituting $r = \xi t^{\alpha}$, $dr = t^{\alpha} d\xi$, since $t = \text{constant}$, we obtain

$$E(t) = 4\pi t^{\delta+5\alpha-2} \int_{\xi_0}^{\xi_1} \left( \frac{P}{\gamma - 1} + \frac{1}{2} U^2 \right) \xi^4 D(\xi) d\xi$$

$$+ 4 \frac{\pi}{3} B t^{3\alpha} (\xi_1^3 - \xi_0^3). \tag{5.9}$$

It is here that we run into the difficulty of making the energy independent of time for a quasi-polytropic medium. This is equivalent to the problem of asymptotic conditions referred to by NEWTON in [4]. If we restrict the problem to a polytropic type variation (equivalent to $B \sim 0$), it is possible to satisfy a constancy condition for the first integral by setting

$$\delta + 5\alpha - 2 = 0 \quad \text{or} \quad \delta = 2 - 5\alpha. \tag{5.10}$$

## 6. Boundary Conditions at Shock Front

The remaining parameters are narrowed down further by making the solution compatible with the basic RANKINE-HUGONIOT conditions across a shock front. These relations apply just as well to a spherical or curved surface as to a plane, since the effect of spherical divergence (the $2u/r$ term) on a finite or sudden jump is of higher order. However, it is not possible to satisfy simultaneously these conditions and an energy condition constant in time for a quasi-polytropic medium. For a polytropic medium we can obtain independence of time by setting

$$\delta = 0. \tag{6.1}$$

The forms for the progressing wave solutions then reduce to

$$u = (r/t)\,U(\xi), \qquad p = (r/t)^2 D(\xi)P(\xi),$$

$$\varrho = D(\xi), \qquad \frac{p}{\varrho} = (r/t)^2 P(\xi) \tag{6.2}$$

with $\xi = rt^{-\alpha}$.

This solution shows that on the shock front or free surface, where $\xi$ is constant, the physical quantities such as velocity, pressure, density, and wave velocity are constant on the rays $r/t = $ constant. This also dimensionalizes the functions (5.2) correctly. The complete set of exponents is now

$$\alpha = 2/5, \qquad \epsilon = -6/5 \tag{6.3}$$

$$\beta = -3/5, \qquad \delta = 0.$$

## 7. Initial Conditions

Since $\xi = \xi_1$ on the shock front, we have, just behind it,

$$U(\xi_1) = \alpha(1 - \mu^2) = 2\alpha/(\gamma + 1), \tag{7.1a}$$

$$D(\xi_1) = \varrho_0/\mu^2, \tag{7.1b}$$

$$P(\xi_1) = \alpha^2\mu^2(1 - \mu^2) = 2\alpha^2\mu^2/(\gamma + 1), \tag{7.1c}$$

with $\alpha = 2/5$ or $1/4$ according as the energy or momentum condition holds.

If we suppose that $p_1 \gg p_0$, the above equations furnish, for a specified material, a definite initial point in the $P-U$ plane, which determines a single solution curve. Note that the constant $\xi_1$, which is later determined from an initial pressure or velocity condition, is not needed for this.

## 8. Momentum Condition

If, as a result of impact, it be supposed that the momentum of an expanding hemispherical cavity remains constant, we obtain a different set of values for the exponents. With $\delta = 0$, these become

$$\alpha = 1/4, \qquad \epsilon = -3/2,$$
$$\beta = 3/4, \qquad \delta = 0.$$

## 9. Equations of State for Metals

Fig. 2 shows log-log pressure-density relationships obtained from [2] for 4 metals in the range 100—400 kb. A straight line fit through these points gives a polytropic relationship

$$p \left( \frac{\varrho}{\varrho_0} \right)^{-\gamma} = A. \qquad 100 \text{ kb} < p < 400 \text{ kb}. \qquad (9.1)$$

Above 400 kb and certainly below 100 kb, where we start to have a transition to elastic-plastic behavior, this type of equation does not apply very well. Within the above range, however, it is convenient to use and does enable the progressing wave analysis to be carried through. At low pressures the quasi- polytropic equation

$$p = A_Q[(\varrho/\varrho_0)^n - 1] \qquad (9.2)$$

permits the pressure to tend to zero without the density doing so.

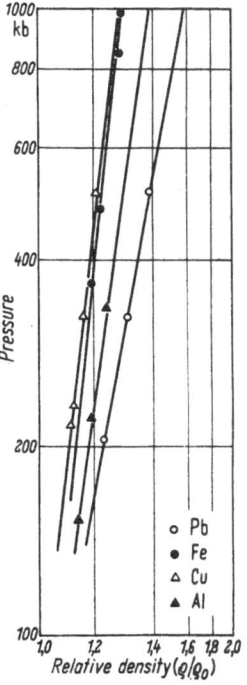

Fig. 2. Equations of state for several metals.

The following table gives some material constants taken from [2] and HUANG's thesis. These were obtained by curve fitting from published experimental data.

Table 1. *Some Equations of State Constants*

| | $\varrho_0$ (g/cc) | $A$ (kb) | $\gamma$ | $A_Q$ | $n$ |
|---|---|---|---|---|---|
| Aluminium (24 ST) | 2.78 | 52.7 | 7.6 | 232 | 3.55 |
| Copper | 8.90 | 15.6 | 8.5 | 404 | 3.65 |
| Lead | 11.34 | 51.07 | 5.9 | 183 | 3.11 |
| Iron | 7.84* | | 16.0* | 480* | 3.53* |
| | 8.38** | 68.45 | 9.0** | 339** | 3.80** |

\* $p < 131$ kb. (b.c.c.).   \*\* $p > 131$ kb. (f.c.c.).

The values for iron are somewhat tentative as the transition pressure is uncertain.

## 10. Further Work on Quasi-Polytropic Media

The basic relations for a shock passing into an undisturbed region in the medium, with $P = 0$, $\varrho = \varrho_0$, $T = T_0$ representing the pressure-free state and $P$, $\varrho$, $T$ the shock-compressed state, are given by

$$p = n A (\varrho/\varrho_0 - 1) \Big/ \Big[ 1 - \frac{1}{2} (n - 1)(\varrho/\varrho_0 - 1) \Big],$$

$$T = \{P/[(\varrho/\varrho_0)^m - 1] - (k_0/m)\}$$
$$\Big\{[\varrho/\varrho_0 - 1] + \frac{1}{m - 1} [(\varrho/\varrho_0)^{m-1} - 1]\Big\} \Big/ \varrho_0 C_v.$$

These equations enable one to study the shock dynamics of metal to metal impact in one-dimension. The compression shock on impact is similar to that in polytropic media but unloading waves behave differently. Whenever such a wave catches up with a shock wave, the state of the medium behind the shock wave soon changes drastically. Further interaction between both waves results in an unsteady shock wave falling behind the rarefaction wave, which propagates with decreasing strength and velocity. Unlike gas dynamics, where isentropic expansion waves occur as continuous simple waves, we have here a rarefaction shock across which the pressure and density of the medium decrease abruptly along with some rise in entropy. Another complication which causes instability of shock waves in metals is that a strong shock may induce b.c.c. → f.c.c. phase transition in certain metals such as iron, steel, bismuth, antimony, which causes in turn the break-up into multiple shock waves. Illustrations of this and other phenomena are given in HUANG's thesis.

## 11. The $P, U$-Diagram

Using the condition of constant energy, the 2/5-power law holds, and the differential equation (5.4) for progressing waves may be solved by numerical procedure (a program in FORTRAN was set up based on the method of increments). A family of integral curves in the diagram for aluminum are shown in Fig. 3. All of them issue from the singular point $A$ enclosed in a rectangle on the diagram and which is located by the condition that (5.4) have the value 0/0. The location of the singular point depends on the material through its value of $\gamma$. For the four metals we obtain:

Table 2

|  | $U_0$ | $P_0$ | $\gamma$ |
|---|---|---|---|
| Aluminum | .0916 | .02884 | 7.6 |
| Copper | .0816 | .0271 | 8.5 |
| Lead | .1199 | .0321 | 5.9 |
| Iron | .0476 | .016609 | 16 |

All the curves issue from the unstable nodal point $A$ along a common tangent. The point $U = P = 0$ is a stable nodal point and $P_0 = 0$, $U_0 = 0.4$ a saddle point. The solution curve for all the 4 metals starts close to the source. This puts an instability into the problem in that the course of the solution curve becomes very sensitive to the location of the initial point. Thus, for aluminum and copper, the solution curve comes out very close to the dividing curve (marked $C$ in Fig. 3) which runs into the saddle. It is not possible to distinguish from the accuracy whether our solution coincides with $C$ or not.

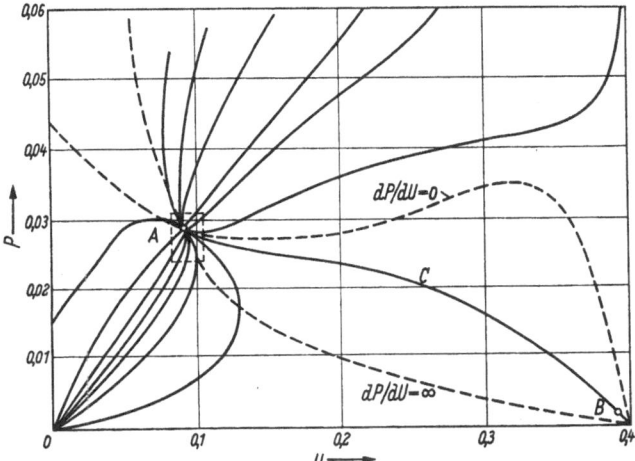

Fig. 3. $P, U$-diagramm, waves: $A L$, $\gamma = 7.6$.

The "source" point itself represents infinite shock strength (but with finite total impulse). The spherical divergence would allow such a shock to decay instantly, unlike the plane case.

The $P, U$-diagram for lead exhibits different behavior in that the solution curve, instead of going to the corner becomes asymptotic to the vertical line $U = \alpha$, approaching $P = \infty$. However, in spite of this, we still obtain a meaningful $R, T$-diagram and a reasonable cavity radius. These are awaiting laboratory tests.

Iron is most unusual in that $\gamma$ changes during the cavity expansion. A previous $P, U$-diagram in [3] based on $\gamma = 16$ only, gave a curve like Al leading into the saddle point. A workable program has been set up to make a $P, U$-diagram taking into account the transition in $\gamma$.

The value $U = 0.4$ represents an asymptotic condition on the particle curves which would occur if the progressing-wave solution were extrapolated to $t = \infty$.

Figs. 4 to 5 show an $R-T$ diagram plotted for an initial cavity radius of 1.698 cm. Pressures in kilobars are also shown on Fig. 4. This figure gives much more detail of the early phase of the expansion up to

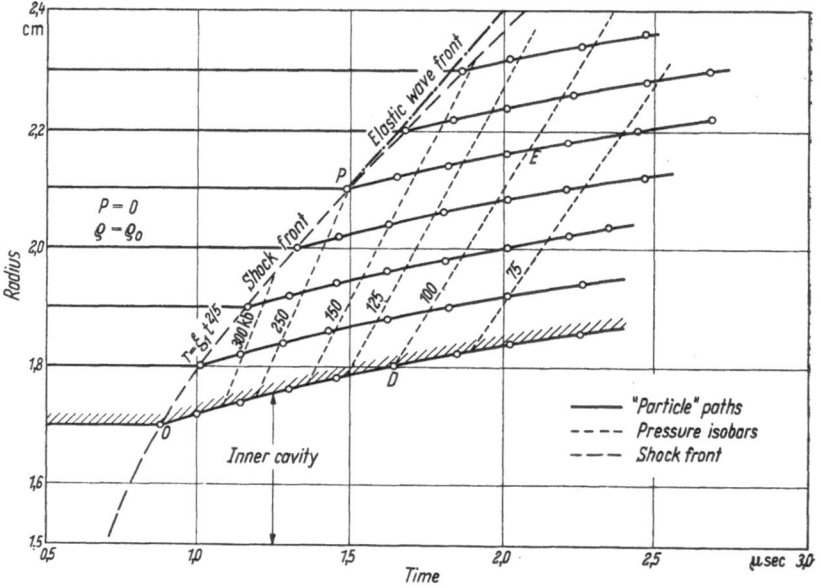

Fig. 4. $R, T$-diagram of shock region for aluminum sphere.

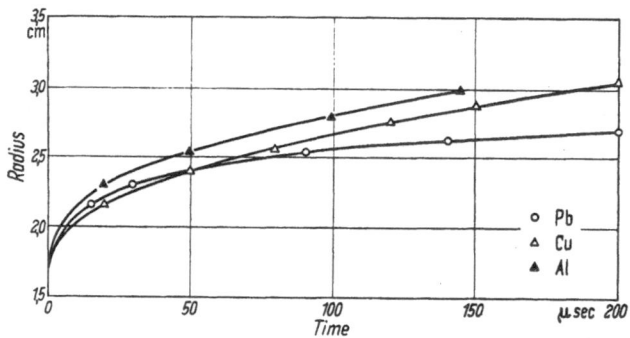

Fig. 5. $R, T$-diagram for cavity formation in several metals.

$t = 2.5$ microsec from its start at $t = 0.9$ μsec. In this elapsed time of 1.6 μsec the inner cavity has only grown to 1.86 cm, which is only 13% of its ultimate change. However, the pressure has already fallen considerably. At the cavity surface it is down to 50 kb.

We also note that the shock velocity at point labelled $P$ on the diagram is equal to the known elastic wave velocity of the material. Beyond this point the 2/5-power law for the shock front starts to deviate from this velocity. Such a condition represents a discrepancy of the progressing wave method from this point on, where it is inevitable because of the equation of state used.

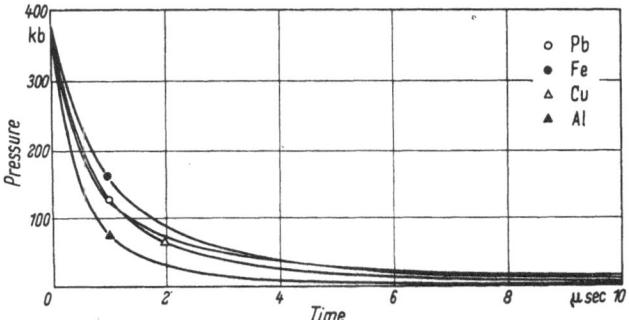

Fig. 6. Pressure decay on cavity surface for several metals.

Fig. 6 shows how the pressure decays with time at the inner cavity surface.

The arbitrary multiplicative constant appearing in the function $\xi(U)$ is determined from the known required value of the density of the material behind the shock front, as given by the RANKINE-HUGONIOT conditions.

## 12. Energy Considerations

The assumption of constant energy is based on an integral extended between a value $\xi = \xi_0$ and the shock front $\xi = \xi_1$, such as curve $BC$ in Fig. 1. The energy in the disturbed part of the solid will only become constant in an asymptotic sense since the cavity surface is not one of the curves $\xi = $ const., except asymptotically. We obtain in the limit a value $E = 1.26 \times 10^{12}$ ergs which agrees favorably with the value available from the detonation of the explosive in the cavity.

## 13. Further Analyses

The "similarity" type solutions discussed above have the drawback that they become singular at time zero and hence cannot supply any information about the initial stage of the motion. This may be unsatisfactory in problems such as hypervelocity impact, where a substantial fraction of the total damage may occur during the initial stage of the motion. The method of expanding the solution for small values of time in a doubly-infinite power series was therefore carried out by HEADING-TON, JAUNZEMIS and DAVIDS. A representation of the exact solution for the initial stage of motion of plane, cylindrical, and spherical waves was determined in the form of a doubly-infinite series and a procedure for computing the leading terms given.

If the space coordinate be denoted by $x$, the material or particle coordinate by $s$, and $t$ the time, we may write the solution to the equations of motion in the form

$$x(s, t) = a_0 + a_1 t + a_2 s + b_1 t^2 + b_2 ts + b_3 s^2 + c_1 t^3 + c_2 t^2 s$$
$$+ c_3 ts^2 + c_4 s^3 + \cdots$$

and the motion of the shock front by a series

$$t^* = t^*(s) = A_1 s + A_2 s^2 + A_3 s^3 + \cdots,$$

where $t^*$ is that value of time $t$ for which the shock wave has arrived at a particle carrying a given label $s$. For initial conditions, the value of velocity $u_i$ was prescribed. The shock front imposes a set of recursion relations, as well as the equation of state of the material, which is assumed polytropic in form.

The details giving the recursion relations for the coefficients are fairly lengthy, and are to be published elsewhere. The scheme for calculating the coefficients $a_i$, $b_i$, $c_i$, $A_i$ of the above infinite series, up to order 3, is presented here. The equation of state is given by the expression

$$p = p_i(\varrho^n - 1)/(\varrho_1^n - 1).$$

We then have

$$a_0 = 0,$$

$$a_1 = u_i,$$

$$a_2 = 1/\varrho_i,$$

$$A_1 = (1 - a_2)/a_1,$$

$$b_2 = a_2(1 - a_2^n)/n - r a_1 a_2,$$

$$b_1 = -\{r n k_1(1 - a_2)/(1 - a_2^n) + [n k_1/(1 - a_2^n)$$
$$+ 2 m a_2 \sigma] b_2 A_1/a_2\}/2 k_3 \sigma,$$

$$c_2 = a_2(1 - a_2^n)[1 - a_2^n(n + 1)]/2 n^2 - r(a_1 b_2 + a_2 b_1)$$
$$- r(r - 1) a_1^2 a_2/2,$$

$$2a_1A_1 = k_{-1}(2b_1A_1^2 + b_2A_1)k_1,$$

$$b_3 = (a_1A_2 + b_1A_1^2 + b_2A_1),$$

$$k_m = \sigma a_2(1 - a_2^n) + mnA_1^2,$$

$$\sigma = s_0^2\alpha^2\varrho_0/p_i,$$

$r = 2$   for spherical waves.

The remaining coefficients $c_1$, $c_3$ and $A_3$ are found by solving a certain set of linear equations, given by HEADINGTON and JAUNZEMIS.

A rigorous proof of the convergence of the series has not yet been given. However, an idea of the rapidity of convergence can be obtained from Fig. 7 where the boundary motion and motion of the shock wave with time have been calculated for a pressure of 500 kb using 1, 2, and 3 terms of the series. The numerical calculations bear out that a fairly rapid convergence is obtained for the range $0 < t < 1$. Other kinematical quantities have also been obtained and will soon be submitted for publication.

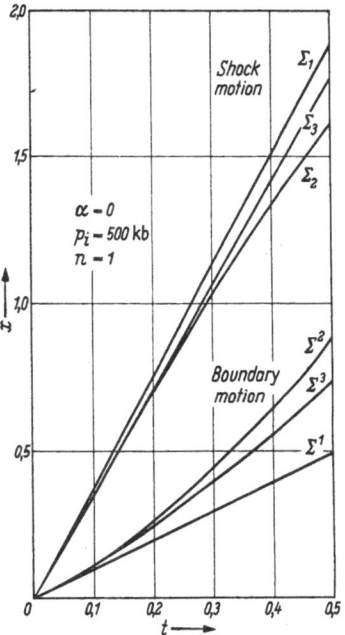

Fig. 7. Convergence of the solutions for the boundary motion and shock motion.

## 14. Summary and Conclusions

In this paper we have attempted to study the cavity formation process in the metal by determining how the important physical variables of cavity radius, velocity, pressure, and density vary with time and position near the cavity. The most prominent general feature of the whole process is the short time of the "shock" regime as compared with the total time of the expansion. One general criterion of the end of the shock is when the supersonic velocity of the shock front drops to sonic, i.e. at the point $P$ of Fig. 4, where the slope attains the value for elastic disturbances in the material. The progressing wave curve cannot be used beyond this point since it would give a subsonic shock velocity. This situation has occured after 0.5 microsec.

The discussion below is based on aluminum but applies equally well to copper. Lead and iron show some peculiarities in the $P, U$-diagram which have not been evaluated as yet.

We note that the highest pressures and densities in the metal are located just behind the shock front, and trail off with decreasing radius to minimum values at the cavity boundary. We note that the equation of state (6.1) which has been used for the calculations has a lower limit of $p = 100$ kb. This could also be used as a criterion for shock termination (point $D$, Fig. 4). It is reached in 0.7 μsec. These conditions thus determine a roughly parallelogram shaped region ODPE in the $R, T$-plane for the validity of the progressing wave region. Note that the cavity has only expanded 0.1 cm during this period, which is 1/30 of the total observed increase in radius. We are thus justified in referring to the shock process as impulsive, i.e., the later stages of the process are insensitive to many features of the shock part. Hence, the progressing wave method of integrations remains valid for the analysis of the shock zone.

The asymptotic characteristics of the progressing waves are thus not of direct physical interest since they do not apply to the problem beyond the region described above. The expansion zone, headed by a wave travelling with the dilatational wave velocity goes on for at least 100 μ seconds, during most of which the metal continues to move by fluid or plastic flow.

The final cavity radius attained is of great interest to the general problem as this value is directly observable on the specimens after blast. This problem is the subject of continued research.

## Summary

We summarize by noting that the following four phenomena are coincident in time:

1. The shock-wave becomes sonic.

2. The pressure at the cavity surface drops to less than 100 kb.

3. The total energy in the disturbed material reaches 90% of its maximum and then levels off asymptotically.

4. The average gas pressure in the cavity (uniform model) equals that in the metal.

All of these occur close to 2 microseconds after initiation of the explosion. This determines a fairly definite time point of changeover of conditions. Up to this time (t = 2,5 microsec for the conditions of this report) we may say the effects of shock predominate. The progressing wave method furnishes an accurate theory for this regime. After this time a relatively long expansion period occurs at constant energy until ultimately terminated by degradation processes.

## 15. Acknowledgement

The work of this paper was sponsored by the Ballistic Research Laboratories, Aberdeen Proving Ground and earlier by the National Aeronautics and Space Administration. Their support is gratefully acknowledged.

### Bibliography

[1] COURANT, R., and K. FRIEDRICHS: Supersonic Flow and Shock Waves, New York: Interscience 1948, Ch. VI.
[2] WALSH, J. M., M. H. RICE, R. G. McQUEEN, and F. L. YARGER: Phys. Rev. 108, No. 2, 196 (1957).
[3] DAVIDS, N., and Y. K. HUANG: J. Aerosp. Sci. 29, No. 5, 550 (1962).
[4] NEWTON, R. G.: J. Appl. Mchs. 19, No. 3, 257 (1952).
[5] SEDOV, L. I.: Similarity and Dimensional Methods in Mechanics, New York: Academic Press 1959.
[6] CHARTERS, A. C., and J. L. SUMMERS: Proc. 3rd Symposium on Hypervelocity Impact, Armour Res. Foundation 1958, p. 101.
[7] DUVALL, G. E., and B. J. ZWOLINSKI: J. Acoust. Soc. Amer. 27, No. 6, 1054 (1955).
[8] DRUMMOND, W. E.: J. Appl. Phys. 28, No. 12 and 29, No. 2, 167 (1958).
[9] ALLEN, W. A., H. L. MORRISON, D. B. RAY, and J. W. ROGERS: Phys. Fluids 2, No. 3, 329 (1959).
[10] GOLDSMITH, W.: Impact, London: Arnold 1960, Ch. V.
[11] RINEHART, J. S., and W. C. WHITE: Am. J. Phys. 20, 14 (1952).
[12] MAURER, W. C., and J. S. RINEHART: J. Appl. Phys. 31, No. 7, p. 1247.
[13] HOPKINS, H. G.: Prog. Solid Mech. I (SNEDDON and HILL ed.), Amsterdam: Amsterdam: North-Holland Publ. Co. 1960, Ch. IV.

# Propagation of Plane Shock Waves in a Stress-Relaxing Medium[1]

## By George E. Duvall

Poulter Laboratories, Stanford Research Institute, Menlo Park, California, U.S.A.

## 1. Introduction

Experimental skill in the production and measurement of shock waves in solids has now progressed to such a stage that it is quite easy to simulate, for a short time, a plane wave driven into a semi-infinite slab by a pressure acting on its surface and to measure the details of stress and strain in the wave front with time resolution the order of $10^{-8}$ sec and with errors of only a few percent in amplitude [1, 2]. Such measurements have demonstrated the existence of an elastic precursor wave preceding the plastic shock in many metals and similar but less understood wave structures in other materials.

Values of yield strength can be readily deduced from elastic precursor amplitudes, and some other details of shock wave structure can also be related to material properties, e.g., pressure-induced phase transitions, if supplemental information about the material is available.

It is the purpose of this paper to show how the decay of an elastic precursor wave preceding a shock can be related to material relaxation from an elastic, nonequilibrium state toward one of equilibrium. We draw upon shock wave measurements in quartzite and, for comparison, consider the theory of shock waves generated by a uniform pressure acting on the surface of a semi-infinite slab.

## 2. Equations of Flow

In Lagrangian coordinates the equations of continuity, motion, and energy conservation in one dimension are, respectively,

$$\frac{\varrho_0}{\varrho^2} \frac{\partial \varrho}{\partial t} + \frac{\partial u}{\partial h} = 0, \tag{1}$$

[1] This work was supported by the Air Force Office of Scientific Research under Contract AF 49 (638)—1086.

$$\frac{\partial u}{\partial t} + \frac{1}{\varrho_0} \frac{\partial p_x}{\partial h} = 0, \tag{2}$$

$$\frac{\partial E}{\partial t} + \frac{p_x}{\varrho_0} \frac{\partial u}{\partial h} = \frac{\partial E}{\partial t} - \frac{p_x}{\varrho^2} \frac{\partial \varrho}{\partial t} = 0, \tag{3}$$

where $\varrho$, $u$, and $E$ are density, particle velocity, and specific internal energy of the medium, $p_x$ is the compressive stress parallel to the direction of wave propagation, $h$ is the undisturbed position of a mass element, and $t$ is the time. Subscripts "0" refer to the undisturbed medium. No heat is supposed to be added to a mass element by conduction or radiation absorption.

Since only linear compression is considered, in which strain parallel to the wave fronts is zero, total strain can be represented by density change, $d\varepsilon_x = d\varrho/\varrho$. The following identity holds between $p_x$, hydrostatic pressure $\overline{p}$, and maximum resolved shear stress $\tau = (p_x - p_y)/2$:

$$p_x = \overline{p} + 4/3\tau. \tag{4}$$

## 3. Stress-Strain Relations

We consider a material with the property that any increment of total strain is the sum of an elastic and an inelastic component. If the latter component is designated $d\varepsilon_x^p$ and assumed to represent no net dilatation, we can derive the following relation [3]:

$$d\varepsilon_x^p = \frac{2}{3}(d\varepsilon_x - d\tau/\mu), \tag{5}$$

in which $d\tau$ is the increment in resolved shear stress accompanying the increment $d\varepsilon_x$ in total strain, and $\mu$ is the shear modulus under conditions that exist when the incremental strain occurs; this modulus is assumed to be a function of density alone. In plastic strain, $d\tau$ may be nonvanishing because of work-hardening; in brittle fracture, it may be nonvanishing because of friction between fractured surfaces. In either case it may be strain-rate dependent.

In the plastic case the dislocation theory of solids leads to a relation between the rate of plastic strain and the density $N$ and velocity $v$ of dislocations responsible for the strain [4]:

$$\partial \varepsilon_x^p / \partial t = Nbv; \tag{6}$$

here $b$ is the BURGERS vector of the material. W. G. JOHNSTON [5] has suggested that the right-hand side of Eq. (6) may for some purposes be considered a function of $\varepsilon_x^p$ and $\tau$ alone; the explicit entrance of time is ignored in the following. Since $\varepsilon_x^p$ is a function of $\tau$ and $\varrho$, and $\tau$ is a function of $p_x$ and $\varrho$, we can write

$$\partial \varepsilon_x^p / \partial t = F(p_x, \varrho)/2\mu. \tag{7}$$

Combining Eqs. (4), (5), and (7), we obtain the expression

$$\frac{\partial p_x}{\partial t} - \left[\frac{4}{3} \frac{\mu}{\varrho} \frac{\partial \varrho}{\partial t} + \frac{\partial \overline{p}}{\partial t}\right] = -F(p_x, \varrho). \tag{8}$$

It is implicit in these considerations that, as $\partial \varepsilon_x^p / \partial t \to 0$, $p_x$ becomes a precise function of $\varrho$, say $p_x^s(\varrho)$. This function exceeds the hydrostatic pressure $\overline{p}$, as indicated in Eq. (4), when the yield stress, $Y = 2\tau$, is non-zero. This relation between $\tau$ and $Y$ for plastic solids is established by use of the von MISES-HENCKY criterion for yield.

It is customary to ignore the thermodynamic properties of elastic-plastic materials when calculating wave effects, and this convention will be adopted here. We note only that if the heating of the medium by compression and plastic work is taken into account and Eq. (3) is properly noted, the function $F$ of Eq. (7) appears in the equivalent of Eq. (8) multiplied by a factor slightly less than unity; there is some uncertainty in the magnitude of this factor because of uncertainty in the amount of irreversible work that is converted into heat. However, the total effect can be neglected to a first approximation at the amplitudes considered here.

Since $\partial \overline{p}/\partial t \equiv c^2 \partial \varrho/\partial t = (K/\varrho)(\partial \varrho/\partial t)$, where $K$ is bulk modulus, Eq. (8) can also be written in the form

$$\frac{\partial p_x}{\partial t} - a^2 \frac{\partial \varrho}{\partial t} = -F(p_x, \varrho), \tag{9}$$

where $a$ is the speed of elastic compression waves at density $\varrho$.

Eqs. (1), (2), and (9) comprise a complete set of relations which can be solved for $p_x(x, t)$, subject to the appropriate boundary conditions, choice of $F$, and knowledge of bulk and shear moduli as functions of density.

In order to cast Eqs. (1), (2), and (9) in characteristic form, Eq. (9) can be solved for $\partial \varrho/\partial t$ and the result substituted into Eq. (1). This yields

$$\frac{1}{\varrho a} \frac{\partial p_x}{\partial t} + \frac{\varrho a}{\varrho_0} \frac{\partial u}{\partial h} = -F(p_x, \varrho)/\varrho a. \tag{10}$$

If Eq. (2) is alternately added to and subtracted from this, we obtain, with Eq. (8), the following characteristic set suitable for numerical integration:

$$C+ : dp_x + \varrho a\,du = -F(p_x, \varrho)dt \quad \text{on} \quad \frac{dh}{dt} = \frac{\varrho a}{\varrho_0}, \tag{11}$$

$$C- : dp_x - \varrho a\,du = -F(p_x, \varrho)\,dt \quad \text{on} \quad \frac{dh}{dt} = -\frac{\varrho a}{\varrho_0}, \tag{12}$$

$$C_0 : \quad dp_x - a^2 d\varrho = -F(p_x, \varrho)dt \quad \text{on} \quad \frac{dh}{dt} = 0. \tag{13}$$

These equations are formally identical to those developed by MAL-VERN [8] to describe wave propagation in a bar. The physical content differs because of lateral restraints. In order to proceed we consider the physical properties of a particular material.

## 4. Compressive Properties of Sioux Quartzite

Sioux quartzite is a pure, fine-grained form of $SiO_2$ for which grain orientation is nearly random and porosity is less than 1%; its natural density is 2.640 g/cc. Extensive measurements have been made of shock waves propagating through it, with results shown in Figs. 1 and 2 [7]. The shock wave typically consists of a first shock traveling at elastic velocity and decaying in amplitude as it travels; this is followed by an ever-widening region of rising stress, terminated by a second shock which is in turn followed and attenuated by a rarefaction that returns the material to an uncompressed state.

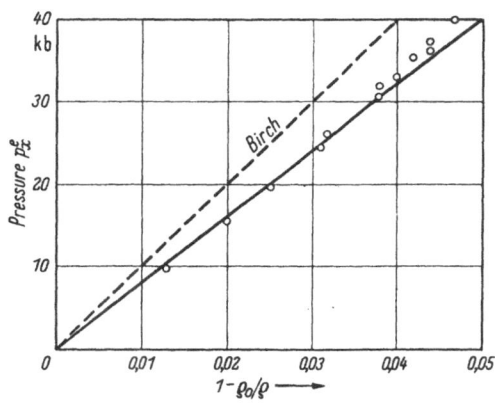

Fig. 1. First wave in Sioux quartzite.

A sequence of amplitudes of the first shock, measured as it receded from the driving source, is shown in Fig. 1. The dotted line is an extrapolation from static measurements made on a high density, "quartzitic sandstone" reported by BIRCH [9]. The difference between static and dynamic data may result from material differences or from flaws in experimental measuring procedures; the latter seems unlikely. For the following calculations, the data of Fig. 1 are represented by the solid straight line:

$$p_x^e = \varrho_0 U_0^2 (1 - \varrho_0/\varrho) = 850 (1 - \varrho_0/\varrho) \text{ kilobars}, \tag{14}$$

where $p_x^e$ is compressive stress corresponding to density $\varrho$ when no deformation or fracture has occurred, and $U_0$ is propagation velocity of the elastic precursor. Since $dp_x^e = a^2 d\varrho$, Eq. (14) leads to the useful relation

$$\varrho a = \varrho_0 U_0 = \text{constant.} \tag{15}$$

This relation will simplify the numerical computation to follow since the $C-$ characteristics will be straight lines and the elastic precursor will lie on a $C+$ characteristic.

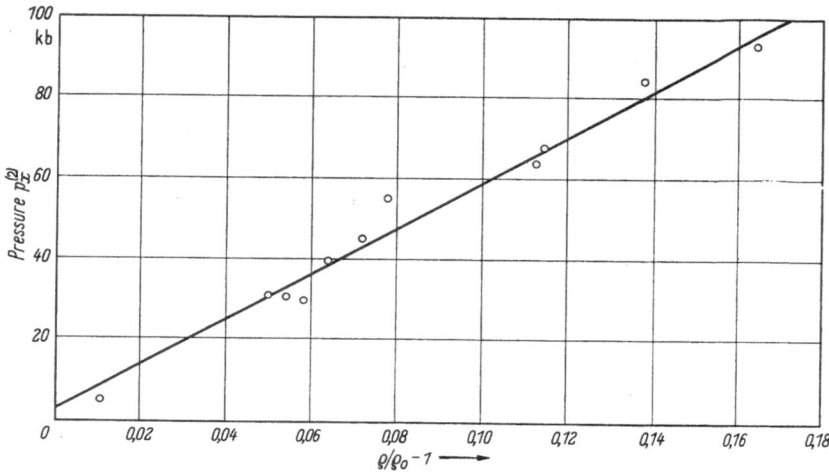

Fig. 2. Second wave in Sioux quartzite.

Measured amplitudes of the second shock are displayed in Fig. 2; they are adequately represented by the solid line

$$p_x^{(2)} = 557 \left(\varrho/\varrho_0 - 1\right) + 2.43 \text{ kilobars;} \tag{16}$$

we shall identify $p_x^{(2)}$ with $p_x^s$.

Since the numerical coefficient in Eq. (14) represents some suitable average of the linear compression modulus, $(K + 4/3\mu)$, and since some part of the variation of $p_x^{(2)}$ with density in Eq. (16) arises from variations of $\bar{p}$ by virtue of Eq. (4), we can derive from the two equations some information about the effect of density on shear strength. From Eqs. (14) and (16), respectively,

$$K + 4/3\mu = 850\varrho_0/\varrho,$$

$$K + 4/3\,\partial\tau/\partial\varepsilon_x = 557\varrho/\varrho_0,$$

so that $\partial\tau/\partial\varepsilon_x \simeq 150$ kilobars.

The nature of the failure process of brittle materials under superposed hydrostatic pressure is not well understood, and it is quite possible that the curve of Eq. (16) does not join continuously with Eq. (14) at the static failure point; this is certainly true in static experiments [10]. In such case $\partial \tau / \partial \varepsilon_x$ may be related to a coefficient of friction: $(1/K) \partial \tau / \partial \varepsilon_x \simeq 0.42$. This is commensurate with values suggested by other geophysical data [6].

## 5. Numerical Integration

We consider explicitly the following problem. A constant pressure, $p_1$, is applied to the surface of a semi-infinite slab of Sioux quartzite and maintained indefinitely. This drives into the quartzite a shock which eventually degenerates to an elastic precursor of constant amplitude followed by a slower wave similar to a plastic shock. The stress-volume relations are illustrated in Fig. 3. The state, $Q$, in the elastic wave is supposed to be reached by a discontinuous jump from undisturbed material and to lie on the curve $p_x^e$, on which the strain of deformation is zero. $Q$ approaches its steady value, $A$, asymptotically in the relaxation process. Relaxation

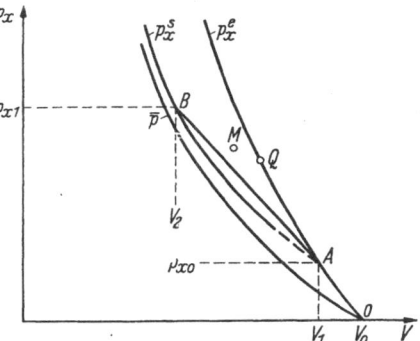

Fig. 3. Stress volume relations in shock.

and deformation occur behind the precursor, and the material passes temporarily through points $M$ lying between $p_x^e$ and $p_x^s$. The state $(V_2, p_{x1} = p_1)$ is the static state ultimately reached by the compressed material.

The details of the distribution of $p_x$ in space and time can be calculated numerically from Eqs. (11) to (13), using the data from Section IV, after a form is given for $F(p_x, \varrho)$. Since the characteristics in the $(h, t)$ plane are straight lines, the problem lies between the elastic precursor and the slab boundary, shown as $OB$ and $OD$, respectively, in Fig. 4.

The elastic precursor is assumed to be a discontinuity in stress, density, and particle velocity propagating with velocity $U_0$. Accordingly, the relation between stress and particle velocity at this front is

$$p_x^e = \varrho_0 U_0 u^e, \tag{17}$$

and if, in an increment of time $dt$, the amplitude of the precursor changes an amount $dp_x$, the accompanying change in $u$ is

$$du^e = dp_x^e/\varrho_0 U_0. \tag{18}$$

Combining Eqs. (11) and (18) leads to the following expression for the rate of change of precursor amplitude with time:

$$\frac{dp_x^e}{dt} = -F(p_x^e, \varrho)/2 \qquad \text{on} \qquad x = U_0 t. \tag{19}$$

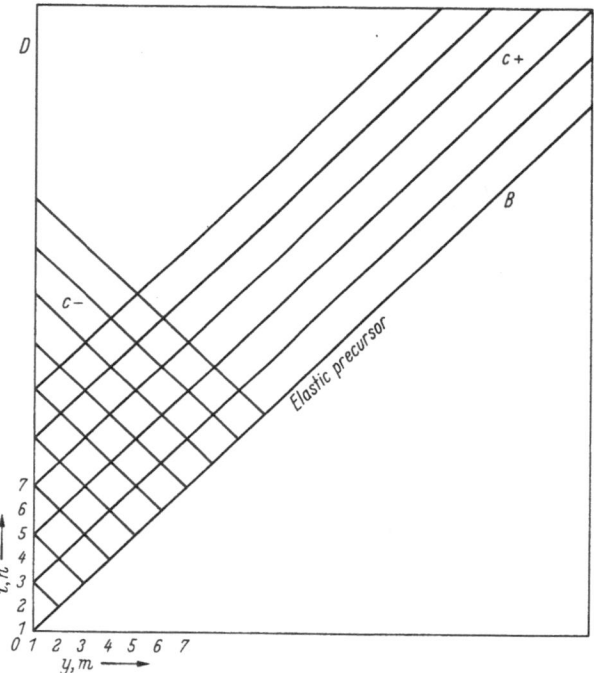

Fig. 4. Characteristics net for transient problem.

The condition at the boundary $h = 0$ is that $p_x \equiv p_1 = $ constant. Then Eq. (13) yields the following expression for density:

$$a^2 \frac{d\varrho}{dt} = F(p_x, \varrho) \qquad \text{on} \qquad h = 0. \tag{20}$$

We introduce the dimensionless variables

$$\tau = t/T, \qquad y = h/U_0 T,$$

where $T$ is a characteristic time in the relaxation process, to be specified later. With these definitions, the $C+$ and $C-$ characteristics become $dy/d\tau = \pm 1$. If $m$ and $n$ are positive integers, the assumption

$$\tau = (n-1)\Delta\tau, \qquad y = (m-1)\Delta\tau$$

produces a characteristic net in the $y,\tau$ plane, as in Fig. 4. In Figs. 5a—c, portions of the net are shown for the diagonal along which the precursor travels, the left boundary ($h=0$), and a general interior point, respectively. Referring to Fig. 5a and Eq. (24), Eq. (22) can be written in difference form:

$$2(p^e_{m+1} - p^e_m) = -F_m T \Delta\tau, \tag{21}$$

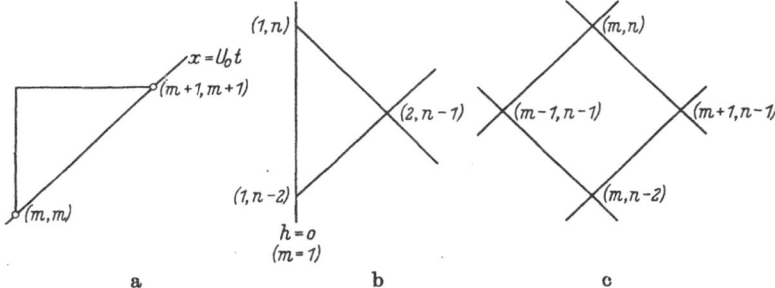

Fig. 5a—c. Net points for transient problem.

where $p^e_m \equiv p^e_{mm}$, etc. The subscript "$x$" on $p$ has been dropped, and it is understood henceforth that $p_{m,n}$ refers to the value of $p_x$ at $y = (m-1)\Delta\tau$, $\tau = (n-1)\Delta\tau$, etc.

With Fig. 5b, the difference form of Eq. (20) becomes

$$a^2_{1n}(\varrho_{1,n+1} - \varrho_{1n}) = F(p_1, \varrho_{1n})T\Delta\tau, \tag{22}$$

where $p_1$ is the constant value of $p_x$ on the left boundary. The values of $u$ on the left boundary are obtained from Eq. (12), referring to Fig. 5b:

$$p_1 - p_{2,n-1} - \varrho_0 U_0(u_{1n} - u_{2,n-1}) = -F_{2,n-1}T\Delta\tau. \tag{23}$$

At an interior point between the left boundary and the precursor diagonal, Eqs. (11) to (13) must be considered simultaneously. Referring to Fig. 5c, the difference equations are, respectively,

$$p_{mn} - p_{m-1,n-1} + \varrho_0 U_0(u_{mn} - u_{m-1,n-1}) = -F_{m-1,n-1}T\Delta\tau, \tag{24}$$

$$p_{mn} - p_{m+1,n-1} - \varrho_0 U_0(u_{mn} - u_{m+1,n-1}) = -F_{m+1,n-1}T\Delta\tau, \tag{25}$$

$$p_{mn} - p_{m,n-2} - a^2_{m,n-2}(\varrho_{mn} - \varrho_{m,n-2}) = -2F_{m,n-2}T\Delta\tau. \tag{26}$$

Adding Eqs. (24) and (25) one obtains

$$2p_{mn} - (p_{m+1, n-1} + p_{m-1, n-1}) + \varrho_0 U_0 (u_{m+1, n-1} - u_{m-1, n-1})$$
$$= -(F_{m-1, n-1} + F_{m+1, n-1}) T \varDelta \tau. \tag{27}$$

Subtraction of Eq. (25) from (24) yields

$$-p_{m-1, n-1} + p_{m+1, n-1} + 2\varrho_0 U_0 u_{mn} - \varrho_0 U_0 (u_{m-1, n-1} + u_{m+1, n-1})$$
$$= -(F_{m-1, n-1} - F_{m+1, n-1}) T \varDelta \tau. \tag{28}$$

In the computation $p_{mn}$ is obtained from Eq. (27), $u_{mn}$ from Eq. (28), and $\varrho_{mn}$ from Eq. (26). The sound velocity is obtained from the equation

$$a^2_{m, n-2} = \varrho_0^2 U_0^2 / \varrho^2_{m, n-2}.$$

A choice of form for $F(p_x \varrho)$ is a matter of some difficulty; in fact it is unlikely that sufficient knowledge of the properties of materials on which to base a correct choice exists, except possibly in very special cases [5, 8]. A simple relaxation relation will illustrate the phenomena to be expected and provide a basis for later refinements. We assume that

$$F(p_x, \varrho) = (p_x - p_x^s)/T, \tag{29}$$

where $T$ is a constant relaxation time. This function ensures that a stressed element at constant density relaxes exponentially toward a state of static compression.

These difference equations have been programmed in ALGOL for the Burroughs 220 automatic digital computer, and the case for quartzite was run with $p_1 = 10^{11}$ dynes/cm², $\varDelta \tau = 0.1$.

Calculated amplitudes of the elastic precursor decay asymptotically according to the relation $p_x^e - 7.13 = 97.87 \exp(-x/5.8 U_0 T)$ and somewhat more slowly than this at first. In Fig. 6 are shown the measured amplitudes of elastic precursors in quartzite as a function of distance from the explosive source. The scatter of data apparently results from irregularities in the wave front which may be produced by small-scale inhomogeneities of the medium. These experiments were performed in two-dimensional steady geometry and are therefore not directly comparable to the calculations. However, if we ignore this and relate the initial slope to the calculated attenuation length of $5.8 U_0 T$, the relaxation time for an element of the material stressed beyond the shear failure point is found to be 2.0 µsec. This may represent the time to fracture under the given stress condition, the rate of movement of fractured surfaces over one another, against friction, or even possibly the rate of plastic deformation if the quartzite is not behaving brittly.

The development of the second wave behind the elastic precursor is illustrated by the stress profiles shown in Fig. 7. Most notable here is that the second wave is beginning to separate distinctly from the precursor and assume a stable profile between 30 and $40\,U_0\,T$. Experiments

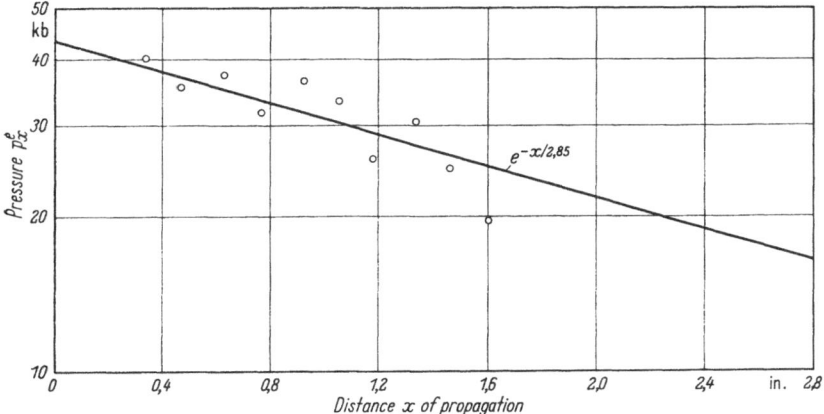

Fig. 6. First wave in Sioux quartzite, amplitude vs. depth (experimental).

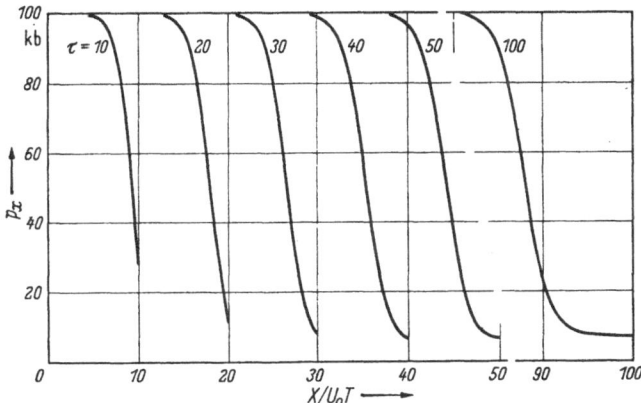

Fig. 7. Shock profiles at successive times in Sioux quartzite (calculated).

do not accurately reveal the second wave profile because of reverberations of the first wave between second wave front and free surface. However, it is clear from the data that a well-defined second shock appears much earlier than indicated by the calculation and that the rise time in the shock is much less than that shown in Fig. 7 [7].

## 6. Comparison with Steady Shock

If we suppose the driving pressure at the slab face, $p_1$, to be maintained for a very long time, the shock wave following the precursor reaches an essentially steady state. The precursor is far ahead of the shock, the boundary far behind, and the shock profile can be discussed on the assumption that it is steady in time. In order to accomplish this, Eqs. (1), (2), and (9) are transformed into Eulerian form, and the origin of coordinates is fixed in the shock front. These equations then become, respectively,

$$d(\varrho u)/dx = 0; \qquad \varrho u = m = \text{constant}, \tag{30}$$

$$\varrho u \frac{du}{dx} + \frac{dp_x}{dx} = 0, \tag{31}$$

$$\frac{dp_x}{dx} - a^2 \frac{dp}{dx} = - \frac{\varrho F(p_x, \varrho)}{m}. \tag{32}$$

The integral of Eq. (30) is

$$p_x + mu = p_x + m^2/\varrho = \text{constant}. \tag{33}$$

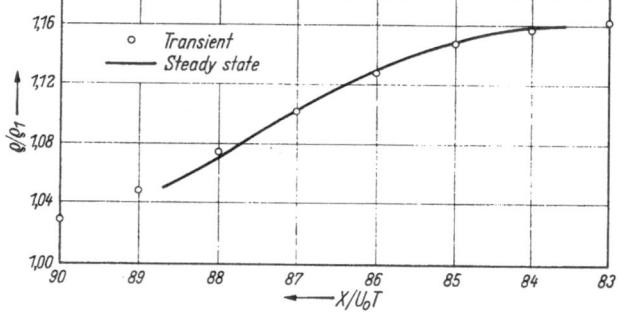

Fig. 8. Comparison of steady and transient profile in second shock, $\varrho_1$ = density before second shock.

Substitution of this into Eq. (31) yields the ordinary differential equation,

$$\left(\frac{m^2}{\varrho^2} - a^2\right) \frac{d\varrho}{dx} = - \frac{\varrho F(\varrho)}{m}, \tag{34}$$

which can be formally integrated by quadrature under the assumption that $\varrho$ approaches steady values as $x \to \pm \infty$.

Substitution of Eqs. (16) and (29) into Eq. (34) yields the solid curve of Fig. 8. The circles shown there are from the numerical integration of the transient equations for $\tau = 100$ (Fig. 7). The transient profile corresponds closely to the stable profile, and from Fig. 7 this is apparently true even after 50 relaxation distances, but not much before.

## 7. Discussion

These calculations and their comparison with experiment indicate clearly that a single relaxation time can be chosen which approximately describes the attenuation of the precursor amplitude in quartzite. However, this relaxation assumption predicts that a steady shock will be established only after about 40 to 50 relaxation times, and that the thickness of the steady shock front will be about four relaxation times. The experimental results indicate that the second shock is established within a very short time of the initiation of the shock, and that the thickness of the shock never exceeds a small fraction of a relaxation time.

While it is true that the experimental conditions do not closely fit the conditions of the calculation, so that a direct comparison is not strictly valid, the order of the discrepancies is so great that some weakness in the physical model is suggested. The most apparent weakness is the assumption of Eq. (29) for relaxation. The data suggest that $T$ decreases very substantially after material has passed through the elastic front; this idea is compatible with theories of dislocation motion and generation in metals and with quasi-static experiments on the failure of brittle solids under static pressure. Other assumptions made here which might lead to discrepancies of the kind described are:

(a) neglect of rise time in the elastic precursor; the quartzite experiments show this to be substantially less than one microsecond and therefore negligible;

(b) variation of precursor velocity with amplitude; this leads to violation of the condition $\varrho a = $ constant; the result would be to accelerate the elastic decay and establishment of the steady state;

(c) assumption of a linear $p-\varrho$ relationship in the second shock; increase of bulk modulus with rise time would lead to a thinner steady shock profile.

The principal value of this calculation appears to be in illustrating that the development of a shock wave in a relaxing medium can be described by conventional theoretical means and that, through comparison of calculations and measurements, there is some promise of understanding the behavior of materials compressed dynamically by large stresses. Further work in both theory and experiment are required to establish the controlling processes.

## Acknowledgment

The author wishes to thank Mr. JOHN O. ERKMAN of Poulter Laboratories for programming the solution of the difference equations.

### References

[1] Solid State Physics, Vol. 6, edited by F. SEITZ and D. TURNBULL, New York: Academic Press 1958, "Compression of Solids by Strong Shock Waves" (M. H. RICE, R. G. McQUEEN, and J. M. WALSH), pp. 1—63.

[2] RICE, M. H.: Rev. Sci. Instr. 32, 449 (1961).

[3] FOWLES, G. R.: J. Appl. Phys. 32, 1475 (1961).

[4] VAN BUEREN, H. G.: Imperfections in Crystals, Amsterdam: North-Holland Publ. Co. 1960.

[5] JOHNSTON, W. G.: J. Appl. Phys. 33, 2716 (1962).

[6] Rock Deformation (A Symposium), edited by DAVID GRIGGS and JOHN HANDIN: Geol. Soc. Am. Memoir 79, 1960, "Mechanism of Seismic Faulting", (E. OROWAN), Ch. 12.

[7] Les Ondes de Détonation, Editions du Centre National de la Recherche Scientifique, Paris 1962, "Shock Wave Stability in Solids", (G. E. DUVALL: p. 351, Discussion by D. R. GRINE). See also V. G. GREGSON and D. R. GRINE: Vela-Uniform: Dynamic Properties of Rocks, Stanford Research Institute, Project PGU-3630, Semiannual Technical Report No. 2, 1962.

[8] MALVERN, L. E.: Quart. Appl. Math. 8, 405 (1951).

[9] Handbook of Physical Constants, edited by BIRCH, SCHAIRER, and SPICER: Geol. Soc. of Am. Special Papers No. 36, January 31, 1942, "Density at High Pressure; Compressibility", (Francis Birch), Section 4.

[10] Rock Deformation (A Symposium), edited by DAVID GRIGGS and JOHN HANDIN: Geol. Soc. Am. Memoir 79, 1960, "Deformation of Rocks at 500° to 800°C", (D. T. GRIGGS, F. J. TURNER, and H. C. HEARD), p. 65ff.

# Précurseurs élastoplastiques d'ondes de choc en milieu solide isotrope

Par Jean Jacquesson

Université de Poitiers, France

## Résumé

La présence de précurseurs d'ondes de choc est liée aux propriétés élastiques et plastiques des solides. Ils modifient l'état du milieu où se propage l'onde de choc et l'on montre que cet état résulte alors d'un équilibre dynamique qui ne dépend que de l'amplitude du choc et des propriétés élastiques et plastiques du milieu.

Si l'on admet ces propriétés indépendantes de la vitesse de compression, et la compression adiabatique, on est amené à définir une courbe de Hugoniot généralisée ou «adiabatique dynamique pratique» dont la connaissance donne la réponse du solide à n'importe quel choc en tenant compte de la présence du précurseur. On montre en outre qu'à son apparition, le précurseur se discerne mal de l'onde de choc, et que pour la même raison un choc faible se discerne mal de son précurseur; enfin on montre que l'existence de ce précurseur entraîne un autoamortissement de l'onde de choc.

## 1. Introduction

Lors des études de compressibilité des métaux par ondes de choc les pressions élevées mises en jeu ont fait schématiser à l'extrême les phénomènes pouvant avoir lieu au pied du choc du fait de leur faible pression, de l'ordre de la limite élastique. On ne peut plus négliger entièrement leur influence pour comprendre le processus physique de compression du solide car ce sont eux qui établissent l'état réel du milieu où se propage l'onde de choc.

On considère d'une manière usuelle une onde plane se propageant en milieu infini. L'absence de dilatation latérale rend le mouvement unidimensionnel et la compression étant supposée adiabatique, deux grandeurs suffisent à définir l'état du milieu: la contrainte $P$ normale à l'onde, qui définira la pression, et le volume spécifique $V$.

Pour un état initial donné, les conditions classiques de HUGONIOT permettent de tracer dans le plan $(P, V)$ la courbe de HUGONIOT donnant les états compatibles avec l'état initial à travers une onde de choc stable, compte tenu de l'équation d'état du milieu.

La forme des courbes de HUGONIOT est donnée qualitativement, fig. 1, pour un solide ne présentant pas de changement de phase thermodynamique; si cette forme est bien définie expérimentalement pour les pressions élevées, assez peu de données existent, par contre, aux pressions inférieures, notamment au voisinage de la limite élastique; la forme de la courbe dans ce domaine a une base essentiellement théorique. On admet la théorie du cisaillement maximum et l'on définit une limite élastique de HUGONIOT pour la pression $P_E$ réalisant ce cisaillement; si la pression dépasse cette valeur, la contrainte de cisaillement reste

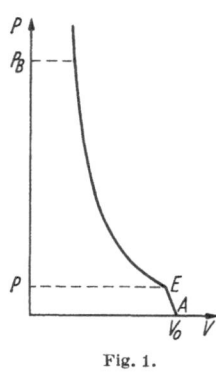

Fig. 1.

constante, on est conduit à admettre alors une brusque variation de la « compressibilité linéaire » du milieu [1] et l'on obtient une courbe présentant un point anguleux à concavité négative qui entraîne une décomposition du front de l'onde de choc (pour une amplitude inférieure à un certain seuil $P_B$).

L'état défini par ce point anguleux est établi devant le choc par une onde précurseur élastique, il est pris à priori comme l'état initial qui définit la courbe de HUGONIOT.

On peut faire deux objections à cette conception des phénomènes. Tout d'abord le point anguleux $P_E$ définit une transition entre, d'une part, des phénomènes élastiques réversibles et, d'autre part, des phénomènes élastiques non réversibles, or la condition de stabilité thermodynamique du choc fait intervenir la concavité de la courbe de compressibilité essentiellement isentropique [2], elle ne pourra donc pas s'appliquer sans quelques restrictions. En second lieu, si l'on considère un milieu polycristallin, pour chaque microcristal la contrainte de cisaillement maximum ne sera pas réalisée pour la même valeur de la pression; on ne pourra plus alors considérer macroscopiquement une brusque variation de la compressibilité pour la pression $P_E$, mais une variation progressive pour une plage plus au moins grande de pressions.

## 2. Hypothèses de définition du milieu

Pour essayer de serrer de plus près la réalité et également la rendre accessible au raisonnement thermodynamique on est amené à délimiter les propriétés du milieu par quelques hypothèses sur son comportement dynamique.

*A.* Nous admettrons tout d'abord qu'aux vitesses élevées de déformation et pour le temps très court que dure la compression, le comportement adiabatique du milieu est régi par une relation $P(V)$ biunivoque, indépendante du temps. Ceci implique en fait deux hypothèses:

a) Nous supposons que la durée des phénomènes de relaxation est nulle ou très faible devant celle de la compression qui leur donne naissance. L'équilibre thermodynamique est alors pratiquement réalisé à tout instant en tout point sans qu'il y ait nécessairement réversibilité du phénomène de compression.

b) Nous admettons en outre que durant le phénomène bref de la propagation, les gradients d'entropie qui peuvent se produire du fait des gradients de pression et de l'irréversibilité de la compression (en phase plastique) ne changent pas le caractère adiabatique local de la compression, c'est à dire n'entraînent pas d'échange de chaleur au sein du milieu. Nous mettons évidemment à part les ondes de choc elles-mêmes. En d'autres termes l'hypothèse d'un phénomène adiabatique est ici plus restrictive que par exemple en acoustique, où l'on peut considérer les compressions et détentes isentropiques. Il y a donc là une approximation qui variera selon la nature du milieu solide étudié, mais qui peut ne pas être négligeable. Elle peut expliquer des écarts entre des résultats théoriques et expérimentaux sans qu'il soit nécessaire de faire intervenir l'influence de la vitesse de déformation.

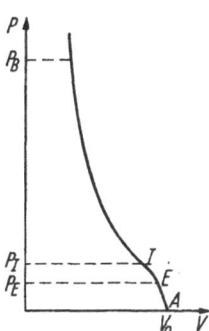

Fig. 2. Courbe de compressibilité unidimensionnelle adiabatique du solide étudié.

Nous avons pensé utile d'expliciter nettement ces hypothèses qui restent implicites dans la théorie de la propagation des ondes plastiques (sans effet de vitesse de déformation); le fait d'admettre directement que la contrainte ne dépend que de la déformation ne fait pas en effet bien apparaître les limites thermodynamiques d'une telle hypothèse.

*B.* La forme de la courbe $P(V)$, fig. 2, résulte de notre discussion d'introduction. Après la partie $AE$ rectiligne élastique la compression entraîne non plus un point anguleux mais une variation telle que $d^2P(V)/dV^2 < 0$ puis aux pressions élevées on aura nécessairement au delà d'une certaine valeur $P_I$ de la pression $d^2P(V)/dV^2 > 0$ sauf anomalies dues à des transformations thermodynamiques que nous n'envisageons pas ici.

Le comportement du solide en compression dynamique linéaire est entièrement défini par cette forme de la courbe $P(V)$. Sur la fig. 3 la courbe $P(V)$ n'est représentée en AEI que pour les pressions inférieures à $P_I$. A chaque point $G_1$ de cette courbe on peut en outre associer une adiabatique de HUGONIOT $H_1$ pour laquelle ce point définit l'état au pied du choc.

### 3. Précurseur élasto-plastique et adiabatique de Hugoniot généralisé

La stabilité des fronts d'onde de choc en milieu solide a déjà été étudiée d'une manière très générale par G. E. DUVALL pour des milieux définis par des courbes de compressibilité isentropique de formes quelconques [3]. Compte tenu des hypothèses que nous avons faites les résultats se transposent à un solide caractérisé par la courbe $P(V)$ définie ci-dessus, cette courbe remplace celle de compressibilité isentropique. Notamment la vitesse du son, définie comme celle d'un ébranlement $dP$ dans le milieu soumis à la pression $P$ est: $c(P) = V(-dP/dV)^{1/2}$, et dans le repère de LAGRANGE défini par le milieu initial (indice 0) au repos: $a(P) = V_0(-dP/dV)^{1/2}$. Les dérivations sont prises le long de la courbe $P(V)$. Ce résultat n'est autre d'ailleurs que celui de la théorie des ondes plastiques [8] pour les solides dont les propriétés sont indépendantes de la vitesse de déformation.

La stabilité du front de choc se retrouve directement sur cette dernière relation, un choc ne peut se détruire lorsque des pressions croissantes ont des vitesses croissantes, c'est à dire tant que $d^2P/dV^2 > 0$. Si, par contre $d^2P/dV^2 < 0$, la vitesse $a(P)$ décroît avec la pression; un front initialement droit s'étalera au cours de sa propagation. C'est le cas notamment pour tout front de pression inférieur à $P_I$ (fig. 3). Seul le front élastique, le plus rapide, gardera sa forme initiale ($d^2P/dV^2 = 0$), mais sans pour autant être un choc au sens thermodynamique. Ce front précèdera une onde centrée de compression. Par contre pour une pression $P_2$ supérieure à $P_I$, la condition de stabilité de choc est satisfaite; un front initialement droit se propagera sous forme d'une onde de choc mais si sa vitesse est inférieure à la vitesse élastique elle pourra être précédée d'une onde précurseur. C'est alors ce précurseur qui établira l'état du solide où se propagera en fait le choc; or la vitesse du choc dépend également de cet état par la relation classique $D = V_1[(P_2 - P_1)/(V_1 - V_2)^{1/2}$ ou les indices 1 et 2 rapportent aux états juste devant et derrière l'onde de choc. On voit donc que l'état amont 1 sera défini par un équilibre de vitesses, celle du choc devant être égale à celle de propagation de l'état 1. Exprimée par rapport à cet état 1, cette *condition de stabilité amont* s'écrit $D = c_1$ soit $(P_2 - P_1)/(V_2 - V_1) = dP/dV_1$ c'est à dire que dans le plan $(P, V)$ la droite de RAYLEIGH joignant les états amont et aval du choc doit être tangente à la courbe de compressibilité en l'état amont.

On peut alors définir géométriquement l'état $G_2$ (fig. 3) correspondant à la pression génératrice $P_2$ comme l'intersection de la tangente à l'état amont $G_1$ et de la courbe de Hugoniot $H_1$ correspondante.

Comme à chaque pression génératrice correspond un état amont différent, d'après la condition de stabilité, la courbe définissant les états stables de choc susceptibles d'être créés dans un solide initialement au

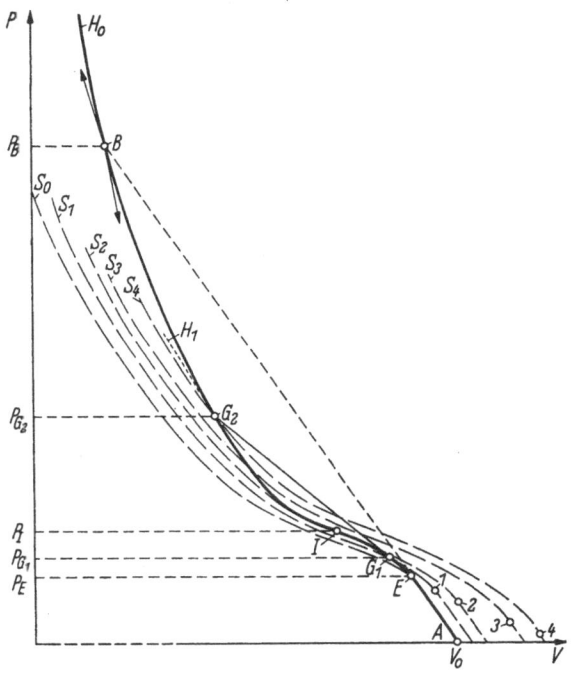

Fig. 3. Courbe de Hugoniot généralisée ou «adiabatique dynamique pratique» du solide étudié.

repos, c'est à dire le lieu du point $G_2$, ne correspond plus à la définition classique de l'adiabatique de Hugoniot. Nous appellerons cette courbe, mise en évidence il y a deux ans [3, 7, 9], *adiabatique de Hugoniot généralisée* ou *adiabatique dynamique pratique* du solide. Elle est évidemment inséparable de l'existence de l'onde précurseur élasto-plastique.

Si la pression croît, l'amplitude $P_1$ maximale du précurseur, au pied du choc, décroît et pour une valeur $P_B$ de la pression génératrice où la droite de Rayleigh atteint la pente de compression élastique $AE$, l'onde de choc devient plus rapide que l'onde élastique.

Le choc est alors sans précurseur et aux pressions supérieures à $P_B$ l'adiabatique pratique se confond avec la courbe de Hugoniot de l'état initial $A$. Au point $B$ correspondant à la pression $P_B$, on peut montrer [5] que l'adiabatique dynamique pratique présente un point anguleux dont les directions des tangentes sont celles des adiabatiques de HUGONIOT correspondant aux états $A$ et $E$. Ce point anguleux n'a pas d'influence sur la stabilité du choc.

Si la pression génératrice $P_2$ décroît, $G_2$ décrit la courbe adiabatique dynamique pratique tandis que $G_1$ s'élève en pression le long de la courbe de compressibilité adiabatique; l'amplitude $P_1$ du précurseur au pied du choc est d'autant plus élevée que la pression génératrice du choc est plus faible, puis il y a disparition du choc pour une pression $P_I$ correspondant à l'état $I$ défini par l'intersection des deux courbes précédentes. Nous allons montrer qu'en ce point l'entropie reste stationnaire et que les tangentes aux deux courbes sont les mêmes, c'est-à-dire que leur jonction se fait suivant un point d'inflexion.

En effet, on sait [6] que l'entropie reste stationnaire en un point situé entre $G_1$ et $G_2$ sur la droite de RAYLEIGH $G_1 G_2$. Lorsque $G_1$ et $G_2$ tendent ensemble vers $I$ la droite $G_1 G_2$ reste tangente à la courbe de compressibilité adiabatique (condition de stabilité amont) et quand ces points se confondent en $I$, la propriété de la droite de RAYLEIGH se retrouve en ce point $I$ le long de la courbe de compressibilité adiabatique. L'entropie y est donc stationnaire et cette courbe est tangente à une isentropique.

D'autre part, une adiabatique de HUGONIOT est osculatrice à l'isentropique passant par le point définissant l'état amont du choc [6]. Si l'on tient compte de la propriété démontrée précédemment et de la construction par point de l'adiabatique dynamique pratique, on voit que lorsque $G_1$ tend vers $I$, cette adiabatique dynamique pratique s'identifie à l'adiabatique de HUGONIOT du point $I$. Par suite adiabatique de HUGONIOT, adiabatique dynamique pratique, isentropique et adiabatique de compressibilité ordinaire sont tangentes en $I$.

D'une manière générale, on appellera *adiabatique dynamique pratique* l'ensemble de la courbe $A E I B H$ (fig. 3). Celle-ci présentera en fait trois parties caractérisant des propriétés différentes du milieu pour la propagation des ondes de compression:

a) De $A$ à $I$ c'est la courbe de compressibilité adiabatique simple et les pressions correspondent aux ondes élastiques et plastiques ordinaires sans possibilité de choc, c'est le domaine des pressions pouvant exister dans les précurseurs.

b) Au-dessus du point d'inflexion $I$ et jusqu'au point $B$, la pression induit un front de choc stable, précédé d'un précurseur élasto-plastique dont l'amplitude maximale $P_1$ est fixée par celle du front de choc $P_2$.

c) Aux pressions supérieures à $P_B$, aucun précurseur ne peut précéder l'onde de choc qui est alors supersonique, on est sur l'adiabatique de HUGONIOT de l'état initial.

La connaissance de l'adiabatique dynamique pratique correspondant à l'état initial du solide permet de connaître exactement la forme des fronts d'ondes qui peuvent s'y propager.

## 4. Evolution de la pression créée par le précurseur

Considérons une pression $P_2$ appliquée à l'instant $t = 0$ et maintenue en un plan origine du milieu, sa grandeur sera telle qu'elle donne naissance à un précurseur ($P_1 < P_2 < P_B$). Nous voulons étudier la pression en fonction du temps $t$ en un plan physique d'abscisse $x$, où par exemple se trouve un capteur de pression. Nous supposerons connue l'adiabatique dynamique pratique du solide.

D'après les résultats établis au paragraphe 3, il se créera une onde de choc précédée d'un précurseur dont l'amplitude maximale $P_1$ sera déterminée par la pression $P_2$. L'équation de ce précurseur peut s'écrire en coordonnées de LAGRANGE:

$$x = a(P)t \qquad \text{pour} \qquad P < P_1$$

avec $a^2(P) = -V_0^2(dP/dV)$, $x$ donnant l'abscisse à l'instant $t$ du plan où règne la pression $P$. La dérivation est faite, comme nous l'avons déjà vu, le long de l'adiabatique élasto-plastique. De ces relations, on tire

$$\frac{V_0^2 t^2}{x^2} = -\frac{dV}{dP},$$

et en dérivant par rapport au temps,

$$\frac{dP}{dt} = -\frac{2t V_0^2}{x^2}\left(\frac{d^2 V}{dP^2}\right)^{-1}.$$

Pour le front élastique, où $d^2V/dP^2 = 0$ on voit que le front reste droit.

Pour l'onde élasto-plastique ($P_E < P < P_1$) on a vu $d^2P/dV^2 < 0$, ce qui revient a $d^2V/d^2P < 0$, d'où $dP/dt > 0$; la pression croît avec le temps suivant la loi:

$$\frac{dP}{dt} = -\frac{2V_0^2}{x a(P)}\left(\frac{d^2 V}{dP^2}\right)^{-1}$$

qui ne dépend que de la courbe de compressibilité. Or suivant l'amplitude du précurseur $d^2V/dP^2$ peut varier d'une valeur nulle (point $E$, fig. 3) à une autre valeur nulle (point $I$, choc évanescent) donc $|dP/dt|$ après le front élastique droit décroît, passe par une valeur minimum pour $P = P_m$, puis croît à nouveau et tend vers l'infini.

Pour une pression génératrice $P_2$, on n'observera de cette évolution que la partie correspondant aux pressions inférieures à $P_1$, pression maximale imposée dans le précurseur par la pression génératrice.

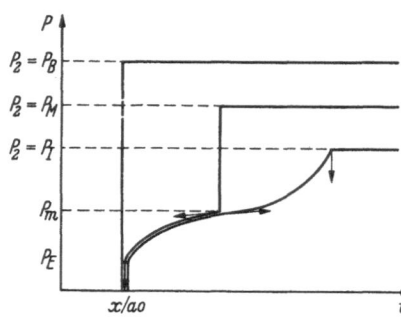

Cette évolution est représentée fig. 4 pour quelques valeurs particulières de cette pression génératrice.

On s'aperçoit alors que suivant son amplitude le choc se dégagera plus ou moins nettement de son précurseur selon que $|dP/dt|$ sera plus ou moins faible au pied du choc. Pour des pressions tendant vers $P_B$, où le choc a une vitesse peu inférieure à celle du front élastique, de même qu'aux pressions tendant vers $P_I$ où le choc est évanescent, $|dP/dt|$ tend vers l'infini; pratiquement le pied du choc se confond avec le choc, on peut dire encore que le pied du choc s'arrondit. C'est pour la pression génératrice $P_M$ imposant la pression amont $P_1 = P_m$ que ce pied sera le plus discernable, et cela d'autant plus que $x$ sera plus grand, c'est-à-dire que l'on sera plus loin du choc initial.

Fig. 4. Pression en fonction du temps en un plan $X$ du solide pour différentes valeurs de la pression $P_2$ génératrice du choc. La pente $|dP/dt|$ au pied du choc passe par un minimum pour $P_2 = P_M$ valeur pour laquelle le «pied» du choc sera le plus net.

## 5. Influence d'un précurseur sur l'amortissement d'une onde de choc

L'énergie existant à un instant donné dans un précurseur d'amplitude donnée $P_1$, se calcule en considérant le travail absorbé par chaque tranche du milieu.

Sous forme potentielle:

$$dE_p = dx \int_1^{\frac{V}{V_0}(P)} - P\,dV.$$

Sous forme cinétique:

$$dE_c = \frac{dx}{V_0} \frac{W^2(P)}{2},$$

où $W(P)$ représente la vitesse d'entraînement matérielle de la tranche et ne dépend que de la pression $\left( W = \int_0^P V_0 \frac{dP}{a(P)} \right)$. Or à l'instant $t$, pour un front initialement droit, on peut exprimer $dx$ en fonction de l'accroissement de pression $dP$ :

$$dx = \frac{da(P)}{dP} t\, dP$$

d'où

$$E = -\int_0^{P_1} (dE_P + dE_c) = -t \int_0^{P_1} \frac{da(P)}{dP} \cdot \left[ \frac{W^2(P)}{2V_0} - \int_1^{\frac{V}{V_0}(P)} P\, dV \right] dP.$$

Comme $da/dP < 0$ l'énergie contenue dans le précurseur est une fonction croissante de la pression maximale $P_1$ qui règne au pied du choc. Mais on voit en outre que pour une pression génératrice $P_2$ maintenue, c'est-à-dire $P_1$ constante, cette énergie est proportionnelle au temps.

Il y a donc transfert d'énergie de l'onde de choc vers le précurseur, cette énergie servant en fait à maintenir l'état amont du choc, c'est-à-dire à réaliser la condition de stabilité amont.

Une autre conséquence importante de l'absorption d'énergie par le précurseur est son rôle dans l'amortissement des ondes de choc réalisées en pratique. Une onde de choc dans un solide est créée soit par un projectile soit par un explosif; l'énergie fournie pour induire le choc est limitée et la pression décroît très rapidement derrière le front de l'onde de choc.

Divers facteurs jouent déjà intrinsèquement pour amortir cette onde de choc, notamment son irréversibilité, ainsi que la détente arrière plus rapide. Parmi ces facteurs on doit ajouter le précurseur puisqu'il absorbe une énergie croissante au cours du temps.

Son influence peut prendre une importance notable car elle s'accentue au cours de la propagation. En effet l'onde de choc qui s'amortit voit l'amplitude du précurseur s'accroître et, par suite, l'effet de ce dernier sur l'amortissement s'accroît. La présence du précurseur entraîne un processus d'autoamortissement de l'onde de choc.

## 6. Conclusions

Lorsque l'on se place dans un milieu ne présentant pas d'effet de vitesse de déformation, la présence, pour une certaine plage de pressions, d'un précurseur élastoplastique oblige à élargir la notion d'adiabatique de HUGONIOT.

Le choc et son précurseur sont entièrement définis par une « courbe de HUGONIOT généralisée » ou « courbe adiabatique dynamique pratique » dont les propriétés résultent d'un équilibre entre la vitesse du choc et celle de l'état amont créé par le précurseur. Cet état amont, défini par la pression génératrice aval, ne se maintient que grâce à un apport d'énergie du front de choc dans le précurseur.

La présence de ce précurseur peut donc jouer un rôle considérable dans l'amortissement d'une onde de choc, d'autant plus que son influence s'accroît par un phénomène de réaction.

Les hypothèses qui nous ont servi à définir le solide sont celles admises pour la propagation des ondes plastiques lorsque les propriétés du milieu sont indépendantes de la vitesse de compression; mais nous avons montré qu'avec ces hypothèses il était plus difficile de considérer comme adiabatiques les compressions dans l'onde plastique, en d'autres termes que cette dernière hypothèse rendait plus restrictive l'application des résultats théoriques aux résultats expérimentaux.

### Bibliographie

[1] RICE, M. H., R. G. McQUEEN, and J. M. WALSH: Solid State Physics, Vol. VI, SEITZ et TURNBULL edit., New York: Academic Press 1958, p. 1—60.
[2] COURANT, R., and K. O. FRIEDRICHS: Supersonic flow and shock waves, New York: Interscience Publishers 1948, p. 142.
[3] DUVALL, G. E.: Les ondes de détonation, C.N.R.S., Paris 1962, pp. 337—353.
[4] COURANT, R., and K. O. FRIEDRICHS: op. cit., p. 243.
[5] JACQUESSON, J.: Thèse, Université de Poitiers, 1962.
[6] COURANT, R., and K. O. FRIEDRICHS: op. cit., p. 141—146.
[7] JACQUESSON, J.: Les ondes de détonation, C.N.R.S., Paris 1962, p. 415—422.
[8] v. KÁRMÁN, TH., and P. DUWEZ: J. Appl. Phys. 21, 987 (1950).
[9] MANDEL, J.: Séminaire de Plasticité, Publ. Scient. et Techn. du Ministère de l'Air No. N.T. 116, p. 167—178.

# Propagation of Magnetoelastic Disturbances in Viscoelastic Bodies

By Sylwester Kaliski and Witold Nowacki

Polish Academy of Sciences, Warsaw, Poland

## 1. Introduction

In the present paper, we formulate the equations of magnetoviscoelasticity and appropriate boundary conditions for a linearly viscoelastic medium with finite conductivity. In particular, we shall consider the boundary conditions for an ideal medium, that is perfectly elastic and perfectly conductive.

In magnetohydrodynamics, the boundary conditions on the surface whose normal is parallel to the vector of the initial magnetic field, assume different forms depending on the manner in which the transition to vanishing mechanical and magnetic viscosities is effected [1, 2].

The same problem arises in magnetoviscoelasticity for the various models of viscoelastic bodies. In particular, the boundary conditions on the contact surface of a liquid and solid body must be discussed.

If, in a boundary problem which concerns the contact of a liquid and a solid body, we start with the equations of the ideal liquid and the elastic solid, and assume perfect conductivity of both media, we obtain a unique form of the boundary conditions. Surface waves on the contact surface of a liquid and a solid were discussed in this manner in [3].

If, on the other hand, we start with the equations for viscous media with finite conductivity and then pass to the limits corresponding to the elastic body and the ideal liquid of perfect conductivity, we obtain different forms of the boundary conditions, which depend on the manner in which the transition to vanishing mechanical and magnetic viscosities has been effected.

In Section 2 of this paper we present general and simplified equations for conductors with finite electric conductivity. Section 3 is devoted to general boundary conditions, while Section 4 deals with the transition to the ideal elastic conductor. In Section 5, we present the forms of the limit passage for different models and for contact between a liquid and a solid when the magnetic field vector is normal to the contact surface.

Finally, in Section 6, we consider a simple example of the transmission
of an elastic wave through the contact surface of the two media under
various types of boundary conditions.

## 2. General Equations

Let us consider a linear viscoelastic medium with finite electric conductivity and suppose that an initial magnetic field exists in this medium.
The action of body forces and external loads produces not only a deformation field but also a coupled electromagnetic field. We assume that
the medium is isotropic and homogeneous and disregard the coupled
thermal effects as well as the effects of the relaxation of the magnetic
and electric induction. These effects have been considered in [4, 5].

The points of departure of our consideration are the equations of
magneto-viscoelasticity which after being linearized [5] assume the
following form:

$$\operatorname{curl} E = -\frac{\mu_0}{c} \frac{\partial h}{\partial t},$$

$$\operatorname{curl} h = \frac{4\pi}{c} j + \frac{\varepsilon}{c} \frac{\partial E}{\partial t} + \frac{\varepsilon \mu_0 - 1}{c^2} \frac{\partial}{\partial t} \left( \frac{\partial u}{\partial t} \times H \right),$$

$$j = \eta \left[ E + \frac{\mu_0}{c} \left( \frac{\partial u}{\partial t} \times H \right) \right], \tag{2.1}$$

$$\operatorname{div} h = 0, \qquad \operatorname{div} D = 0,$$

$$D = \varepsilon \left[ E + \frac{1}{c} \frac{\mu_0 \varepsilon - 1}{\varepsilon} \left( \frac{\partial u}{\partial t} \times H \right) \right].$$

$$\int_0^t \left\{ a(t - \tau) \frac{\partial}{\partial \tau} \nabla^2 u + [a(t - \tau) + b(t - \tau)] \frac{\partial}{\partial \tau} \operatorname{grad} \operatorname{div} u \right\} d\tau$$

$$+ \frac{\mu_0}{c} (j \times H) + X = \varrho \frac{\partial^2 u}{\partial t^2}. \tag{2.2}$$

The system comprises the equations of electrodynamics of slowly
moving media. The vectors $h$ and $E$ indicate the magnetic and electric
fields, respectively, $j$ is the vector of current density, $D$ the vector of
electric induction, $H$ the vector of the initial constant magnetic field,
$c$ the velocity of light in vacuum; $\mu_0$ and $\varepsilon$ are the magnetic and electric
permeabilities, and $\eta$ is the electric conductivity.

Eqs. (2.2) are the equations of motion of a viscoelastic medium.
Here, $u$ denotes the displacement vector, $X$ the body force per unit
volume, while $a(t)$ and $b(t)$ are relaxation functions of the viscoelastic

medium, and $\varrho$ is its density. For $\boldsymbol{H} = 0$, Eqs. (2.2) reduce to the well-known equations of viscoelasticity [6].

Restricting our considerations to good conductors we disregard the displacement currents. Thus, for $\varepsilon = \mu_0 = 1$, neglecting the term $(\mu_0/c) \partial \boldsymbol{E}/\partial t$ and eliminating the magnitudes of $\boldsymbol{E}$ and $\boldsymbol{j}$, we obtain the following set of equations

$$\frac{\partial \boldsymbol{h}}{\partial t} - \operatorname{curl}\left(\frac{\partial \boldsymbol{u}}{\partial t} \times \boldsymbol{H}\right) - \lambda_m \nabla^2 \boldsymbol{h} = 0, \qquad \lambda_m = \frac{c^2}{4\pi\mu_0\eta}$$

$$\int_0^t \left\{ a(t-\tau) \frac{\partial}{\partial t} \nabla^2 \boldsymbol{u} + [a(t-\tau) + b(t-\tau)] \frac{\partial}{\partial t} \operatorname{grad} \operatorname{div} \boldsymbol{u} \right\} d\tau$$

$$+ \boldsymbol{X} + \frac{\mu_0}{c} (\boldsymbol{j} \times \boldsymbol{H}) = \varrho \frac{\partial^2 \boldsymbol{u}}{\partial t^2}, \qquad (2.3)$$

where

$$\boldsymbol{E} = \frac{\lambda_m}{c} \operatorname{curl} \boldsymbol{h} - \frac{\mu_0}{c}\left(\frac{\partial \boldsymbol{u}}{\partial t} \times \boldsymbol{H}\right), \qquad \boldsymbol{j} = \frac{c}{4\pi} \operatorname{curl} \boldsymbol{h},$$

$\lambda_m$ is the magnetic viscosity.

As in problems of viscoelasticity, an elastic-viscoelastic analogy can be formulated here. Suppose that external loads and body forces have been applied at the moment $t = 0+$, and that all load and body forces are proportional to the same function of time. After performing the one-sided LAPLACE transform, we may write Eqs. (2.3) in the form

$$p[\overline{\boldsymbol{h}} - \operatorname{curl}(\overline{\boldsymbol{u}} \times \boldsymbol{H})] - \lambda_m \nabla^2 \overline{\boldsymbol{h}} = 0,$$

$$\overline{\mu} \nabla^2 \overline{\boldsymbol{u}} + (\overline{\mu} + \overline{\lambda}) \operatorname{grad} \operatorname{div} \overline{\boldsymbol{u}} + \frac{\mu_0}{c} (\overline{\boldsymbol{j}} \times \boldsymbol{H}) + \overline{\boldsymbol{X}} = \varrho p^2 \overline{\boldsymbol{u}}, \qquad (2.4)$$

where

$$\overline{\mu}(p) = p\,\overline{a}(p), \qquad \overline{\lambda}(p) = p\,\overline{b}(p),$$

$\overline{a}$, $\overline{b}$ being the LAPLACE transformation of the relaxation functions $a(t)$ and $b(t)$.

Let us compare Eqs. (2.4) with the corresponding equations of magnetoelasticity:

$$p[\overline{\boldsymbol{h}}_0 - \operatorname{curl}(\overline{\boldsymbol{u}}_0 \times \boldsymbol{H})] - \lambda_m \nabla^2 \overline{\boldsymbol{h}}_0 = 0,$$

$$\mu \nabla^2 \overline{\boldsymbol{u}}_0 + (\mu + \lambda) \operatorname{grad} \operatorname{div} \overline{\boldsymbol{u}}_0 + \frac{\mu_0}{c} (\overline{\boldsymbol{j}}_0 \times \boldsymbol{H}) + \overline{\boldsymbol{X}} = \varrho p^2 \overline{\boldsymbol{u}}_0, \qquad (2.5)$$

where $u_0$, $h_0$, $j_0$, etc., refer to perfectly elastic bodies of finite electric conductivity. The comparison shows that instead of solving a magneto-viscoelastic problem we may solve the "associated" magnetoelastic problem. When its solution has been obtained, the Lamé constants $\mu$, $\lambda$ are replaced by the quantities $\bar{\mu}$, $\bar{\lambda}$, which are the functions of the parameter $p$; the inverse Laplace transform of $u_0$ then furnishes the solution of the magnetoviscoelastic problem. This analogy, however, is only valid when the load and body forces as well as the displacements in the boundary conditions of the problem are proportional to the same function of time.

## 3. Boundary Conditions

Let us now discuss the boundary conditions for two viscoelastic media of finite electric conductivity with a plane contact surface. Suppose that the vector of the initial magnetic field is arbitrarily oriented with respect to the plane of the contact. Taking into account the homogeneous initial conditions assumed in Section 2, we establish the boundary conditions for the Laplace transforms of the mechanical and electromagnetic quantities.

Two types of boundary conditions will be considered:

(a) boundary conditions for two viscoelastic media in contact, and

(b) boundary conditions for a viscoelastic medium in contact with a vacuum.

In case (a) we obtain a set of twelve boundary conditions (see [7]). The indices (1) and (2) denote the two media, and $n$ the unit vector normal to the plane of contact. We have

$$\bar{u}^{(1)} = \bar{u}^{(2)}, \qquad \bar{\sigma}_{ni}^{(1)} + \bar{T}_{ni}^{(1)} = \bar{\sigma}_{ni}^{(2)} + \bar{T}_{ni}^{(2)},$$

$$\bar{D}_n^{(1)} = \bar{D}_n^{(2)}, \qquad \bar{b}_n^{(1)} = \bar{b}_n^{(2)},$$

$$[n \times (\bar{E}^{(2)} - \bar{E}^{(1)})] = \frac{p\bar{u}_n}{c} (\mu_0^{(2)} - \mu_0^{(1)})H_t,$$

$$[n \times (\bar{h}^{(2)} - \bar{h}^{(1)})] = 0, \tag{3.1}$$

where

$$\bar{\sigma}_{ik} = \bar{\mu}(\bar{u}_{i,k} + \bar{u}_{k,i}) + \delta_{ik}\bar{u}_{k,k},$$

$$\bar{T}_{ik} = \frac{\mu_0}{4\pi} [H_i\bar{h}_k + H_k\bar{h}_i - \delta_{ik}\bar{h} \cdot H],$$

$$B = B_0 + b = \mu_0(H + h). \tag{3.2}$$

The system (3.1) contains the conditions of continuity of the transforms of the displacements, the conditions of the equality of the transforms of the sums of the mechanical stresses and the components of Maxwell's tension tensor, and the conditions of continuity of the transforms of the normal inductions and tangential fields for the moving boundary surface. The last two conditions are given in linearized form using an approximation for the terms in $v/c$ and without introducing two-sided values of the tangent field $H_t$. If the quantities $\mu^{(\alpha)}$, $\alpha = 1.2$ of the media are approximately equal, then the term connected with the motion of the contact surface may be disregarded in (3.1). If, in the boundary conditions of the type (b), the medium characterized by subscript 1 is the vacuum, we must set

$$\mu_0^{(1)} = 1, \quad \sigma_{ik}^{(1)} = 0, \quad b_n^{(1)} = h_n^{(1)}, \quad D_n^{(1)} = E_n^{(1)} \quad \text{in} \quad (3.1).$$

## 4. Limit Passage to the Ideal Conductor

Lettering $\lambda_m \to 0$ in (2.3) and replacing the relaxation functions $a(t)$, $b(t)$ by the LAMÉ constants $\mu$, $\lambda$, we pass from a viscoelastic body with finite electric conductivity to the perfectly elastic and perfectly conducting body. Eliminating the quantities $E$ and $h$ from (2.3), we then obtain the following equations:

$$\mu \nabla^2 u + (\lambda + \mu)\, \text{grad div}\, u + \frac{\mu_0}{4\pi}\, [\text{curl curl}\, (u \times H)] \times H$$

$$+ X = \varrho\, \frac{\partial^2 u}{\partial t^2}, \tag{4.1}$$

where

$$E = -\frac{\mu_0}{c} \left( \frac{\partial u}{\partial t} \times H \right), \qquad h = \text{curl}\, (u \times H). \tag{4.2}$$

For a perfectly conducting viscoelastic body, it is sufficient to assume that $\lambda_m \to 0$. We then obtain the equation of motion

$$\int_0^t \left\{ a(t - \tau)\, \frac{\partial}{\partial \tau}\, \nabla^2 u + [a(t - \tau) + b(t - \tau)]\, \frac{\partial}{\partial \tau}\, \text{grad div}\, u \right\} d\tau$$

$$+ \frac{\mu_0}{4\pi}\, [\text{curl curl}\, (u \times H)] \times H + X = \varrho\, \frac{\partial^2 u}{\partial t^2}, \tag{4.3}$$

the boundary conditions (4.2) remaining valid.

Note that with $\mu = 0$ and $\partial u/\partial t \not\equiv 0$ Eq. (4.1) reduces to the acoustic equation of an ideal fluid. Similarly, taking coefficients $a(t)$, $b(t)$ in (4.3) that correspond to VOIGT's model with $\bar{a} = 0$, we obtain the acoustic equation of a viscous liquid.

Passing from a body with mechanical and magnetic viscosities to a body that is perfectly conductive and elastic, we must also make the appropriate transitions to the limit in the boundary conditions. Here, the manner in which the ratio of the magnetic and mechanical viscosities in the liquid and in the solid tends to zero is essential. This concerns the coefficients of viscosity when the initial field $H$ is perpendicular to the contact surface (see Section 5).

In the present section, we give the boundary conditions for the case when the vector of the initial magnetic field is parallel to the contact plane. We consider two basic types of contact

(a) The displacements of the media are continuous across the contact surface. We then have the following boundary conditions:

$$\overline{u}^{(1)} = \overline{u}^{(2)}, \qquad \overline{\sigma}_{ni}^{(1)} + \overline{T}_{ni}^{(1)} = \overline{\sigma}_{ni}^{(2)} + \overline{T}_{ni}^{(2)}, \tag{4.4}$$

where the quantities $\overline{T}_{ni}$ are given by (3.2) and the quantities $\overline{h}_i$ by (4.2). The quantities $\overline{T}_{ik}$ can be then expressed explicitly in terms of the displacements $\overline{u}_i$.

(b) In the contact plane there appears the ideal tangential slip. The boundary conditions now take the form

$$\overline{\sigma}_{33}^{(1)} + \overline{T}_{33}^{(1)} = \overline{\sigma}_{33}^{(2)} + \overline{T}_{33}^{(2)}, \qquad \overline{u}_n^{(1)} = \overline{u}_n^{(2)},$$

$$\overline{\sigma}_{ni}^{(1)} = 0, \qquad \overline{\sigma}_{ni}^{(2)} = 0, \qquad i = 1,2. \tag{4.5}$$

The second condition (4.5) can be replaced by the relation between the tangential components of the vectors $h^\alpha$ ($\alpha = 1, 2$) and the surface currents [7].

## 5. Passage to the Limit in Boundary Conditions for $H = H_3$

If the initial magnetic field is perpendicular to the contact surface then the boundary conditions assume one of two forms depending on how the mechanical and magnetic viscosities tend to zero. These boun-

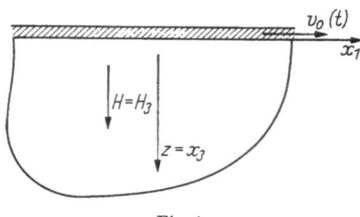

Fig. 1.

dary conditions will be discussed for a particular one-dimensional problem with shear strains since these influence the alternative forms of the boundary conditions.

Let us consider a rigid plate on the surface of a viscoelastic semi-space with finite electric conductivity. This plate is set in motion in a tangential

direction, the initial conditions being assumed homogeneous (Fig. 1). The equations of motion (2.3), then contain the function $a(t)$ and the dependent variables $u_1(z, t) = u$ and $h_1(z, t) = h$. The remaining quantities either are equal to zero or do not enter the problem. Thus, the system (2.3) takes the form

$$\lambda_m \frac{\partial^2 h}{\partial z^2} - \frac{\partial h}{\partial t} + H \frac{\partial^2 u}{\partial t \partial z} = 0,$$

$$\int_0^t a(t - \tau) \frac{\partial^3 u}{\partial z^2 \partial \tau} \, d\tau + \frac{H}{4\pi} \frac{\partial h}{\partial z} = \varrho \frac{\partial^2 u}{\partial t^2}, \qquad H_3 = H.$$

(5.1)

Neglecting the inertia term in the equations of motion, we treat the problem as quasi-static. This does not influence the physical conditions on the boundary. After performing the Laplace transform on Eqs. (5.1), the initial conditions being assumed homogeneous, and after solving these equations we obtain the following relations:

$$\bar{u} - \bar{u}_\infty = C_1 e^{-\bar{\varkappa} z},$$

$$\bar{h} - \bar{h}_\infty = \frac{H\bar{\varkappa}}{a_0^2 \bar{v}^2} C_1 e^{-\bar{\varkappa} z},$$

(5.2)

where

$$\bar{\varkappa}^2 = \frac{p}{\lambda_m} (1 + a_0^2 \bar{v}^2), \qquad a_0^2 = \frac{H^2}{4\pi\varrho}, \qquad \bar{v}^2 = \frac{\varrho}{\bar{\mu}},$$

$$\bar{\mu} = \bar{a} \, p, \qquad \bar{u}_\infty = \bar{u}(z, p)|_{z=\infty}, \qquad \bar{h}_\infty = \bar{h}(z, p)|_{z=\infty}.$$

Making use of the boundary condition for $z = 0$,

$$\bar{v}(0, p) = \bar{v}_0,$$

(5.3)

we obtain

$$\bar{v} - \bar{v}_\infty = (\bar{v}_0 - \bar{v}_\infty) e^{-\bar{\varkappa} z},$$

(5.4)

$$\bar{h} - \bar{h}_\infty = \frac{H\bar{\varkappa}}{a_0^2 \bar{v}^2} (\bar{v}_0 - \bar{v}_\infty) e^{-\bar{\varkappa} z}.$$

(5.5)

Here, $\bar{v} = p\bar{u}$ is the transform of the velocity.

It follows from relations (5.4) and (5.5) that

$$\bar{v}_0 - \bar{v}_\infty = \frac{a_0^2 \bar{v}^2}{H} \left[ \frac{\lambda_m p}{1 + a_0^2 \bar{v}^2} \right]^{\frac{1}{2}} (\bar{h}_0 - \bar{h}_\infty).$$

(5.6)

Eq. (5.6) is essential for the discussion of possible combinations of boundary conditions when the mechanical and magnetic viscosities tend to zero. For a solid body with $\bar{\mu} \neq 0$, when the magnetic viscosity $\lambda_m$ as well as the parameters characterizing the mechanical viscosity tend simultaneously to zero, condition (5.6) always yields

$$\bar{v}_0 = \bar{v}_\infty,$$

(5.7)

that is the continuity of tangential velocities. For the Voigt model with $\bar{\mu} = \mu(1 + p\tau)$, $\tau = \beta/\mu$, where $\beta$ is the coefficient of viscosity, Eq. (5.6) takes the form

$$\bar{v}_0 - \bar{v}_\infty = \frac{a_0^2 v_0^2}{H} \sqrt{\frac{p\lambda_m}{(1 + p\tau)(1 + p\tau + a_0^2 v_0^2)}} (\bar{h}_0 - \bar{h}_\infty), \quad v_0^2 = \frac{\varrho}{\mu}. \quad (5.8)$$

It is apparent from (5.8) that in the limit we obtain relation (5.7) when $\lambda_m$ and $\tau$ tend to zero in such a way that their ratio remains constant. For the MAXWELL model, we have $\bar{\mu} = \frac{p\tau}{1 + p\tau}$, where $\tau = \beta/\mu$. Eq. (5.6) therefore furnishes

$$\bar{v}_0 - \bar{v}_\infty = \frac{a_0^2 v_0^2}{H} \sqrt{\frac{\lambda_m(1 + p\tau)^2}{\tau[p\tau(1 + a_0^2 v_0^2) + a_0^2 v_0^2]}} (\bar{h}_0 - \bar{h}_\infty). \quad (5.9)$$

If the process is assumed, in the limit, to be stationary, (5.9) becomes

$$v_0 - v_\infty = \sqrt{\frac{\lambda_m}{4\pi\beta}} (h_0 - h_\infty). \quad (5.10)$$

It is apparent that when $\lambda_m \to 0$ while $\beta \neq 0$, in the limit we obtain the boundary condition (5.7). If $\beta \to 0$, then according to the three possibilities

$$\frac{\lambda_m}{\beta} \to 0, \quad \frac{\lambda_m}{\beta} \to \infty, \quad \frac{\lambda_m}{\beta} \to s, \quad (5.11)$$

where $s$ is a constant, we obtain the following boundary conditions:

$$\bar{v}_0 = \bar{v}_\infty, \quad \bar{h}_0 = \bar{h}_\infty, \quad \bar{v}_0 - \bar{v}_\infty = \sqrt{\frac{s}{4\pi}} (\bar{h}_0 - \bar{h}_\infty) \quad (5.12)$$

respectively. The variant $\beta \to 0$ is physically artificial, because it implies a vanishing time of relaxation, and consequently excludes elastic stresses. For the stationary laminar flow of a viscous liquid, (5.8) yields

$$v_0 - v_\infty = \sqrt{\frac{\lambda_m}{4\pi\beta'}} (h_0 - h_\infty). \quad (5.13)$$

Here, $\beta'$ denotes the coefficient of viscosity of the liquid. For $\beta' \to 0$ the liquid becomes ideal. Depending on the manner in which $\lambda_m$ and $\beta'$ tend to zero [see (5.11)], we obtain three variants of the boundary conditions corresponding to the physically significant cases.

When two media, for instance a viscous liquid and a viscoelastic solid are in contact, we obtain the following equations for the MAXWELL model (here the process may also be stationary)

$$v(z) - v_\infty = \sqrt{\frac{\lambda_m}{4\pi\beta}} \, (h(z) - h_\infty),$$

$$v(z) - v_\infty = \sqrt{\frac{\lambda_m}{4\pi\beta'}} \, (h(z) - h_\infty). \tag{5.14}$$

The first of these equations concerns the solid with the viscosity $\beta$, and the second the viscous liquid with the viscosity $\beta'$. It follows from these equations that on the contact surface $z = 0$ of these media the variants (5.12) of the boundary conditions may occur.

We conclude that a condition of the type $h^{(1)} = h^{(2)}$ may obtain, provided the surface currents vanish in the transition from the visco-elastic to the perfect conductor. The disappearance of the surface currents is identical with the disappearance of the tangential component of the stress tensor. This case occurs for a liquid, for $\beta'/\lambda_m \to 0$. It also appears for the MAXWELL model for $\beta/\lambda_m \to 0$; this case, however, should be regarded as unrealistic.

## 6. Example

The simple example that will now be discussed, illustrates the extreme alternatives of the boundary conditions. Consider two perfectly conductive media: an ideal liquid and a perfectly elastic solid, and suppose the initial magnetic field to be perpendicular to their plane of contact. Assume moreover that a tangential force, varying harmonically with time, acts in this plane. We consider the one-dimensional problem in the $x, z$ plane.

The liquid in the negative half-space $z < 0$ is denoted by the index (2) and the elastic solid in the half-space $z \geq 0$ by the index (1). The magnetic permeability of both media is assumed to be 1.

The equations of motion of the media take the form

$$\frac{\partial^2 u}{\partial x^2} - \frac{1}{a^2} \frac{\partial^2 u}{\partial t^2} = 0, \tag{6.1}$$

$$h = H \frac{\partial u}{\partial z}. \tag{6.2}$$

In these equations, the symbols

$$u = u^{(2)}, \qquad h = h^{(2)}, \qquad a^2 = {}_2a_0^2 = H^2/(4\pi\varrho_2)$$

should be used for the liquid, and the symbols

$$u = u^{(1)}, \qquad h = h^{(1)}, \qquad a^2 = {}_1a_0^2 + c^2, \qquad {}_1a_0^2 = \frac{H^2}{4\pi\varrho_1}, \qquad c^2 = \frac{\mu}{\varrho_1}$$

for the elastic solid. According to the discussion in the preceding section we consider two extreme variants of the boundary conditions corresponding to the transition to the limit in the liquid:

$$1. \ \frac{\lambda_m}{\beta'} = 0; \qquad 2. \ \frac{\lambda_m}{\beta'} = \infty.$$

For the first boundary condition, we have the following relations for $z = 0$:

$$u^{(1)} = u^{(2)}, \qquad \sigma_{31}^{(1)} + T_{31}^{(1)} = T_{31}^{(2)} + P e^{i\omega t}$$

$$\text{or} \qquad \varrho_1 a^2 \frac{\partial u^{(1)}}{\partial z} - \varrho_{22} a_0^2 \frac{\partial u^{(2)}}{\partial z} = P e^{i\omega t}. \tag{6.3}$$

For the second boundary condition we find for $z = 0$

$$h^{(1)} = h^{(2)} \qquad \text{or} \qquad \frac{\partial u^{(1)}}{\partial z} = \frac{\partial u^{(2)}}{\partial z},$$

$$\sigma_{31}^{(1)} = P e^{i\omega t} \qquad \text{or} \qquad \varrho_1 c^2 \frac{\partial u^{(1)}}{\partial z} = P e^{i\omega t}. \tag{6.4}$$

Setting

$$u(z, t) = u^*(z) e^{i\omega t}, \tag{6.5}$$

we obtain

$$u^{*(1)} = A_1 e^{-i\alpha_1 z}, \qquad u^{*(2)} = A_2 e^{i\alpha_2 z}, \tag{6.6}$$

where

$$\alpha_1^2 = \frac{\omega^2}{a^2}, \qquad \alpha_2^2 = \frac{\omega^2}{{}_2 a_0^2}.$$

Taking into account the boundary conditions (6.3), we obtain

$$u^{(1)} = -\frac{P}{\Delta} e^{i(\omega t - \alpha_1 z)} \qquad u^{(2)} = -\frac{P}{\Delta} e^{i(\omega t + \alpha_2 z)},$$

$$h^{(1)} = \frac{H P \alpha_1}{\Delta} e^{i(\omega t - \alpha_1 z)} \qquad h^{(2)} = -\frac{H P \alpha_2}{\Delta} e^{i(\omega t + \alpha_2 z)}, \tag{6.7}$$

where

$$\Delta = \varrho_1 a^2 \alpha_1 + \varrho_{22} a_0^2 \alpha_2.$$

For the boundary conditions (6.4) we find

$$u^{(1)} = -\frac{P}{\mu \alpha_1} e^{i(\omega t - \alpha_1 z)} \qquad u^{(2)} = \frac{P}{\mu \alpha_2} e^{i(\omega t + \alpha_2 z)},$$

$$h^{(1)} = \frac{P H}{\mu} e^{i(\omega t - \alpha_1 z)} \qquad h^{(2)} = \frac{P H}{\mu} e^{i(\omega t + \alpha_2 z)}. \tag{6.8}$$

As is readily seen from (6.7) and (6.8), we have radiation of waves of identical displacement or velocity amplitudes in the first case and identical amplitudes of the field $h$ or the derivative of the displacement in the second case. Thus, depending on the manner in which the mechanical and magnetic viscosities in the liquid tend to zero, we have obtained different solutions to the contact problem between a liquid and a solid.

The manner in which the viscosities tend to zero in the solid has no influence on the solution for all physically realistic cases.

### References

[1] KULIKOVSKIJ, A. G., and G. A. LJUBIMOV: Magnetohydrodynamics (in Russian), Moscow: Fizmatgiz, 1962.

[2] STEWARTSON, K.: Fluid Mech. 8, No. 1 (1960).

[3] KALISKI, S.: Proc. Vibr. Probl. 3, No. 1 (1962).

[4] KALISKI, S., and W. NOWACKI: Bull. Acad. Polon. Sciences 10, No. 4 (1962).

[5] KALISKI, S., and J. PETYKIEWICZ: Proc. Vibr. Probl. 1, No. 4 (1959).

[6] FREUDENTHAL, A. M., and H. GEIRINGER: Encyclopedia of Physics, Vol. VI, Berlin/Göttingen/Heidelberg: Springer 1958.

[7] KALISKI, S.: Problems of Continuum Mechanics, Soc. Ind. Appl. Math., Philadelphia, 1961.

# On the Numerical Solution of a Wave Propagation Problem in the Theory of Dislocation Motion

By **John D. Campbell** and **David B. Taylor**

Oxford University Engineering Laboratory, Oxford, England

## Summary

Partial differential equations are derived governing the motion of a dislocation line in an anelastic crystalline material, including the effects of inertia, line tension and a velocity-dependent dissipative force. The equations employ Lagrangian co-ordinates and consist of a non-linear second-order equation and a quasi-linear first-order equation, with two dependent and two independent variables.

It is shown that by the introduction of a third dependent variable the equations may be transformed to three quasi-linear first-order equations of hyperbolic type, with three sets of characteristics. The differential relations holding along these characteristics are derived, and a numerical technique employing back-interpolation is used to integrate them for several values of the parameters involved.

The calculations were performed on a digital computer and the results are plotted to show the influence of source length and resistive force on the shape of the dislocation as it expands into the slip plane. The effect of these variables on the time required to form a closed loop is also discussed.

## 1. Introduction

The study of plastic wave propagation in metals has been severely hampered by the lack of reliable and comprehensive data on the fundamental properties of crystalline materials subjected to very rapidly applied stresses. In the absence of such data, theoretical work on plastic waves has been based on the assumption that the behaviour can be described by a constitutive law relating stress, strain and strain rate.

The earliest investigators [1, 2, 3, 4] assumed that this relation was independent of strain rate, so that the behaviour could be described by a single stress-strain curve (not necessarily that obtained in a 'static' test). However, certain experimental results [5, 6] indicate that, for some metals at least, this assumption is probably incorrect.

An alternative hypothesis has been proposed [7, 8], in which the plastic strain rate is assumed to be governed by the instantaneous values of stress and strain. Results of wave analyses based on this assumption are found to differ radically from those based on the strain-rate independent relation. A conclusive experimental test of the two hypotheses has, however, not yet been performed, since the particular types of test so far made cannot be analysed by three-dimensional wave theory, but only by an approximate treatment the validity of which has been questioned [9].

It is therefore a matter of some interest to determine whether any light can be thrown on the matter by consideration of the actual mechanisms of plastic deformation. It is known that in crystalline materials the most important of these mechanisms is the movement of dislocations. A complete derivation of dynamic plastic behaviour must await the development of the theory of distributions of dislocations, but it seems likely that a prerequisite is a knowledge of the behaviour of a single dislocation source.

An analysis of the constant-speed motion of a straight dislocation [10] gives the energy of a moving dislocation in terms of its velocity. This result has been used [11] to obtain an approximate solution to the problem of the motion of a dislocation forming a FRANK-READ source, under the action of a suddenly-applied stress. The solution showed that, in the absence of any dissipative force opposing the motion of the dislocation, the time required to form the first loop from the source is extremely small — typically of the order of $10^{-9}$ sec.

However, it is known that, in some materials at least, there is a considerable frictional drag on dislocations, and that this drag increases with increasing velocity [12, 13, 14]. It is evident that the operation of a FRANK-READ source, and in particular its nucleation time, will be considerably affected if the drag force becomes comparable with the force due to the applied stress.

The macroscopic plastic flow of the metal under rapid loading depends on the rate at which fresh dislocations can be generated, as well as on the motion of the dislocations which are already present [14]. It is therefore of interest to analyse the behaviour of a FRANK-READ source under shock loading, taking account of a velocity-dependent force opposing the motion. In the earlier paper, it was necessary to take into account the relativistic behaviour of dislocations, since in the absence of dissipative forces the dislocation velocity rapidly approaches that of sound. In the present analysis, however, it is assumed that relativistic effects may be neglected; this assumption will be valid if the dissipative force is large enough to limit the dislocation speed to less than say one-third of the sonic speed.

## 2. Energy of a Moving Dislocation

The energy per unit length ($E_s$) of a straight screw dislocation moving at a constant velocity $v$ can be calculated from its stress field [10]. If $v^2 \ll c^2$, where $c$ is the elastic shear wave velocity, it is given by

$$E_s = E_0 \left(1 + \frac{1}{2} \frac{v^2}{c^2}\right), \tag{1}$$

where $E_0$ is the strain energy of the stationary dislocation.

The corresponding analysis for an edge dislocation [15] gives the energy per unit length as

$$E_e = \frac{E_0}{1 - v} + \frac{1}{2} E_0 \frac{v^2}{c^2} \left(1 + \frac{c^4}{c_1^4}\right), \tag{2}$$

where $v$ is POISSON's ratio and $c_1$ is the elastic dilatation wave velocity. Since $c/c_1$ is of the order of $1/2$ for metals, the second term in (2) differs from that in (1) by less than 10%. It is therefore a good approximation to take the velocity-dependent part of the energy of a dislocation to be independent of the angle between the dislocation and its BURGERS vector. Eqs. (1) and (2) show that the rest energy of a dislocation segment varies by about $\pm 16\%$ with its orientation; however, it seems probable that this will have only a second-order effect on the behaviour of moving dislocations. The energy of the dislocation per unit length will therefore be taken as

$$E = \Gamma \left(1 + \frac{1}{2} \frac{v^2}{c^2}\right), \tag{3}$$

where $\Gamma$ is a constant.

The energy of a curved dislocation has not been calculated, but is expected to be lower than that of a straight one due to the interaction of stress fields. In the present analysis this can only be taken into account by using a smaller (constant) value of $\Gamma$.

When a dislocation is accelerated, the change in its stress field is propagated at elastic wave speed. This means that the velocity-dependent energy takes a finite time to reach the value given by the steady-state solution. The importance of this time lag will depend on the effective radius of the dislocation's stress field and on its core energy, as well as on the accelerations involved. The effect is neglected in the present treatment of the problem.

A review of theories of the energy dissipated by a moving dislocation has been given by WEERTMAN [16]; he considers three proposed damping mechanisms, each of which results in a linear relationship between the applied stress and the dislocation velocity. On the other hand, experimental rezults [*12, 13*], indicate that the relation is non-linear, the velocity increasing much more rapidly with stress at low velocities. At velocities greater than about $c/100$, however, the relationship may be approximated by a linear one; further, at these velocities the resistance to motion is approximately the same for edge and screw dislocations. In the present paper, therefore, a linear relation is assumed for the numerical calculations, though the basic analysis is applicable to any arbitrary relation.

### 3. Equations Governing the Motion

The motion of the dislocation will be described by using Lagrangian coordinates $(\alpha, t)$, where $\alpha$ defines an arbitrary point on the dislocation line at time $t = 0$ (Fig. 1). Each point on the line is taken to move at right angles to the line with speed $v$, thus tracing out a 'trajectory'. The trajectories $\alpha =$ const. thus form a family of curves orthogonal to the dislocation lines $t =$ const.

The length of dislocation line between adjacent trajectories $\alpha, \alpha + d\alpha$, at time $t$ is denoted by $A\,d\alpha$, where $A$ is a function of $\alpha$ and $t$.

The energy equation for this segment as it moves forward a distance $v\,dt$ is

$$\sigma b A\,d\alpha\,v\,dt = \frac{\partial}{\partial t}\,(EA\,d\alpha)\,dt + wA\,d\alpha\,dt,$$

where $\sigma$ is the applied shear stress and $w$ is the rate of dissipation of energy per unit length of dislocation when it is moving at speed $v$.

Substituting from (3) and assuming that $v^2 \ll c^2$, this becomes

$$\sigma b = \frac{\Gamma}{c^2}\,\frac{\partial v}{\partial t} + \frac{\Gamma}{Av}\,\frac{\partial A}{\partial t} + \frac{w}{v}. \qquad (4)$$

Assuming that $\sigma$ is constant, (4) may be written in non-dimensional form as

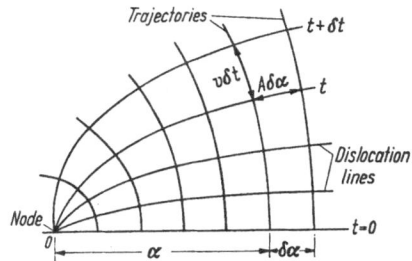

Fig. 1. Illustrating the Lagrangian coordinates $(\alpha, t)$ and the dependent variables $(A, v)$.

$$\frac{1}{Av'}\,\frac{\partial A}{\partial t'} + \frac{\partial v'}{\partial t'} = p, \qquad (5)$$

where $v' = v/c$, $t' = \sigma bct/\Gamma$, and $p = 1 - w/\sigma bv$, a given function of $v$.

In addition to Eq. (5), the quantities $A$ and $v$ are related by the geometrical compatibility condition for plane motion. This condition is derived in the earlier paper, and is

$$\frac{\partial}{\partial \alpha}\left[\frac{1}{A}\,\frac{\partial v}{\partial \alpha}\right] + \frac{\partial}{\partial t}\left[\frac{1}{v}\,\frac{\partial A}{\partial t}\right] = 0. \tag{6}$$

This equation is unchanged in form if the non-dimensional variable $\alpha' = \sigma b \alpha/\Gamma$ is used together with $v'$ and $t'$ as previously defined. In the remainder of the paper, the primes are omitted, and $\alpha$, $t$, $v$ used to denote the non-dimensional variables.

## 4. Boundary Conditions

The solution obtained in the earlier paper [11] is applicable to the motion of a dislocation with one fixed point (single mill); it holds for a dislocation with two fixed points (double mill or Frank-Read source) only until the disturbances propagated from the two ends interact with each other. The present analysis will however be applied to the general motion of a double mill.

The length of the source, between the fixed points or nodes, is denoted by $2\lambda\Gamma/\sigma b$, so that if $\alpha = 0$ defines one node, the other is defined by $\alpha = 2\lambda$. It is assumed that the dislocation is straight and has zero velocity at time $t = 0$; further, $A = 1$ by definition at $t = 0$.

The first term in Eq. (5) represents the normal force on a dislocation element due to its line tension, and this is evidently zero when the line is straight. The term $p$ represents the applied stress minus the resistive force, and the latter is zero at $t = 0$ since the velocity is then zero. Thus Eq. (5) shows that $\partial v/\partial t = 1$ at $t = 0$.

The boundary conditions can therefore be stated as

$$\left.\begin{array}{l} A = 1, \quad v = 0, \quad \dfrac{\partial v}{\partial t} = 1 \quad \text{on} \quad t = 0, \quad 0 \leq \alpha \leq 2\lambda, \\[2mm] \text{and } v = 0 \quad \text{on} \quad \alpha = 0 \quad \text{and} \quad \alpha = 2\lambda, \quad t \geq 0. \end{array}\right\} \tag{7}$$

The problem is thus properly posed; the conditions on $t = 0$ constitute Cauchy conditions for Eqs. (5) and (6), and those on $\alpha = 0$ and $\alpha = 2\lambda$ are satisfactory secondary boundary conditions.

It can be shown [11] that the angle made by the dislocation line with the $x$-axis at time $t$ is given by

$$\varphi = \int^{t} \frac{1}{A}\,\frac{\partial v}{\partial x}\,dt, \tag{8}$$

where the integration is performed along a trajectory $\alpha = \text{const}$. Since distance $s$ along the dislocation is defined by $ds = A\,d\alpha$, the coordinates

$(x, y)$ of a point on the dislocation may be obtained by integrating the equations

$$dx = A \cos \varphi \, d\alpha, \qquad dy = A \sin \varphi \, d\alpha \qquad (9)$$

along the dislocation line at any given time $t$.

Alternatively, we may use the angle made by a trajectory $\alpha$ with the negative $x$-axis, given by

$$\theta = \int^{\alpha} \frac{1}{v} \frac{\partial A}{\partial t} \, d\alpha, \qquad (10)$$

the integration being performed along a line $t = $ const. The coordinates $(x, y)$ are then found by integrating the equations

$$dx = -v \cos \theta \, dt, \qquad dy = v \sin \theta \, dt \qquad (11)$$

along a trajectory $\alpha$.

## 5. The Choice of Equations for a Numerical Solution

A numerical procedure could be developed to obtain a solution of Eqs. (5) and (6) using the method of characteristics, but a method involving the solution of a smaller number of non-linear equations can be devised if Eq. (6) is replaced by two first-order equations through the introduction of a third dependent variable. Two possible ways of effecting this will be discussed.

The new dependent variable is defined as

$$f = \frac{1}{v} \frac{\partial A}{\partial t} - \frac{\partial v}{\partial \alpha}$$

or $$\qquad (12)$$

$$g = \frac{1}{v} \frac{\partial A}{\partial t} + \frac{\partial v}{\partial \alpha}.$$

Then from (5) and (12) we obtain

$$\frac{1}{A} \frac{\partial v}{\partial \alpha} = p - \frac{f}{A} - \frac{\partial v}{\partial t} = -p + \frac{g}{A} + \frac{\partial v}{\partial t}. \qquad (13)$$

Using (12) and (13), second-order derivatives may be eliminated from (6), giving

$$\frac{f}{A^2} \frac{\partial A}{\partial \alpha} + p' \frac{\partial v}{\partial \alpha} - \frac{1}{A} \frac{\partial f}{\partial \alpha} + \frac{\partial f}{\partial t} = 0$$

or $$\qquad (14)$$

$$\frac{g}{A^2} \frac{\partial A}{\partial \alpha} + p' \frac{\partial v}{\partial \alpha} - \frac{1}{A} \frac{\partial g}{\partial \alpha} - \frac{\partial g}{\partial t} = 0,$$

where $p' = dp/dv$.

Eqs. (5), (12) and (14) now constitute a system of three first-order equations for $A$, $v$, and $f$ or $g$. The boundary conditions (7) are modified by substituting $f = 0$ or $g = 0$ for $\partial v/\partial t = 1$ on $t = 0$, $0 \leq \alpha \leq 2\lambda$; this follows from Eqs. (5) and (12).

A further advantage obtained from the introduction of $f$ or $g$ is that the expressions for $\varphi$ and $\theta$ can be written in a form which does not involve the integration of derivatives. Thus (8) becomes

$$\varphi = \int_0^t (p - f/A)dt - v = -\int_0^t (p - g/A)dt + v. \qquad (15)$$

The corresponding expressions for $\theta$ are less useful, since $\theta$ becomes indeterminate on $\alpha = 0$, so that the integrals must be evaluated from $\alpha = \lambda$, where $v$ is unknown.

The characteristic directions of the quasi-linear hyperbolic equations (5), (12) and (14) are found to be given by

$$\frac{d\alpha}{dt} = 0, \quad \pm \frac{1}{A}. \qquad (16)$$

The first of these corresponds to the trajectories $\alpha = $ const., and the other two to waves propagated along the dislocation line at the sonic speed. Along these characteristics the following differential relations hold:

On the trajectories $\alpha = $ const.,

$$\frac{1}{A}\, dA + vdv = pvdt. \qquad (17)$$

On the curves $dt = A d\alpha$ ($F_1$ family),

$$dv = (p - f/A)dt. \qquad (18)$$

On the curves $dt = -A d\alpha$ ($F_2$ family),

$$\frac{2f}{A}\, dA + (vf + A p')dv - 2df$$

$$= \left[ p(vf + A p') + \frac{f}{A}\, (vf - A p') \right] dt. \qquad (19)$$

When $g$ is used, equation (18) with $f$ replaced by $g$ appears instead of Eq. (19), and likewise (19) with $f$ replaced by $g$ appears instead of (18).

It may be noted that when there is no dissipative force, that is when $p = 1$, $p' = 0$, the first of (14) can be satisfied identically by taking $f = 0$; the family $F_2$ does not then exist and it is possible to integrate the equations to give an exact solution to the problem of the half-mill [11].

## 6. The Analytic Solution in a Limited Region

In the region $R$ (Fig. 2) between the $F_1$ characteristic through $\alpha = 0$, $t = 0$, the $F_2$ characteristic through $\alpha = 2\lambda$, $t = 0$, and the boundary $t = 0$, the problem can be solved analytically since the solution is independent of $\alpha$. Equating all $\alpha$-derivatives to zero and integrating with respect to $t$ we obtain $A = 1$, $f = g = 0$, and $v$ is given by

$$t = \int_0^v (1/p)\,dv. \qquad (20)$$

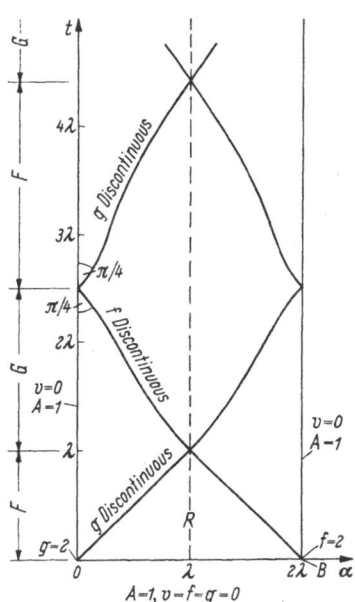

Fig. 2. The $(\alpha, t)$ plane, showing boundary conditions and special characteristics.

Since $A = 1$ in this region, the bounding characteristics have slopes $\pm 1$ and hence are given by the lines $\alpha = t$, $\alpha = 2\lambda - t$.

The motion of the dislocation in this region is therefore straight-line motion at a velocity $v$ given by (20).

## 7. Symmetry

The motion is evidently symmetrical about the line $x = \lambda$. This implies that in the $(\alpha, t)$ plane, $A$ and $v$ are symmetric about the line $\alpha = \lambda$.

From (12) it follows that

$$g - f = 2\frac{\partial v}{\partial \alpha}, \qquad (21)$$

and hence that $(g - f)$ is asymmetric about $\alpha = \lambda$. It is clear that the numerical solution can be confined to the strip $t \geq 0$, $0 \leq \alpha \leq \lambda$.

## 8. Discontinuities

The boundary conditions (7) show that $\partial v/\partial t$ is discontinuous at the points $O\,(0, 0)$ and $B\,(2\lambda, 0)$ in the $(\alpha, t)$ plane. Discontinuities in the first derivatives of both $A$ and $v$ are propagated along the member of $F_1$ passing through $O$, the member of $F_2$ passing through $B$, and their reflections; these characteristics will be referred to as special characteristics. Since $f$ and $g$ are functions of the first derivatives of $A$ and $v$, we may expect discontinuities in their values also across the special characteristics; discontinuities in $A$ and $v$, however, can be excluded on physical grounds.

It follows from (19) that since $A$ and $v$ are continuous, $df/dt$ is defined and finite along any member of $F_2$; hence if $f$ has a discontinuity it must occur across a member of $F_2$ and propagate along this characteristic. By similar reasoning a discontinuity of $g$ can only occur across a member of $F_1$ and it must propagate along this characteristic.

## 9. The Numerical Technique

It is usually convenient when integrating along characteristics in the presence of discontinuities to use the characteristics themselves as the mesh on which to carry out the computation. Since two families of characteristics define such a mesh, the presence of a third family makes interpolation necessary and the advantage of the method is lost.

A suitable alternative is available in the method proposed by HARTREE [17] in which a fixed rectangular mesh is used. The use of this mesh simplifies the computation of the integrals (9) and (15) involved in obtaining the solution in the $(x, y)$ plane; it also makes possible the transformation from $f$ to $g$ or $g$ to $f$ by the numerical evaluation of $\partial v/\partial \alpha$.

In the computer program written, the variables $(A, v, f)$ are used in regions in which $g$ has a discontinuity, and $(A, v, g)$ are used in regions in which $f$ has a discontinuity; these regions are indicated by $F$ and $G$ respectively in Fig. 2.

A square mesh of side $\Delta \alpha = \Delta t$ is used, and the transformation from $f$ to $g$ is made on the first line $t = T$ of the mesh after a special characteristic intersects $\alpha = \lambda$ or $\alpha = 0$. Eq. (21) is used, $\partial v/\partial \alpha$ being evaluated from a formula using values of $v$ at four mesh points.

The nature of the discontinuities at $O$ (Fig. 2) will now be examined; those at $B$ follow immediately from the symmetry conditions.

Since $v$ and $\partial v/\partial t$ vanish on $\alpha = 0$, $p$ is unity and it is seen from (5) that $\partial A/\partial t$ is zero along $\alpha = 0$; hence $A = \text{const.} = 1$, as continuity of $A$ at $O$ must be maintained. It therefore follows from (5) that $v^{-1}\partial A/\partial t = 1$ along $\alpha = 0$ and hence, from the first of (12), that

$$(f)_{\alpha=0} = 1 - \left(\frac{\partial v}{\partial x}\right)_{\alpha=0}. \tag{22}$$

The fact that $f = 0$ on $t = 0$ and $f$ is continuous along any member of $F_2$ implies that $f$ is continuous at $O$, and its value there is zero. Eq. (22) then shows that

$$\lim_{t \to 0} \left[\left(\frac{\partial v}{\partial x}\right)_{\alpha=0}\right] = 1. \tag{23}$$

The value of $\partial v/\partial \alpha$ therefore has a discontinuity of unity at $O$. It then follows from (21) and (22) that

$$\lim_{t \to 0} [(g)_{\alpha=0}] = 2.  \tag{24}$$

The pattern of the special characteristics, along which the discontinuities of $f$ and $g$ are propagated, is sketched in Fig. 2.

The numerical technique used is as follows (see Fig. 3). $S$, $C$ and $T$ represent mesh points at which the dependent variables are already known. $PL$ and $PR$ are segments of the $F_1$ and $F_2$ characteristics which pass through the mesh point $P$. Since in the present problem $A \geq 1$, $L$ and $R$ lie in the interval $ST$ and the use of numerical approximations with a truncation error of order $\Delta \alpha^3$ can be expected to yield a convergent method giving an error of order $\Delta \alpha^2$.

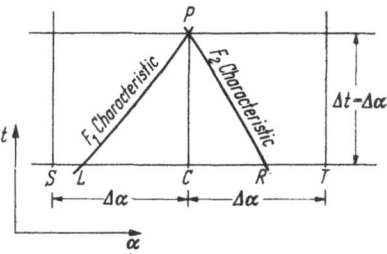

Fig. 3. Illustrating the numerical technique.

If the derivatives of $A, v$ and $f$ or $g$ are suitably smooth on $ST, CP$, $LP$ and $RP$, the solution at $P$ can be obtained from approximations to the integrals of (17), (18) and (19). The trapezoidal rule is used for these and all other integrations in the program; the error involved here is of order $\Delta \alpha^3$.

Using the notation

$$\mu_{AB}[\Phi(x)] \equiv 1/2 [\Phi(x_A) + \Phi(x_B)]  \tag{25}$$

Eqs. (17), (18) and (19) become

$$\mu_{CP}[1/A](A_P - A_C) + \mu_{CP}[v](v_P - v_C) = \mu_{CP}[pv]\Delta t + O(\Delta \alpha^3),  \tag{26}$$

$$v_P - v_L = \mu_{LP}[p - f/A]\Delta t + O(\Delta \alpha^3),  \tag{27}$$

and

$$\mu_{RP}[2f/A](A_P - A_R) + \mu_{RP}[vf + A p'] - 2(f_P - f_R)$$
$$= \mu_{RP}\left[p(vf + A p') + \frac{f}{A}(vf - A p')\right]\Delta t + O(\Delta \alpha^3).  \tag{28}$$

Values at $L$ and $R$ are obtained by interpolation based on three points; this yields six equations of the form

$$\psi(\alpha) = \psi_S \frac{(\alpha - \alpha_C)(\alpha - \alpha_T)}{2\Delta \alpha^2} - \psi_C \frac{(\alpha - \alpha_T)(\alpha - \alpha_S)}{\Delta \alpha^2}$$
$$+ \psi_T \frac{(\alpha - \alpha_S)(\alpha - \alpha_C)}{2\Delta \alpha^2} + O(\Delta \alpha^3),  \tag{29}$$

where $\psi$ represents $A, v, f$ or $g$.

The values of $\alpha_L$ and $\alpha_R$ required by these interpolations are obtained from

$$\alpha_L = \alpha_C - \frac{\Delta t}{\mu_{IP}[A]} + O(\Delta \alpha^3), \tag{30}$$

$$\alpha_R = \alpha_C + \frac{\Delta t}{\mu_{RP}[A]} + O(\Delta \alpha^3). \tag{31}$$

The system (26) to (31) consists of eleven non-linear equations for the values of the three dependent variables at $L$, $R$ and $P$, and the values of $\alpha$ at $L$ and $R$. These equations are solved iteratively in the program written, the iteration being terminated when successive estimates of the values of the dependent variables at $P$ differ by less than $\Delta t/1000$; this requires 3 to 5 iterations at each point.

In the neighbourhood of the boundaries $\alpha = 0$, $\alpha = \lambda$, this procedure is modified by the boundary conditions. These are simplified by the use of both $f$ and $g$, which are related on $\alpha = 0$ by

$$f + g = 2 \tag{32}$$

and on $\alpha = \beta$ by

$$f - g = 0. \tag{33}$$

Eq. (32) follows from (12) and (22), and (33) from (21) using symmetry.

The use of $f$ and $g$ avoids discontinuities in the dependent variables, but discontinuities in their derivatives remain. These are propagated along the special characteristics, so that no differentiation, interpolation or integration may be carried out across these characteristics. It is therefore necessary to compute the position of the special characteristics at each stage, and in their neighbourhood to calculate additional values of the main variables at points other than normal mesh points.

The mapping to the $(x, y)$ plane is carried out numerically by using the trapezoidal rule approximations to the integrals of (9) and (15). In these integrations, special steps are taken to subdivide any interval containing a special characteristic. As a check, the value of $y$ is also computed from (11) for the straight trajectory $\alpha = \lambda$, yielding the value $Y^*$, say. Denoting by $(X, Y)$ the values for the trajectory $\alpha = \lambda$ obtained in the main computation, the quantities $|X - \lambda|$, $|Y - Y^*|$, may be used as a measure of the accuracy obtained.

## 10. Numerical Results

The program was written for the 'Mercury' computer in the Computing Laboratory, Oxford. A maximum of 96 mesh points for each value of $t$ was allowed for. The machine completed the calculations for such a row in about one minute.

The function $p(v)$ was taken as

$$p = 1 - kv. \tag{34}$$

This corresponds to a viscous-type resisting force on the dislocation, the limiting velocity of which, under the given applied stress, is $1/k$ when the curvature is zero.

Fig. 4 shows results for the case $\lambda = 1$, $k = 0$, i.e. a source to which the static breakout stress is suddenly applied, the energy dissipation being zero. It is seen that the dislocation passes through a position which is approximately semicircular at a time given by $t = 1.6$, and continues to expand into the slip plane with increasing velocity. Since at

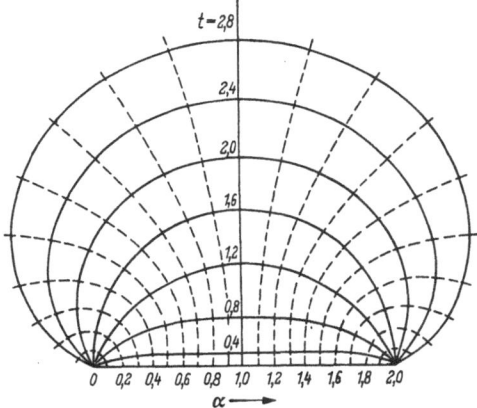

Fig. 4. Solution for $\lambda = 1$, $k = 0$.

$t = 2.8$ the solution gives a maximum velocity somewhat greater than the sonic speed, it is evident that relativistic effects are already important in this case; the solution is therefore not carried any further.

Fig. 5 shows results for the case $\lambda = 1$, $k = 0.5$. Here the dislocation passes through an approximately semicircular configuration at $t \simeq 2.0$. A continuation of the computation shows that the loop closes at a time estimated as $t \simeq 14$, and this time may be called the nucleation time for the source. In this solution also the maximum velocities are found to be somewhat greater than the sonic speed, so that relativistic effects should be taken into account.

Fig. 6 shows results for $\lambda = 10$, $k = 3$; it is seen that in this case the dislocation does not pass through an approximately semicircular configuration. The computed shapes are similar to those obtained in the approximate treatment of the relativistic motion of a half-mill with zero resistance [11]. In this computation, it was necessary to use 80 points per row to obtain a sufficiently small spacing of the calculated points where $A$ is large. Since the velocities are fairly small a large number of time steps was required (384 to obtain the results shown). At $t = 48$ the values of $|X - \lambda|/\lambda$ and $|Y - Y^*|/Y^*$ are less than 4%, but it is possible that larger errors may exist elsewhere in the computed values; it seems probable that this type of solution cannot usefully be carried much further by the present technique. It would be preferable in such cases to neglect the inertia force and to treat the equations as parabolic.

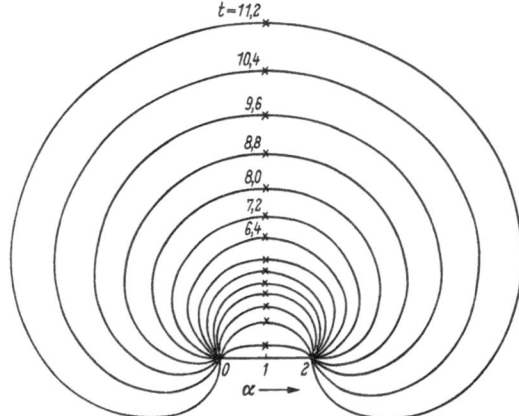

Fig. 5. Solution for $\lambda = 1$, $k = 0.5$.

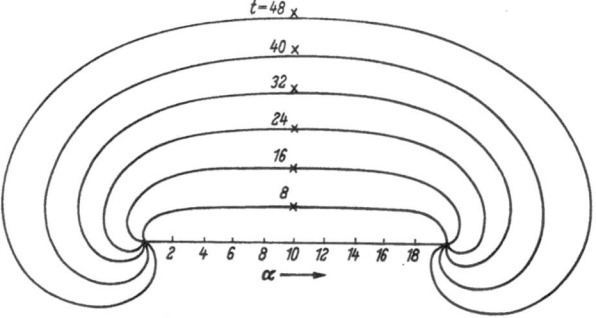

Fig. 6. Solution for $\lambda = 10$, $k = 3$.

## 11. Discussion and Conclusions

The results show that the presence of a velocity-dependent dissipative force may have a considerable effect on the motion of a dislocation source. It increases the nucleation time in two ways: by decreasing the maximum velocity reached, and by altering the shape of the dislocation.

It is clear that for large $\lambda$ the nucleation time will become large, as each end of the dislocation 'winds up' into a tight spiral around its node. For some finite value of $\lambda$ in the neighbourhood of unity, however, the nucleation time must become infinite, when the work done by the applied stress is insufficient to supply the line energy and energy dissipated.

It follows that for a given resistance the nucleation time is a minimum at a certain value of $\lambda$, say $\lambda^*$, which is probably of the order of, but greater than, unity. We may therefore expect that when a stress $\sigma$ is suddenly applied to a crystal containing sources whose lengths are distributed over a wide range, the first sources to operate will be those whose length is

$$2l = 2\lambda^* \Gamma / \sigma b. \tag{35}$$

A large suddenly-applied stress thus causes short sources to operate before long ones. If the stress is applied slowly, however, the long sources will operate first at low stress levels. It is expected therefore that the density of operative slip planes will increase with speed of loading, as will the stress at which a given number of sources operate.

As noted in section 10, if the dissipative forces are small, relativistic effects become important. These can in principle be treated by methods similar to those given in the paper. The equation of motion may be taken (in non-dimensional terms) as

$$\frac{\beta}{Av} \frac{\partial A}{\partial t} + \frac{1}{\beta} \frac{\partial v}{\partial t} = p, \tag{36}$$

where

$$\beta = (1 - v^2)^{1/2}.$$

Second derivatives may be eliminated from (6) by using a modified variable $f$ or $g$, defined by

$$\left.\begin{aligned} f &= \frac{\beta}{v} \frac{\partial A}{\partial t} - \frac{\partial v}{\partial \alpha}, \\ g &= \frac{\beta}{v} \frac{\partial A}{\partial t} + \frac{\partial v}{\partial \alpha}, \end{aligned}\right\} \tag{37}$$

instead of those defined by (12).

The characteristics are found to be given by

$$\frac{d\alpha}{dt} = 0, \quad \pm \frac{\beta}{A}, \tag{38}$$

and differential relations similar to (17), (18), (19) can be derived for the three families of characteristics.

A comparison of Fig. 6 with the curves obtained from the approximate relativistic solution for $k = 0$ suggests that energy dissipation and relativistic behaviour have a similar effect on the shape of the dislocation loop. We may therefore expect the nucleation time to increase by a factor of order $k$ due to viscous-type resistance, since the limiting velocity is then $1/k$.

In Fig. 6, it is seen that each side of the line must travel a further horizontal distance of about 9 units before the loop closes. With $k = 3$, this requires a time greater than 27 units, and the nucleation time may therefore be estimated as about 80 units. This result may be compared with the value 15.7 obtained from Eq. (68) of [11], in the approximate relativistic solution for $\lambda = 10$.

Since the ratio $80/15.7 \simeq 5 > k = 3$, it appears that the increase in the nucleation time due to energy dissipation is appreciably greater than is accounted for by the reduction in the limiting velocity.

## Acknowledgment

The authors are indebted to the Director and Staff of the Computing Laboratory, Oxford University, for helpful suggestions during the development of the program, and for the use of the Ferranti 'Mercury' computer.

### References

[1] DONNELL, L. H.: Trans. A.S.M.E. 52, 153 (1930).
[2] TAYLOR, G. I.: Ministry of Supply Report, 1942 (see Scientific Papers, Vol. 1, Cambridge University Press 1958, p. 467).
[3] VON KÁRMÁN, TH.: N.D.R.C. Report, 1942, also J. Appl. Phys. 21, 987 (1950).
[4] RAKHMATULIN, K. A.: Prikl. Mat. Mekh. 9, 91 (1945).
[5] STERNGLASS, E. J., and D. A. STUART: J. Appl. Mech. 20, 427 (1953).
[6] ALTER, B. E. K., and C. W. CURTIS: J. Appl. Phys. 27, 1079 (1956).
[7] SOKOLOVSKY, V. V.: Prikl. Mat. Mekh. 12, 261 (1948).
[8] MALVERN, L. E.: J. Appl. Mech. 18, 203 (1951).
[9] CRAGGS, J. W.: J. Mech. Phys. Solids 5, 115 (1957).
[10] FRANK, F. C.: Proc. Phys. Soc. 62, 131 (1949).
[11] CAMPBELL, J. D., J. A. SIMMONS and J. E. DORN: J. Appl. Mech. 28, 447 (1961).
[12] JOHNSTON, W. G., and J. J. GILMAN: J. Appl. Phys. 30, 129 (1959).
[13] STEIN, D. L., and J. R. LOW: J. Appl. Phys. 31, 362 (1960).
[14] GUARD, R. W.: Acta Met. 9, 163 (1961).
[15] ESHELBY, J. D.: Proc. Phys. Soc. 62, 307 (1949).
[16] WEERTMAN, J.: Response of metals to high velocity deformation, Interscience Press 1961, p. 237.
[17] HARTREE, D. R.: Calculating instruments and machines, University of Illinois Press 1949.

# Vibrations d'une structure lamellaire sous fluage localisé

Par Robert Mazet

Université de Paris et Office National d'Etudes et de Recherches Aéronautiques, France

## Introduction

Cette communication n'entre peut-être pas tout à fait dans le cadre du Colloque, car, loin de traiter le cas d'une onde de contrainte se propageant dans un milieu élasto-plastique, elle suppose que les contraintes qui provoquent la déformation plastique sont localisées dans une partie bien définie et très étroite de la structure. J'ai montré ailleurs [6] que les hypothèses habituelles de la plasticité, jointes à une répartition statistique des seuils de glissement (ou seuils de fluage), rendaient compte d'une manière satisfaisante du comportement d'une structure lamellaire sous des forces extérieures quelconques tant que les déformations restent suffisamment petites. L'idée d'une répartition statistique des seuils de fluage existe déjà dans une Note de C. DE CARBON [1]. Elle a été reprise par B. PERSOZ [2, 3] qui l'a utilisée dans des modèles rhéologiques. Elle n'a pas encore donné lieu, à ma connaissance, à des applications à l'échelle des structures intervenant dans la construction mécanique (on trouvera un exposé préliminaire dans [4]).

## 1. Définition de la structure

Rappelons qu'une structure lamellaire est caractérisée par un déplacement dominant de chacun de ses points $P$ à partir de la position d'équilibre non chargée. Par rapport à une base fixe $0, xyz$, ce déplacement est défini par des expressions de la forme

$$u = a(P)\zeta, \qquad v = b(P)\zeta, \qquad w = c(P)\zeta, \tag{1}$$

les $a, b, c$ étant des fonctionnelles opérant linéairement sur $\zeta$, déplacement dominant de référence (supposé petit) qui définit la déformée moyenne à l'instant $t$.

Lorsque la structure flue, nous admettons:

1. que le domaine non élastique $D_0$ est localisé (éventuellement par une entaille provoquant un affaiblissement de la structure dans cette

région) dans la « tranche » associée par les formules (1) à la partie de la déformée moyenne comprise entre $x = 0$ et $x = \varDelta$ ($\varDelta$ petit vis-à-vis de $L$, longueur de référence caractérisant l'étendue de la structure);

2. que la structure reste lamellaire et que la direction du déplacement dominant ne change pas.

On peut, dans ces conditions, définir un état quelconque de la déformée moyenne à partir d'une base de déformées normées dont l'une figure la déformée de la structure lorsque $D_0$ est en fluage pur; nous admettrons qu'une seule autre suffit pour figurer la déformée purement élastique qui se superpose à la première au-delà de $D_0$. On montre que la déformée moyenne est alors définie par la fonction suivante (en admettant qu'elle ne dépend que de $x$):

$$[0 \leq x \leq \varDelta] \quad \zeta = Lg\left(\frac{x}{L}\right) c_0 \frac{\varDelta}{L} q_2, \tag{2}$$

$$[\varDelta \leq x \leq L] \quad \zeta = L\left\{\left[f\left(\frac{x}{L}\right) - g\left(\frac{x}{L}\right) c_0 \frac{\varDelta}{L}\right] q_1 + g\left(\frac{x}{L}\right) c_0 \frac{\varDelta}{L} q_2\right\}. \tag{3}$$

$\zeta = Lf\left(\frac{x}{L}\right)$ est la forme normée purement élastique $[f(0) = 0,\ f(1) = 1]$,
$\zeta = Lg\left(\frac{x}{L}\right) c_0 \frac{\varDelta}{L}$ la forme normée de fluage pur $\left[g(0) = 0,\ g'\left(\frac{\varDelta}{L}\right) = 1,\right.$
$\left. c_0 = f'\left(\frac{\varDelta}{L}\right) \frac{L}{\varDelta}\right]$, $q_1$ *est le paramètre de déformation élastique*, $q_2$ *le paramètre de déformation élasto-plastique* (Fig. 1).

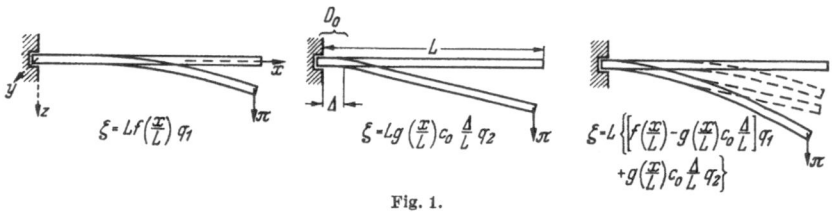

Fig. 1.

Dans le cas où le seuil de fluage $s_M$ varie statistiquement d'un point à l'autre de la « section » $S_0$ [ensemble des points $P$ associés par (1) au point $x = \varDelta$ de la déformée moyenne], il est commode d'introduire un nouveau paramètre $\alpha$. Ce paramètre, dit *paramètre de déformation plastique vraie*, est lié à $q_1$ et à $q_2$ par

$$q_2 = (1 - \theta)q_1 + \theta\alpha, \tag{4}$$

$\theta$ désignant une fonction bien définie de $q_2$ dépendant de la distribution statistique des $s_M$.

## 2. Définition du materiau et des forces intérieures

Nous supposons que le matériau est à écrouissage irréversible sans recouvrance. Il possède donc, en tout point, un seuil de fluage défini par la condition de von MISES, séparant deux états dans lesquels les forces intérieures obéissent à des lois différentes: un état élastique (dont nous supposerons la viscosité interne négligeable) et un état plastique sans variation de volume (où la viscosité sera, au contraire, rapidement croissante avec la vitesse de déformation). Comme lois des forces intérieures, nous adoptons celles de l'élément élasto-plastique simple de B. NAYROLES [5] légèrement modifiées pour tenir compte de la non-variation du volume.

L'état élastique correspond à $s \leq s_M$ ($s$: module du déviateur des contraintes); les lois des forces intérieures y sont, en supposant nul l'écrouissage actuel, les lois classiques de l'Elasticité

$$[s \leq s_M] \quad \sigma_m = \frac{E}{1-2\nu}\, \varepsilon_m, \quad s = \frac{E}{1+\nu}\, e \tag{5}$$

($\sigma_m$: moyenne des contraintes principales; $\varepsilon_m$: moyenne des déformations principales; $e$: module du déviateur des déformations; $E$: module de YOUNG, $\nu$: coefficient de POISSON).

D'autre part, l'état plastique correspond à $s > s_M$; les lois des forces intérieures y sont celles de LEVI-MISES et de NADAÏ-HOFF que nous écrirons en supposant $\dot{e} > 0$ [pas de retour en arrière dans les régions plastiques de la structure]:

$$[s > s_M] \quad \dot{\varepsilon}_m = 0, \quad \dot{e} = \frac{1+\nu}{E}\, \dot{s} + f\left[\frac{1+\nu}{E}\,(s - s_M)\right] \tag{6}$$

la fonction $f(u)$ [telle que $f(0) = 0$] peut s'approximer, selon NADAÏ-HOFF, par un monome de la forme $\frac{u^n}{\tau}$, $n$ et $\tau$ étant, au même titre que $E$ et $\nu$, des constantes caractéristiques du matériau ($n > 1$).

Au lieu du seuil de fluage $s_M$, il sera commode d'introduire la limite élastique $e_M$ définie par

$$s_M = \frac{E}{1+\nu}\, e_M. \tag{7}$$

## 3. Equations du mouvement dans le cas général

Nous avons montré [6] que, dans le cas où la limite élastique $e_M$ varie statistiquement d'un point à l'autre de $S_0$, $q_2$ (ou, si l'on préfère, $\alpha$) ne suffit pas pour caractériser l'état de fluage de la structure à l'instant $t$. Il faut lui adjoindre un nouveau paramètre $\beta$ dont dépend la viscosité de fluage et qui est lié à $\alpha$ par l'équation différentielle

$$\beta + \varkappa(\theta)\varphi(\dot{\beta}) = \theta\alpha - \frac{H}{D_1}, \qquad (8)$$

$\varkappa[\theta)$ étant une fonction bien définie de $\theta$, $\theta$ et $H$ des fonctions bien définies de $q_2$ [dépendant de la distribution statistique des $e_M$], $\varphi(v)$ l'inverse de la fonction $f(u)$ de la formule (6) et $D_1$ une constante numérique finie (Fig. 2). $\beta$ est le *paramètre d'étirement moyen*.

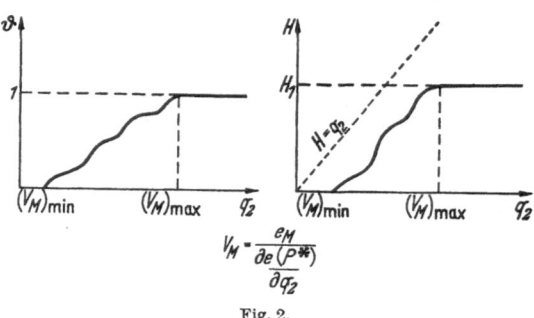

Fig. 2.

Les paramètres étant au nombre de 4: $q_1$ (déformation élastique), $q_2$ (déformation élasto-plastique), $\alpha$ (déformation plastique vraie) et $\beta$ (étirement moyen), nous devons écrire 4 équations: les deux équations de LAGRANGE relatives, par exemple, à $q_1$ et à $\alpha$ et les relations (8) et (4). Il vient, pour la structure soumise à la force $\pi(t)$ agissant au point de référence $L$ dans la direction du déplacement dominant:

$$m_{11}\ddot{q}_1 + m_{12}\frac{\theta\Delta}{L}\ddot{\alpha} = -\gamma\left[1 + \frac{\sigma_0 K_2\Delta}{n_0 L}(1-\theta)\right]q_1 + \pi L\left(1 - g_1 c_0\frac{\theta\Delta}{L}\right)$$

$$m_{12}\theta\ddot{q}_1 + m_{22}\frac{\theta^2\Delta}{L}\ddot{\alpha} = -\gamma\frac{K_2}{n_0}(\theta\alpha - \beta) + \pi L g_1 c_0\theta, \qquad (9)$$

$$\beta + \varkappa(\theta)\varphi(\dot{\beta}) = \theta\alpha - \frac{H}{D_1},$$

$$(1-\theta)q_1 + \theta\alpha = q_2.$$

$m_{11}$, $m_{12}$, $m_{22}$ sont les coefficients de l'énergie cinétique $T$ calculée à partir de (1) et de (3) après expression de $q_2$ en fonction de $q_1$ et de $\alpha$ au moyen de (4):

$$2T = m_{11}\dot{q}_1^2 + 2m_{12}\theta\frac{\Delta}{L}\dot{q}_1\dot{\alpha} + m_{22}\theta^2\frac{\Delta^2}{L^2}\dot{\alpha}^2, \qquad (9\,\mathrm{A})$$

$$\gamma = E\int_{D-D_0}\left[\frac{3}{1-2\nu}\left(\frac{\partial\varepsilon_m}{\partial q_1}\right)^2 + \frac{1}{1+\nu}\left(\frac{\partial e}{\partial q_1}\right)^2\right]dD.$$

[$D$: volume total de la structure non chargée];

$\gamma$ est la raideur généralisée (élastique) relative à $q_1$; $\sigma_0$, $n_0$, $K_2$, $g_1$ sont des constantes numériques finies dépendant de la structure; $n_0$, qui fait intervenir le point $P^*$ de $S_0$ le plus exposé au fluage, joue le rôle d'un coefficient d'entaille.

Le système (9) est valable pour

$$q_2 > \frac{(e_M)\min}{\partial e(P^*)/\partial q_2}, \quad \dot{q}_2 > 0.$$

Nous avons spécialement en vue aujourd'hui l'application aux vibrations sous fluage.

## 4. Première application: Fluage localisé établi sous charge constante

Supposons $\pi = $ Constante, $q_2 > \dfrac{(e_M)\max}{\partial e(P^*)/\partial q_2}$ et cherchons une solution du système (9) de la forme

$$\dot{q}_1 = c^{te}, \qquad \dot{q}_2 = c^{te}, \qquad \dot{\alpha} = c^{te}, \qquad \dot{\beta} = c^{te}.$$

On a: $\theta = 1$, $H = H_1$, $\varkappa = $ Constante $\varkappa_1$. Les équations deviennent:

$$0 = -\gamma q_1 + \pi L \left( 1 - g_1 c_0 \frac{\Delta}{L} \right),$$

$$0 = -\gamma \frac{K_2}{n_0} (\alpha - \beta) + \pi L g_1 c_0, \tag{10}$$

$$\beta + \varkappa_1 \varphi(\dot{\beta}) = \alpha - \frac{H_1}{D_1},$$

$$\alpha = q_2.$$

Il s'en suit que

$$q_1 = \frac{\pi L}{\gamma} \left( 1 - g_1 c_0 \frac{\Delta}{L} \right) = \text{Constante},$$

$$\alpha - \beta = \frac{\pi L}{\gamma} \frac{g_1 c_0 n_0}{K_2} = \frac{H_1}{D_1} + \varkappa_1 \varphi(\dot{\beta}). \tag{11}$$

Les relations (11) font connaître $\dot{\beta}$ $[= \dot{\alpha}]$ en fonction de $\pi$, ainsi que la différence constante $\alpha - \beta$.

## 5. Deuxième application:
## Vibrations autour du fluage localisé établi

Affectant de l'indice $s$ (stationnaire) la solution précédente, nous nous proposons de linéariser les petits mouvements de la structure autour de cette solution. Nous devons faire $\theta = 1$, $H = H_1$, $\varkappa = \varkappa_1$, $\alpha = q_2$ dans les équations (9). Posons:

$$q_1 = (q_1)_s + \xi, \qquad q_2 = (q_2)_s + \eta, \qquad \beta = \beta_s + \chi.$$

Il vient, en tenant compte de (10):

$$m_{11}\ddot{\xi} + m_{12}\frac{\Delta}{L}\ddot{\eta} = -\gamma\,\xi,$$

$$m_{12}\ddot{\xi} + m_{22}\frac{\Delta}{L}\ddot{\eta} = -\gamma\,\frac{K_2}{n_0}\,(\eta - \chi), \tag{12}$$

$$\chi + \varkappa_1[\varphi(\dot{\beta}_s + \dot{\chi}) - \varphi(\dot{\beta}_s)] = \eta.$$

Considérons le terme $\varphi(\dot{\beta}_s + \dot{\chi}) - \varphi(\dot{\beta}_s)$. Malgré la proximité du point singulier $v = 0$ de la fonction $\varphi(v)$, on peut, en présumant que $\dot{\chi}$ oscillera entre 2 valeurs sensiblement constantes et opposées telles que $\dot{\beta}(= \dot{\beta}_s + \dot{\chi})$ reste $> 0$, linéariser ce terme par la formule des accroissements finis:

$$\varphi(\dot{\beta}_s + \dot{\chi}) - \varphi(\dot{\beta}_s) \simeq U_m\dot{\chi}, \tag{13}$$

$U_m$ étant la valeur moyenne de $d\varphi/\partial\dot{\beta}$ dans l'intervalle de variation de $\dot{\beta}$.

Il est facile de rattacher $U_m$ à une grandeur physique. Désignons en effet par $v_s$ la vitesse de déplacement du point $L$ dans le fluage établi sous charge constante $\pi$. Elle est donnée par:

$$v_s = (\dot{\zeta}_L)_s = Lg_1 c_0\frac{\Delta}{L}\dot{\alpha}_s = g_1 c_0\frac{\Delta}{L}\dot{\beta}_s,$$

d'où:

$$\frac{dv_s}{d\pi} = g_1 c_0\frac{\Delta}{L}\frac{d\dot{\beta}_s}{d\pi},$$

Mais $\dot{\beta}_s$ est lié à $\pi$ par la relation (11). Décrivons celle-ci par rapport à $\pi$:

$$\frac{L}{\gamma}\frac{g_1 c_0 n_0}{K_2} = \varkappa_1\frac{d\varphi}{d\dot{\beta}_s}\frac{d\dot{\beta}_s}{d\pi}$$

d'ou:

$$\frac{d\varphi}{d\dot{\beta}_s} = \frac{L}{\gamma}\frac{g_1 c_0 n_0}{K_2\varkappa_1}\frac{1}{d\dot{\beta}_s/d\pi} = \frac{L\,g_1^2 c_0^2 n_0\Delta}{\gamma K_2\varkappa_1}\frac{d\pi}{dv_s}$$

Prenons les valeurs moyennes pour une variation de $\dot{\beta}_s$ égale à celle de $\dot{\beta}$ lors de la variation de $\dot{\chi}$:

$$U_m = \frac{L g_1^2 c_0^2 n_0 \Delta}{\gamma K_2 \varkappa_1} \left(\frac{d\pi}{dv_s}\right)_m. \tag{14}$$

Le système (12) linéarisé s'écrit maintenant:

$$m_{11}\ddot{\xi} + \gamma\xi + m_{12}\frac{\Delta}{L}\ddot{\eta} = 0,$$

$$m_{12}\ddot{\xi} + m_{22}\frac{\Delta}{L}\ddot{\eta} + \frac{\gamma K_2}{n_0}(\eta - \chi) = 0, \tag{15}$$

$$\chi + \varkappa_1 U_m \dot{\chi} = \eta.$$

Cherchons l'effet d'accompagnement de $\chi$ sur $\eta$ à la fréquence $\omega$:

$$\frac{(\chi)}{(\eta)} = \frac{1}{1 + \varkappa_1 U_m j\omega} = \frac{1 - \varkappa_1 U_m j\omega}{1 + \varkappa_1^2 U_m^2 \omega^2};$$

d'ou:

$$\eta - \chi = \frac{\varkappa_1^2 U_m^2 \omega^2}{1 + \varkappa_1^2 U_m^2 \omega^2}\eta + \frac{\varkappa_1 U_m}{1 + \varkappa_1^2 U_m^2 \omega^2}\dot{\eta}.$$

En supposant $\varkappa_1^2 U_m^2 \omega^2$ grand devant l'unité, on peut écrire approximativement:

$$\eta - \chi = \eta + \frac{1}{\varkappa_1 U_m \omega^2}\dot{\eta}.$$

Les deux premières équations (15) deviennent alors:

$$m_{11}\ddot{\xi} + \gamma\xi + m_{12}\frac{\Delta}{L}\ddot{\eta} = 0,$$

$$m_{12}\ddot{\xi} + m_{22}\frac{\Delta}{L}\ddot{\eta} + \gamma\frac{K_2}{n_0}\left(\frac{1}{\varkappa_1 U_m \omega^2}\dot{\eta} + \eta\right) = 0. \tag{16}$$

En l'absence du terme en $\dot{\eta}$, le système serait conservatif (nous avons négligé l'amortissement interne dans le domaine élastique). La présence de ce terme introduit visiblement un amortissement sur les deux paramètres $\xi$ et $\eta$.

Posons:

$$\frac{1}{\varkappa_1 U_m \omega^2} = \frac{L}{\Delta}f. \tag{17}$$

Nous verrons que nous pouvons traiter $f$ comme un infiniment petit du premier ordre. L'équation caractéristique de (16) s'écrit:

$$\begin{vmatrix} m_{11} p^2 + \gamma & m_{12} \dfrac{\varDelta}{L} p^2 \\[2ex] m_{12} p^2 & m_{22} \dfrac{\varDelta}{L} p^2 + \dfrac{\gamma K_2}{n_0} \left( \dfrac{L}{\varDelta} f p + 1 \right) \end{vmatrix} = 0.$$

Elle admet une racine voisine de $j \sqrt{\gamma/m_{11}}$. Désignons cette racine par $p = j\omega + \varepsilon$ et calculons les parties principales de $\omega$ et de $\varepsilon$ en fonction de l'infiniment petit principal $f$. Il vient, après un calcul simple:

$$\omega^2 \simeq \frac{\gamma}{m_{11}} - \frac{m_{12}^2 \gamma n_0}{m_{11}^3 K_2} \frac{\varDelta}{L},$$

$$\varepsilon \simeq \frac{L}{\varDelta} f \frac{m_{12}^2 \gamma^2 \dfrac{\varDelta}{L}}{2 \, m_{11}^3 \dfrac{\gamma K_2}{n_0}} = -f \frac{m_{12}^2 \gamma n_0}{2 \, m_{11}^3 K_2}.$$

En remplaçant $f$ par sa valeur tirée de (17) et de (14) et limitant, d'autre part, $\varepsilon$ à sa partie principale en $\dfrac{\varDelta}{L}$, on trouve:

$$\varepsilon \sim - \frac{m_{12}^2}{m_{11}} \frac{\omega^2}{2 g_1^2 c_0^2 L^2} \left( \frac{d v_s}{d \pi} \right)_m. \tag{18}$$

Il est remarquable que la partie principale de $\varepsilon$ ne dépende de $\dfrac{\varDelta}{L}$ que par l'intermédiaire de $\dfrac{d v_s}{d \pi}$ $\left(\text{qui est, toutes choses égales d'ailleurs,} \right.$ infiniment petit comme $\left. \dfrac{\varDelta}{L} \right)$.

Le taux d'amortissement apporté par le fluage est:

$$\frac{|\varepsilon|}{\omega} = \frac{m_{12}^2}{m_{11}} \frac{\omega}{2 g_1^2 c_0^2 L^2} \left( \frac{d v_s}{d \pi} \right)_m. \tag{19}$$

$\Big[ L$: longueur de référence; $\omega$: fréquence de la vibration (libre ou entretenue à la résonance); $\pi$: force appliquée en $L$; $\underline{v}_s$: vitesse en $L$ du fluage établi sous $\pi$; $m_{11}$, $m_{12}$, $g_1$, $c_0$: constantes caractéristiques de la structure et de la base de référence définies comme suit: $m_{11}$ et $m_{12}$ selon (9 A) $\Big\{$avec $\dfrac{\varDelta}{L} = 0$ et $\theta = 1\Big\}$, $g_1 = g(1)$, $c_0 = \dfrac{L}{\varDelta} f' \left( \dfrac{\varDelta}{L} \right) \Big]$.

Dans le cas particulier où la force $\pi$ est produite par une masse $M$ fixée au point $L$ de la structure et agissant verticalement, et où la masse de la structure est négligeable devant $M$, la formule (19) se simplifie considérablement:

$$m_{11} \simeq M L^2, \qquad m_{12} \simeq M L^2 g_1 c_0, \qquad \pi = M g$$

($g$: intensité de la pesanteur), d'où:

$$\frac{|\varepsilon|}{\omega} \simeq \frac{M \omega}{2g} \left( \frac{d v_s}{d M} \right)_m.$$

### Bibliographie

[1] DE CARBON, C.: C.R. Acad. Sc. Paris **215**, 241 (1942).

[2] PERSOZ, B.: Rheologica Acta, No. 2—3, 90 (1958).

[3] PERSOZ, B.: Cahiers de Rhéologie **IV**, No. 3, 31 (1959).

[4] MAZET, R.: Creep in Structures, Berlin/Göttingen/Heidelberg: Springer 1962, p. 326.

[5] NAYROLES, B.: C.R. Acad. Sc. Paris **254**, 3485 (1962).

[6] MAZET, R.: C.R. Acad. Sc. Paris **256**, 2971 (1963).

# Dynamics of an Anelastic String of Variable Length

By G. N. Savin

Institute of Mechanics, Kiev, Ukr. S.S.R.

This paper briefly surveys current methods of solving problems concerning the motion of elastic or anelastic strings of variable length such as cables of mine cages. Particular attention will be given to the propagation of stress waves in these cables when the law of winding is prescribed.

The investigations described in Reference [2] show that the methods of solution developed for strings of variable length may be used to solve dynamic problems concerning other one-dimensional continua of variable length such as rods or beams (see Section 11 in Reference 2).

It was shown in Reference [1] that the motion of an anelastic string of variable length is described by a very intricate system of functional-differential equations in a variable domain with equally complicated boundary conditions. Accordingly, an approximate solution of the problem was emphasized in Reference [1], which consists in the substitution of a system of initial equations into an infinite number of integro-differential relations. By a suitable choice of the function that is to be determined, the integro-differential equations can be reduced to a system of ordinary differential equations of the second order with variable coefficients. These equations were investigated in a number of papers by the author and his students [2]. Methods devised by N. M. KRYLOV and N. N. BOGOLUBOV [5], YU. A. MITROPOLSKY [7], S. F. FESHCHENKO [8], and YU. D. SOKOLOV [9] turned out to be most effective.

With the aid of electronic computers, these methods have yielded the following interesting results concerning the motion of anelastic strings of variable length: (a) when a weight attached to the lower end of the string is lowered at constant speed, the stress waves in the string are rapidly damped (the same is true for an elastic string); (b) depending on the degree of energy dissipation, there exists a critical value of the speed of ascent such that the stress waves in the string are damped; (c) on account of dissipation of energy, the higher harmonics are damped more rapidly than the lower harmonics. This is important because it enables us to solve the problem in terms of a single ordinary differential equation with variable coefficients.

The possibility of slip between string and drum must be considered in formulating the appropriate boundary conditions for a string of variable length.

There exists a velocity of ascent at which the stress in the string is within 2 to 3% of the static value corresponding to the weight attached to its lower end. Moreover, there exist velocities of ascent for which there is no slip between string and drum and the stress in the string wound on the drum equals, within 2 to 3%, the weight that is being raised. The investigation of the motion of an anelastic string of variable length[1] have led to the formulation of new problems in the mechanics of one-dimensional continua of variable length. These concern hyperbolic equations where the domain under consideration is fixed in neither variable. No general formulation or solution has as yet been found, and the authors had to adapt existing methods to the solution of the problem on hand.

It should be noted that methods of solving problems of this kind necessarily differ from the classical methods of mathematical physics. For instance, the concepts of natural modes and frequencies become meaningless because with the length of the string the natural frequencies become time-dependent, and the independence of the natural modes of oscillations is lost.

The necessity of treating time-dependent phenomena of this kind has led to the development of special asymptotic methods by the Ukrainian mathematicians N. M. KRYLOV and N. N. BOGOLUBOV [5] and YU. A. MITROPOLSKY [6]. These prove to be particularly convenient for the treatment of problems concerning the motion of one-dimensional continua of varying length.

In our opinion, the most convenient and most general method of treating problems of this kind is the method of integro-differential equations with variable limits of integration indicated in Reference [2], although this reference deals primarily with the motion of an anelastic string (mine winding rope), some generalizations are given in Sections 11, 12, and 23 through 25. These generalizations concern rods, beams, and shafts of variable length as well as systems subjected to moving loads.

While in principle the motion of an unelastic string of variable length can be described to any desired degree of approximation by an infinite system of integro-differential equations, the mathematical difficulties usually become prohibitive for all but the first two orders of approximation. In this case, however, the problem can be reduced to a single integro-differential equation.

---

[1] Stresses in elastic winding ropes were first investigated by J. PERRY [3] and A. N. DINNIK [4].

The basic idea of the method used to derive this kind of equation is readily discussed for the following simple example of a string of variable length. Suppose that the upper end of the string is wound on a rotating drum the surface of which has the given linear velocity $V_c(t)$ (Fig. 1). To facilitate the discussion, we imagine the length $CO'$ of string that is wound on the drum to be unwound and straightened out along the vertical $CO$ without change in longitudinal strains.

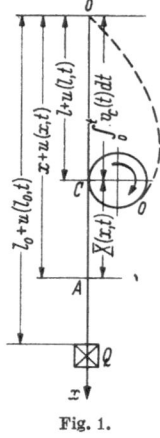

Fig. 1.

Choosing the positive $x$ direction vertical and downward, and denoting the Lagrangian and Eulerian coordinates of the typical cross section $A$ (Fig. 1) by $x$ and $X(x, t)$, respectively, we have

$$X(x, t) = x + u(x, t) - \int_0^t V_c(t)\,dt, \qquad (1)$$

where $u(x, t)$ is the elongation which the length $O'CA$ of the string has experienced.

For an elastic string, the relation between the force $P(x, t)$ and the strain $\varepsilon = \partial u/\partial x$ is

$$P(x, t) = EA\,\partial u/\partial x, \qquad (2)$$

where $E$ is the modulus of elasticity and $A$ the cross sectional area. Denoting the weight per unit length of the string by $q$, we therefore have the equation of motion

$$\frac{q}{g}\,\frac{\partial^2 X}{\partial t^2} - \frac{\partial P}{\partial x} = q. \qquad (3)$$

In view of (1) and (2), Eq. (3) can be written as follows:

$$\frac{q}{g}\,\frac{\partial^2 u}{\partial t^2} = EA\,\frac{\partial^2 u}{\partial x^2} + q\left(1 + \frac{\dot{V}_c}{g}\right), \qquad (4)$$

where $\dot{V}_c = dV_c/dt$.

The equation of motion of the weight $Q$ at the lower end of the string furnishes the following boundary condition at that end:

$$\left(\frac{Q}{g}\,\frac{\partial^2 u}{\partial t^2} + EA\,\frac{\partial u}{\partial x}\right)_{x=l_0} = Q\left(1 + \frac{\dot{V}_c}{g}\right) \qquad (5)$$

where $l_0$ is the original length of the string.

If the original length of the part $CO$ of the string is denoted by $l$, we have the material derivative

$$\frac{du(l, t)}{dt} = \left(\frac{\partial u}{\partial x}\right)_{x=l} \cdot \frac{dl}{dt} + \left(\frac{\partial u}{\partial t}\right)_{x=l}. \qquad (6)$$

If there is no slip between string and drum, we have $(\partial u/\partial t)_{x=l} = 0$ and (6) reduces to

$$\frac{du(l, t)}{dt} = \left(\frac{\partial u}{\partial x}\right)_{x=l} \cdot \frac{dl}{dt}. \tag{7}$$

On the other hand, if at $C$ the string slips with respect to the drum (Fig. 1), the relation (7) must be replaced by

$$\frac{du(l, t)}{dt} = \left(\frac{\partial u}{\partial x}\right)_{x=l} \cdot \frac{dl}{dt} + b\left(\frac{\partial u}{\partial t}\right)_{x=l}, \tag{8}$$

where $b$ is a slip coefficient.

Integration of (7) and (8) furnishes the following boundary conditions at $C$:

$$u(l, t) = \int_0^t \left(\frac{\partial u}{\partial x}\right)_{x=l} \cdot \frac{dl}{dt} \cdot dt$$

or $\qquad\qquad\qquad\qquad\qquad\qquad\qquad\qquad\qquad\qquad\qquad\qquad$ (9)

$$u(l, t) = \int_0^t \left(\frac{\partial u}{\partial x} \frac{dl}{dt} + b\frac{\partial u}{\partial t}\right)_{x=l} \cdot dt.$$

In general, the Eqs. (9) cannot be integrated.

The original length of that part of the string that is wound on the drum is found from the relation

$$l(t) = \int_0^t \frac{V_c(t)dt}{1 + \partial u(l, t)/\partial x}, \tag{10}$$

which may be replaced by the simpler relation

$$l(t) = \int_0^t V_c(t)dt \tag{11}$$

as $(\partial u(l, t)/\partial x)_{x=l} \ll 1$.

The initial conditions of the problem are

$$u(x, 0) = f_1(x), \qquad \frac{\partial u(x, 0)}{\partial t} = f_2(x), \tag{12}$$

where $f_1(x)$ and $f_2(x)$ are known functions characterizing the elongation and velocity of the string at the instant $t = 0$.

The problem under consideration is therefore reduced to that of integrating Eq. (4) under the boundary conditions (5) and (9) with the initial conditions (12) with due regard to relations (10) and (11). To obtain

the integro-differential equation equivalent to the differential Eq. (4)
with the boundary conditions (5) and (9), we introduce the function

$$K(x,\, s,\, t) = \begin{cases} (s - l)/(EA) & \text{for } s \le x, \\ (x - l)/(EA) & \text{for } s \ge x. \end{cases} \tag{13}$$

Note that the time enters into this function through $l(t)$ in accordance
with (10) or (11). For $l(t) = \text{const}$, the function (13) may be interpreted
as an influence function (Fig. 2).

Multiplying both sides of (4) by $K(x,\, s,\, t)$, integrating from $l(t)$ to $l_0$,
and using (5) and (9), we obtain

$$u(x,\, t) = -\int_{l(t)}^{l_0} K(x,\, s,\, l)\, \frac{q}{g} \left[ \frac{\partial^2 u}{\partial t^2} - g - \dot{V}_c \right] ds$$

$$- \frac{x - l(t)}{EA}\, \frac{Q}{g} \left[ \frac{\partial^2 u}{\partial t^2} - g - \dot{V}_c \right]_{x = l_0} \tag{14}$$

$$+ \int_0^t \frac{\partial u(l,\, t)}{\partial x}\, \dot{l}\, dt.$$

If we set

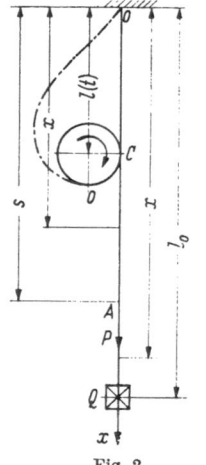

Fig. 2

$$\varrho(x) = \frac{1}{g}\, [q + Q\delta(x - l_0)], \tag{15}$$

where $\delta(x - l)$ is the Dirac function, Eq. (14)
becomes

$$u(x,\, t) = -\int_{l(t)}^{l_0} K(x,\, s,\, l)\varrho(s) \left[ \frac{\partial^2 u}{\partial t^2} - g - \dot{V}_c \right] ds \tag{16}$$

$$+ \int_0^t \frac{\partial u(l,\, t)}{\partial x}\, \dot{l}\, dt,$$

where $\dot{l} = dl/dt$.

Introducing

$$v(x,\, t) = u(x,\, t) - \int_0^t \frac{\partial u(l,\, t)}{\partial x}\, \dot{l}\, dt,$$

and noting that $\partial u/\partial x = \partial v/\partial x$ we find

$$u(x,\, t) = v(x,\, t) + \int_0^t \frac{\partial v(l,\, t)}{\partial x}\, \dot{l}\, dt. \tag{17}$$

Substituting (17) into (16), we finally obtain the following integro-differential equation for the considered motion of an elastic string of variable length:

$$v(x, t) = -\int_{l(t)}^{l_0} K(x, s, l)\varrho(s)\left[\frac{\partial^2 v(s, t)}{\partial t^2} - g - \dot{V}_c + \frac{d}{dt}\left(i\,\frac{\partial v(l, t)}{\partial x}\right)\right]ds. \quad (18)$$

This equation is equivalent to the original system of Eqs. (4) through (7) (see Section 11 of Reference [2]).

It follows from (17) and (12) that the initial conditions for the function $v(x, t)$ have the form

$$v(x, 0) = f_1(x), \qquad \frac{\partial v}{\partial t} = f_2(x) - f_1'[l(0)]\dot{l}(0). \quad (19)$$

The integro-differential equation for an anelastic string will naturally differ from (18) on account of the difference in the relations between stresses and strains.

For example, for a visco-elastic string with the stress-strain relation

$$P(x, t) = EA\,\frac{\partial}{\partial x}\left(u + \mu\,\frac{\partial u}{\partial t}\right), \quad (20)$$

the integro-differential equation has the form

$$v(x, t) + \mu\,\frac{\partial v(x, t)}{\partial t} = -\int_{l(t)}^{l_0} K(x, s, l)\,\frac{q}{g}\left[\frac{\partial^2 v}{\partial t^2} - g - \dot{V}_c\right]ds$$

$$- K(x, l_0, l)\,\frac{Q}{g}\left(\frac{\partial^2 v}{\partial t^2} - g - \dot{V}_c\right)_{x=l}$$

$$- \frac{d}{dt}\left(i\,\frac{\partial v}{\partial x}\right)_{x=l}\left[K(x, l_0, l)\,\frac{Q}{g} + \int_{l(t)}^{l_0} K(x, s, l)\,\frac{q}{g}\,ds\right]. \quad (21)$$

Appropriate initial conditions must be added to this equation. No exact solution of (18) or (21) seems to be known. Approximate solutions were therefore emphasized in References [1] and [2]. Even finding an approximate solution is in general a very difficult problem.

When the kernel parameters in (18) or (21) and the limits of integration depend weakly on time, that is, when they vary only slightly during one natural period, the asymptotic methods of nonlinear vibration theory developed by YU. A. MITROPOLSKY [7] for systems with a finite number of degrees of freedom may be adapted to the solution of these equations.

6*

The main time-dependent parameter in (18) or (21) is the function $l(t)$, which will be assumed to depend on $\tau = \epsilon t$, where $\epsilon$ is small in comparison to unity. The derivative

$$\frac{dl}{dt} = \epsilon \frac{dl}{d\tau}$$

is then proportional to $\epsilon$. This condition is fulfilled in all examples considered in Reference [2].

To find the numerical value of this small parameter $\epsilon$, one must introduce dimensionless coordinates and time

$$x' = x/l_0, \qquad t' = \omega t, \tag{22}$$

where $l_0$ is the total length of the string (Fig. 2) and $\omega$ the corresponding lowest natural frequency.

Considering $l$ as a function of $\tau = \epsilon t$, using dimensionless coordinates and time, and omitting primes for the sake of brevity, we obtain the equation

$$v(x, t) = -\int\limits_{l(\tau)}^{l_0} K(x, s, t) \left[\frac{\partial^2 v}{\partial t^2} - g - \dot{V}_c\right] \varrho(s)\,ds$$

$$+ \epsilon \left[\int\limits_{l(\tau)}^{l} K(x, s, t)\varrho(s)\,ds\right] \frac{d}{dt}\left(\frac{dl}{d\tau} \cdot \frac{\partial v(l, t)}{\partial x}\right). \tag{23}$$

For this kind of equation, the conventional definitions of natural modes and frequencies are meaningless. It is however reasonable to assume that with the slowly varying parameters in (23) the modes and frequencies determined by this integro-differential equation do also vary slowly.

For a string of fixed length with a weight $Q$ at the lower end, it is found (Section 24 of Reference [2]) that over a wide range of values of

$$\alpha = q(l_0 - l)/Q$$

the natural modes which are the eigenfunctions of the integral equation

$$v(x, t) = -\int\limits_{l=\text{const}}^{l_0} K(x, s)\varrho(s) \frac{\partial^2 v(s, t)}{\partial t^2}\,ds \tag{24}$$

are very close to the functions

$$\varphi_1(x) = x - l,$$

$$\varphi_2(x) = (x - l)\left(Q + \frac{q(l_0 - l)}{4} - \frac{(x - l)^2}{l_0 - l}\left(Q + \frac{q(l_0 - l)}{3}\right),\right.$$

$$\varphi_3(x) = (x - l)\begin{vmatrix} b_4 & b_5 \\ b_5 & b_6 \end{vmatrix} - \frac{(x - l)^2}{l_0 - l}\begin{vmatrix} b_3 & b_5 \\ b_4 & b_6 \end{vmatrix} + \frac{(x - l)^3}{(l_0 - l)^2}\begin{vmatrix} b_3 & b_4 \\ b_4 & b_5 \end{vmatrix}, \quad (25)$$

and so on, where

$$b_n = Q + \frac{q(l_0 - l)}{n} \qquad (n = 3, 4, 5, \ldots).$$

The functions (25) constitute a full system of orthogonal functions with respect to the weighting function

$$\varrho(x) = \frac{1}{g}\left[q + Q\delta(x - l_0)\right].$$

In Reference [2], these functions have been called "rope functions".
As is readily seen, Eq. (24) is the homogeneous part of Eq. (23) with $l = \text{const}$ and hence $dl/dt = 0$, $V_c = 0$ and $\dot{V}_c = 0$.
The asymptotic solution of Eq. (23) is written in the form

$$v(x, t) = \sum \varphi_i(x, l)[a_i \cos \psi_i + \varepsilon u_i^{(1)}(a, \psi, \tau)]$$

$$+ \varepsilon^2 u_i^{(2)}(a, \psi, \tau) + \cdots + \varepsilon^n u_i^{(n)}(a, \psi, \tau)]$$

$$+ \int_{l(t)}^{l_0} K(x, s, l)\varrho(s)(g - \dot{V}_c)ds, \qquad (26)$$

where $\varphi_i(x, l)$ are the functions (25), and $a_i(t)$ and $\psi_i(t)$ are functions found from a corresponding system of ordinary differential equations of the first order in accordance with the methods of nonlinear theory of vibrations developed by N. N. BOGOLUBOV and YU. A. MITROPOLSKY (see References [6] and [7]). For the problem on hand, these methods were discussed in detail in Section 25 of Reference [2]. Chapters II through VII of this reference deal with waves of stress and strain in both elastic and anelastic strings of variable length throughout the entire cycle of raising and lowering the weight when the time-variation of the speed of winding or unwinding is represented by trapezoids. The integro-differential Eqs. (18) and (23), and the initial conditions (19) are used in this investigation. The asymptotic method of S. F. FESHCHENKO [10] is found to be the most convenient one for the solution of the integro-differential equations (see Eq. (11) in Reference [1] and Section 25 of Reference [2]).

## References

[1] SAVIN, G. N.: Some Dynamic Problems of Non-Elastic Thread of Variable Length, Proceedings, 9th International Congress of Applied Mechanics, Brussels, 1958.
[2] SAVIN, G. N., and O. A. GOROSHKO: Dinamika niti peremennoy dliny, (Russian), Izdatelstvo Akad. Nauk Ukr. SSR, Kiev 1962.
[3] PERRY, J.: Phil. Mag., ser. 6, 11, 107 (1906).
[4] DINNIK, A. N.: Yuzhny Inzhener (Russian), No. 11—12 (1916).
[5] KRYLOV, N. M., and N. N. BOGOLUBOV: Vvedenie v nelineinuyu mehaniku (Russian), Kiev, 1937.
[6] BOGOLUBOV, N. N., and YU. A. MITROPOLSKY:. Asimptoticheskie metody v teorii nelineinyh kolebaniy (Russian), Fizmatigiz, 1958.
[7] MITROPOLSKY, YU. A.: Nestazionarnye prozessy v nelineinyh kolebatelnyh sistemah (Russian), Izdatelstvo Akad. Nauk Ukr. SSR, 1955.
[8] SAVIN, G. N., and S. F. FESHCHENKO: Dopovidi Akademii Nauk SSR (Ukrainian), No. 6 (1958).
[9] SOKOLOV, YU. D.: Prykladna Mehanika (Ukrainian) 1, No. 1 (1955).
[10] FESHCHENKO, S. F., and M. I. SHKIL: Prykl. Meh. (Ukrainian) 4, No. 3 (1958).

# Anelastic Waves in Thin Plates

## By Henryk Zorski[1]

Polish Academy of Sciences, Warsaw, Poland

## Introduction

In investigating small displacements and the resulting state of stress in thin plates or shells obeying a linear stress-strain law, the three-dimensional problem is reduced to a two-dimensional one by means of a hypothesis concerning the distribution of the relevant geometric and mechanical quantities along the thickness of the shell or plate. The simplest hypothesis in bending is the KIRCHHOFF hypothesis, while for the motion "in the $xy$ plane" it is usually assumed that a plane state of stress prevails. However, in the case of high frequencies and short wave lengths, particularly in the case of anelastic materials the equations of which contain higher time derivatives, the above assumptions are incorrect and it is necessary to establish a system of equations taking into account additional phenomena, such as rotary inertia, shear displacement, etc. This is done by changing the hypothesis or employing a method of successive approximations.

In this paper an attempt is made to derive the required system of equations on a different basis. The reduction to a two-dimensional problem is carried out by means of the exact solution for an infinite layer in an operational form.

We shall mainly be concerned with the problem of bending.

## 1. The Fundamental Equations

The LAMÉ equations for displacements of the form[2] $e^{i\omega t} u_i (x^m)$

$$\mu \nabla^2 u_i + (\lambda + \mu) \nabla_i \nabla_p u^p = -\omega^2 u_i \qquad (1.1)$$

are satisfied (in rectangular coordinates) when

$$u_i = \partial_i \varphi + \varepsilon_i^{pq} \partial_p \psi_q \qquad (1.2)$$

---

[1] In 1962–63 Visiting Professor, University of Kansas, Lawrence, Kansas, U.S.A.
[2] $i, j, k, \ldots = 1, 2, 3$, while the Greek indices $\alpha, \beta, \mu, \ldots$ run from 1 to 2.

provided the potentials satisfy the HELMHOLTZ equations

$$(\nabla^2 + K_1^2)\varphi = 0, \qquad (\nabla^2 + K_2^2)\psi_i = 0, \qquad (1.3)$$

where $K_1^2 = \dfrac{\omega^2}{c_1^2}$, $K_2^2 = \dfrac{\omega^2}{c_2^2}$, $C_1^{-2} = \dfrac{\sigma}{\lambda + 2\mu}$, $C_2^{-2} = \dfrac{\sigma}{\mu}$. We assume that in general the body is anelastic and the quantities $\mu$, $\lambda$ depend on frequency $\omega$. The deformations and stresses are given by the formulae

$$\varepsilon_{ij} = \partial_i \partial_j \varphi + \varepsilon_{(i}^{pq} \partial_{j)}, \partial_p \psi_q, \qquad (1.4a)$$

$$\sigma_{ij} = \lambda \delta_{ij} \nabla^2 \varphi + 2\mu(\partial_i \partial_j \varphi + \varepsilon_{(i}^{pq} \partial_{j)} \partial_p \psi_q). \qquad (1.4b)$$

Without loss of generality we assume that $\psi_3 \equiv 0$. Applying the double FOURIER transform $(x^\alpha) \to (\xi_\alpha)$, $\alpha = 1, 2$, and denoting the images of the potentials by $\tilde{\varphi}(\xi_\alpha, z)$ and $\tilde{\psi}_\alpha(\xi_\alpha, z)$ we have the solutions of (1.3) $(\xi^2 = \xi_1^2 + \xi_2^2, z = x^3)$:

$$\tilde{\varphi}(\xi_\alpha, z) = A(\xi_\alpha) \cosh \gamma_1 z + B(\xi_\alpha) \sinh \gamma_1 z, \qquad \gamma_1 = \sqrt{\xi^2 - K_1^2},$$

$$\tilde{\psi}_\alpha(\xi_\alpha, z) = A_\alpha(\xi_\alpha) \cosh \gamma_2 z + B_\alpha(\xi_\alpha) \sinh \gamma_2 z, \qquad \gamma_2 = \sqrt{\xi^2 - K_2^2}. \qquad (1.5)$$

Consider now an infinite layer bounded by planes $z = \pm h$ and loaded on its surfaces by forces $q_i^\pm(x^\alpha)$; then the boundary conditions take the form

$$\tilde{\sigma}_{33}(\xi_\alpha, \pm h) = \tilde{q}_3^\pm(\xi_\alpha), \qquad \sigma_{\alpha 3}(\xi_\alpha, \pm h) = \tilde{q}_\alpha^\pm(\xi_\alpha). \qquad (1.6)$$

We confine ourselves in this paper to forces $q_\alpha^\pm(x_\alpha)$ equal at corresponding points $x^\alpha$ and acting in the same directions; generalization to arbitrary forces presents no difficulties. Thus, we have $q_i^+(x^\alpha) + q_i^-(x^\alpha) = 0$. We assume moreover that the forces are concentrated at point $x^\alpha = 0$. Introducing the notations $\tilde{Q}_i = \tilde{Q}_i^0 \delta(x)\delta(y)$ $(x = x^1, y = x^2, \delta$ is the Dirac function) we obtain the following expressions for constants $A, \cdots, B_2$;

$$A = -\frac{i}{4\mu h}(\xi_1 \tilde{Q}_1^0 + \xi_2 \tilde{Q}_2^0) \frac{\cosh \gamma_2 h}{\mathcal{M}^{(1)}}, \quad B = \frac{1}{4\mu h} \frac{1}{\gamma_1}\left(\xi^2 - \frac{1}{2}K_2^2\right)\tilde{Q}_3^0 \frac{\cosh \gamma_2 h}{\mathcal{M}^{(2)}},$$

$$A_1 = \frac{i}{4\mu h}\xi_2 \tilde{Q}_3^0 \frac{\cosh \gamma_1 h}{\mathcal{M}^{(2)}}, \qquad A_2 = -\frac{i}{4\mu h}\xi_1 \tilde{Q}_3^0 \frac{\cosh \gamma_1 h}{\mathcal{M}^{(2)}},$$

$$B_1 = -\frac{1}{2\mu h}\frac{1}{\gamma_2^3}\frac{1}{\mathcal{M}^{(1)}}\left\{\tilde{Q}_1^0\left[\frac{1}{2}\left(\xi^2 - \frac{1}{2}K_2^2\right)\xi_1\xi_2 \cosh \gamma_1 h\right.\right.$$

$$\left. - \xi_1\xi_2\gamma_1\gamma_2 \frac{\sinh \gamma_1 h}{\sinh \gamma_2 h}\cosh \gamma_2 h\right] - \tilde{Q}_2^0\left[\frac{1}{2}\left(\xi^2 - \frac{1}{2}K_2^2\right)\left(\gamma_2^2 + \xi_1^2\right)\right.$$

$$\left.\left.\cosh \gamma_1 h - \xi_1^2\gamma_1\gamma_2 \frac{\sinh \gamma_1 h}{\sinh \gamma_2 h}\cosh \gamma_2 h\right]\right\},$$

$$B_2 = -\frac{1}{2\mu h}\frac{1}{\gamma_2^3}\frac{1}{m^{(1)}}\left\{\tilde{Q}_1^0\left[\frac{1}{2}\left(\xi^2-\frac{1}{2}K_2^2\right)(\gamma_2^2+\xi_2^2)\cosh\gamma_1 h\right.\right.$$

$$-\xi_2^2\gamma_1\gamma_2\frac{\sinh\gamma_1 h}{\sinh\gamma_2 h}\cosh\gamma_2 h\Big] - \tilde{Q}_2^0\left[\frac{1}{2}\left(\xi^2-\frac{1}{2}K_2^2\right)\xi_1\xi_2\cosh\gamma_1 h\right.$$

$$\left.\left.-\xi_1\xi_2\gamma_1\gamma_2\frac{\sinh\gamma_1 h}{\sinh\gamma_2 h}\cosh\gamma_2 h\right]\right\},$$

where

$$\mathcal{M}^{(1)}(\xi_\alpha) = \left(\xi^2-\frac{1}{2}K_2^2\right)\frac{\sinh\gamma_2 h}{\gamma_2 h}\cosh\gamma_1 h - \xi^2\gamma_1^2\frac{\sinh\gamma_1 h}{\gamma_1 h}\cosh\gamma_2 h, \quad (1.7)$$

$$\mathcal{M}^{(2)}(\xi_\alpha) = \left(\xi^2-\frac{1}{2}K_2^2\right)\frac{\sinh\gamma_1 h}{\gamma_1 h}\cosh\gamma_2 h - \xi^2\gamma_2^2\frac{\sinh\gamma_2 h}{\gamma_2 h}\cosh\gamma_1 h.$$

Calculating the displacements we find that the expression for $\tilde{w}(\xi_\alpha,z)$ contains $\tilde{Q}_3^0$ only, while $\tilde{u}^\alpha(\xi_\alpha,z)$ contain both $\tilde{Q}_1^0$ and $\tilde{Q}_2^0$; taking linear combinations of the latter so that the right-hand sides contain $\tilde{Q}_1^0$ and $\tilde{Q}_2^0$ only, we arrive at the required expressions. In what follows we choose as characteristic the displacements of the middle surface[1]. Thus, setting $z=0$ we have

$$\mathcal{L}_{\alpha\beta}^{(u)}(\xi_\alpha)\tilde{u}^\beta(\xi_\alpha,0) = \frac{1}{2\mu}\mathcal{K}^{(1)}(\xi_\alpha)\tilde{Q}_\alpha^0, \qquad (1.8)$$

$$\mathcal{L}^{(w)}\tilde{w}(\xi_\alpha,0) = -\frac{K_2^2}{8\mu h}\mathcal{K}^{(2)}(\xi_\alpha)\tilde{Q}_3^0,$$

where

$$\mathcal{L}_{\alpha\beta}^{(u)}(\xi_\alpha) = \frac{2h}{K_2^2}\left\{-\mathcal{K}^{(1)}\gamma_2^2\frac{\sinh\gamma_2 h}{\gamma_2 h}\delta_{\alpha\beta} + \left[\gamma_2^2\frac{\sinh\gamma_2 h}{\gamma_2 h}\cosh\gamma_2 h\right.\right. \qquad (1.9)$$

$$\left.\left.+\left(\xi^2-\frac{1}{2}K_2^2\right)\frac{\sinh\gamma_2 h}{\gamma_2 h}\cosh\gamma_1 h - 2\gamma_1^2\frac{\sinh\gamma_1 h}{\gamma_1 h}\cosh\gamma_2 h\right]\xi_\alpha\xi_\beta\right\},$$

$$\mathcal{L}^{(w)}(\xi_\alpha) = \mathcal{M}^{(2)}(\xi_\alpha).$$

On the basis of relations (1.8) we proceed to derive the required partial differential equations of equilibrium in two variables $x$ and $y$, for the displacements of the middle plane of the plate. The following reasoning is essential for the reduction of the problem to two dimensions.

The system of equations (1.8) (a system of two equations and a separate equation for $w$ — the problem is split up into "bending" and "plane state") is a system for the images (in the sense of the Fourier double transform) of characteristic displacements of the plate; the

---

[1] If we choose, for instance, the displacements of one of the bounding planes, the resulting equations would be different.

functions of $\xi_\alpha$, $\mathscr{L}_{\alpha\beta}^{(u)}$ and $\mathscr{L}^{(w)}$, $\mathscr{K}^{(1)}$, $\mathscr{K}^{(2)}$ are analytic in the arguments $\xi_\alpha$ and moreover, are analytic functions in parameter $h$ which can be regarded as a small parameter. Multiplication of, say, $\tilde{u}^\beta$ by $(-i\xi_\alpha)$ corresponds to the differentiation of the original $u^\beta(x, y)$ with respect to $x_\alpha$, i.e.

$$(-i\xi_\alpha)\,\tilde{u}^\beta \doteqdot \frac{\partial u^\beta}{\partial x^\alpha} \tag{1.10}$$

or, more generally, in curvilinear coordinates to the covariant differentiation, i.e. $(-i\xi_\alpha)\tilde{u}^\beta \doteqdot \nabla_\alpha u^\beta$. Consequently, Eqs. (1.8) may be regarded as operational equations (in the spirit of MIKUSIŃSKI's operational calculus, [1], with analytic operators $\mathscr{L}_{\alpha\beta}^{(u)}(\xi_\alpha)$, $\mathscr{L}^{(w)}(\xi_\alpha)$, $\mathscr{K}^{(1)}(\xi_\alpha)$, $\mathscr{K}^{(2)}(\xi_\alpha)$ containing moreover (analytically) the small parameter $h$. Functions $\cosh\gamma_1 h$, $\gamma_2^2$, $\dfrac{\sinh\gamma_2 h}{\gamma_2 h}$ etc. contain only $\xi^2$ corresponding to $-\nabla^2$ and are consequently tensorially invariant operators; thus all considered operators have tensorial properties indicated by their indices. The general form of the expansion of the operators into the power series of $h$ is the following:

$$\mathscr{L}_{\alpha\beta}^{(u)}(\xi_\alpha) = -h\left[l_{\alpha\beta}^{(u)}(0,0) + \sum_{n=1}^{\infty}\sum_{m=n-1}^{n} l^{(u)}(m,n)h^{2m}\xi^{2n}\cdot\delta_{\alpha\beta}\right.$$

$$\left. + \sum_{n=1}^{\infty}\sum_{m=n-1}^{n} l^{1(u)}(m,n)h^{2m}\xi^{2n}\cdot\xi_\alpha\xi_{\bar{\beta}}\right], \tag{1.11}$$

$$\mathscr{L}^{(w)}(\xi_\alpha) = l^{(w)}(0,0) + \sum_{n=1}^{\infty}\sum_{m=n-1}^{n} l^{(w)}(m,n)h^{2m}\xi^{2n},$$

$$\mathscr{K}^{(\alpha)}(\xi_\alpha) = K(0,0) + \sum_{n=1}^{\infty}\sum_{m=n-1}^{n} K(m,n)h^{2m}\xi^{2n}.$$

Furthermore all coefficients of the expansion are analytic functions of frequency $\omega$. To estimate the magnitude of the terms of the operators introduce the following dimensionless parameters:

$$\frac{1}{\lambda_x^2} = h^2\xi_\alpha, \qquad \nu_\alpha^2 = h^2 K_\alpha^2 = \frac{h^2\omega^2}{c_x^2}, \qquad \frac{1}{\lambda^2} = \frac{1}{\lambda_1^2} + \frac{1}{\lambda_2^2} \tag{1.12}$$

In the case of plane waves the first parameter is equal to $\dfrac{h^2}{\lambda^2}$ where $\lambda$ is the wave length; in this case Eqs. (1.8) (provided of course the external forces are of the type necessary to engender a plane wave) are simple algebraic equations for the amplitudes. In terms of these parameters, retaining terms of order $\nu_\alpha^4$, $\frac{1}{\lambda^4} \cdot \frac{\nu_\alpha^2}{\lambda^2}$ we have (in $\mathscr{L}^{(w)}$ we retain some terms of a higher order)

$$\mathscr{L}_{11}^{(u)} = -\frac{1}{h}\left\{\nu_2^2\left[1 - \frac{1}{6}(\nu_2^2 + 3\nu_1^2)\right] - \frac{4}{\lambda^2}\left[\left(1 - \frac{\nu_1^2}{\nu_2^2}\right) - \frac{1}{6}\left(2\nu_2^2 - \frac{\nu_1^4}{\nu_2^2}\right)\right]\right.$$

$$- \frac{1}{\lambda_2^2}\left[1 + \frac{1}{6}(\nu_2^2 - g\nu_1^2)\right] - \frac{4}{3}\frac{1}{\lambda_1^4}\left(1 - \frac{\nu_1^2}{\nu_2^2}\right) - \frac{1}{3}\frac{1}{\lambda_1^2\lambda_2^2}\left(3 - \frac{\nu_1^2}{\nu_2^2}\right)$$

$$\left. + \frac{1}{3}\frac{1}{\lambda_2^4}\left(1 - 3\frac{\nu_1^2}{\nu_2^2}\right)\right\},$$

$$\mathscr{L}_{12}^{(u)} = \frac{1}{h}\frac{1}{\lambda_1\lambda_2}\left\{\left[\left(3 - 4\frac{\nu_1^2}{\nu_2^2}\right) - \frac{1}{6}\left(g\nu_2^2 + 15\nu_1^2 + 4\frac{\nu_1^4}{\nu_2^2}\right)\right] - \frac{1}{3}\frac{1}{\lambda^2}\left(1 - \frac{\nu_1^2}{\nu_2^2}\right)\right\},$$

$$\mathscr{L}^{(w)} = \frac{1}{h^4}\left\{\frac{1}{2}\nu_2^4\left[1 - \frac{1}{6}(\nu_1^2 + 3\nu_2^2)\right]\right.$$

$$+ \frac{1}{6}\frac{\nu_2^2}{\lambda^2}\left[(3\nu_2^2 - 2\nu_1^2) - \frac{1}{20}(9\nu_2^4 - 4\nu_1^4 + 3\nu_1^2\nu_2^2)\right]$$

$$\left. - \frac{1}{3}\frac{1}{\lambda^4}\left[(\nu_2^2 - \nu_1^2) - \frac{1}{10}(4\nu_2^4 - \nu_1^4 - 2\nu_1\nu_2^2)\right] - \frac{1}{15}\frac{1}{\lambda^6}(\nu_2^2 - \nu_1^2)\right\},$$

$$\mathscr{K}^{(1)} = \left(1 - \frac{\nu_1^2}{2} + \frac{\nu_1^4}{24}\right) + \frac{1}{2}\frac{1}{\lambda^2}\left[\left(3 - 2\frac{\nu_2^2}{\nu_1^2}\right) - \frac{1}{6}\left(2\nu_1^2 - \frac{\nu_2^4}{\nu_1^2}\right)\right]$$

$$+ \frac{1}{24}\frac{1}{\lambda^4}\left[\left(5 - 4\frac{\nu_2^2}{\nu_1^2}\right) - \frac{1}{10}\left(3\nu_1^2 - 2\frac{\nu_2^4}{\nu_1^2}\right)\right]. \tag{1.13}$$

$\mathscr{L}_{22}^{(u)}$ is obtained by replacing in $\mathscr{L}_{11}^{(u)}$ $\xi_1^2$ by $\xi_2^2$ and $\xi_2^2$ by $\xi_1^2$. Obviously for smaller wave lengths and higher frequencies (i.e. for larger values of parameters $\frac{1}{\lambda_\alpha^2}$, $\nu_a^2$) more terms in the expansion should be retained. With this degree of accuracy we may finally write the differential equations (for purely technical reasons we change the notation for the coefficients)

$$\mathscr{L}_{\alpha\beta}^{(u)}u^\beta = \frac{1}{2\mu}\mathscr{K}^{(1)}Q_\alpha, \qquad \mathscr{L}^{(w)}w = -\frac{h}{4\mu}\mathscr{K}^{(2)}Q_3, \tag{1.14}$$

where the differential operators are given by the formulae ($g_{\alpha\beta}$ is the metric tensor of $x^\alpha$)

$$\mathscr{L}_{\alpha\beta}^{(u)} = -\frac{1}{h}\left[(l^{(1)} + h^2l^{(2)}\nabla^2 + h^4l^{(3)}\nabla^4)\cdot g_{\alpha\beta} + h^2(l^{(4)} + h^2l^{(5)}\nabla^2)\nabla_\alpha\nabla_\beta\right],$$

$$\mathscr{L}^{(w)} = l^{(6)} + h^2l^{(7)}\nabla^2 + h^4l^{(8)}\nabla^4 + h^6l^{(9)}\nabla^6, \tag{1.15}$$

$$\mathscr{K}^{(1)} = K^{(1)} + h^2K^{(2)}\nabla^2 + h^4K^{(3)}\nabla^4.$$

The coefficients and the forces are the following:

$$l^{(1)} = \nu_2^2 \left[ 1 - \frac{1}{6} (\nu_2^2 + 3\nu_1^2) \right],$$

$$l^{(2)} = 1 + \frac{1}{6} (\nu_2^2 - 9\nu_1^2),$$

$$l^{(3)} = \frac{1}{3} \left( 1 - 3 \frac{\nu_2^2}{\nu_1^2} \right),$$

$$l^{(4)} = \left( 3 - 4 \frac{\nu_1^2}{\nu_2^2} \right) - \frac{1}{6} \left( 9\nu_2^2 - 9\nu_1^2 - 4 \frac{\nu_1^4}{\nu_2^2} \right),$$

$$l^{(5)} = \frac{1}{3} \left( 5 - 7 \frac{\nu_1^2}{\nu_2^2} \right),$$

$$l^{(6)} = \nu_2^2 \left[ 1 - \frac{1}{6} (\nu_1^2 + 3\nu_2^2) \right],$$

$$l^{(7)} = - \frac{2}{3} \left[ (3\nu_2^2 - 2\nu_1^2) - \frac{1}{20} (9\nu_2^4 - 4\nu_1^4 + 3\nu_1^2\nu_2^2) \right],$$

$$l^{(8)} = - \frac{4}{3} \left[ \left( 1 - \frac{\nu_1^2}{\nu_2^2} \right) - \frac{1}{10} \left( 4\nu_2^2 - \frac{\nu_1^4}{\nu_2^2} - 2\nu_1^2 \right) \right],$$

$$l^{(9)} = \frac{2}{15} \left( 1 - \frac{\nu_1^2}{\nu_2^2} \right),$$

$$Q_\alpha = Q_\alpha^0 \delta(x^1) \delta(x^2), \qquad\qquad Q_3 = Q_3^0 \delta(x^1) \delta(x^2). \qquad (1.16)$$

**Remarks.** 1. The derived differential equations describe the behaviour of the middle surface of the plate. In exactly the same way equations for, say, $u^a(x, y, h)$ and $w(x, y, h)$ may be derived; the nature of the differential operators remains the same but the coefficients change which can lead to essentially different properties of the motion.

2. We based our treatment on analytic operators. In principle, we could however consider operators of type $\frac{1}{\mathcal{K}^{(1)}} \mathcal{L}_{\alpha\beta}^{(u)}$, etc. This has the advantage of leaving the right-hand side in the original form but then the problem (much more essential) of convergence of the resulting series arises. In fact, it can be shown that in the simplest cases of plane waves there always exists a combination of $\lambda^2$ and $\omega^2$ for which $\mathcal{K}^{(1)} = 0$ and similarly $\mathcal{K}^{(2)} = 0$. We shall therefore deal exclusively with analytic operators.

The coefficients $l$ and $K$ in (1.11) and (1.15) contain frequency $\omega$; similarly to the case of operators $(-i\xi_\alpha)$ expression $(i\omega)$ is an operator of differentiation with respect to time. Thus, in the case of an elastic material

$$l^{(1)} = - \frac{h^2}{c_2^2} \frac{\partial^2}{\partial t^2} \left[ 1 + \frac{h^2}{6} \left( \frac{1}{c_2^2} + \frac{3}{c_1^2} \right) \frac{\partial^2}{\partial t^2} \right], \qquad l^{(2)} = 1 - \frac{h^2}{6} \left( \frac{1}{c_2^2} - \frac{9}{c_1^2} \right) \frac{\partial^2}{\partial t^2},$$

etc. In the case of anelastic material $c_1^2$ and $c_2^2$ also contain $(i\omega)$ and are therefore operators. For instance for the MAXWELL material

$$c_1^2 = \frac{1}{\sigma}\left(K + \frac{4G}{3}\frac{i\omega}{t_0^{-1}+i\omega}\right), \qquad c_2^2 = \frac{G}{\sigma}\frac{i\omega}{t_0^{-1}+i\omega}$$

while for the KELVIN-VOIGT material $c_1^2 = \frac{1}{\sigma}\left[K + \frac{4G}{3}(1 + t^*i\omega)\right]$, $c_2^2 = \frac{G}{\sigma}(1 + t^*i\omega)$. In this case the whole equation has to be multiplied by the common denominator containing $i\omega$, to obtain analytic operators.

**Example.** Consider transverse vibrations of an infinite plate, assuming $Q_\alpha^0 = 0$. Writing the second equation (1.14) in full (it is to be borne in mind that $\lambda^2 \ll 1$, $\nu_{1,2}^2 \ll 1$) we have for an elastic material

$$\left\{\frac{h^6}{15(1-\nu)}\nabla^6 - \frac{2h^4}{3(1-\nu)}\left(1 + \frac{h^2}{10}\frac{5-6\nu}{1-\nu}\frac{1}{c_2^2}\frac{\partial^2}{\partial t^2}\right)\nabla^4\right.$$

$$+ \frac{2h^4}{3(1-2\nu)}\frac{1}{c_1^2}\frac{\partial^2}{\partial t^2}\left[(5-4\nu) + \frac{h^2}{20}\frac{19-31\nu+16\nu^2}{1-\nu}\frac{1}{c_2^2}\frac{\partial^2}{\partial t^2}\right]\nabla^2$$

$$\left. - \frac{h^2}{c_2^2}\frac{\partial^2}{\partial t^2}\left(1 + \frac{h^2}{6}\frac{7-8\nu}{1-2\nu}\frac{1}{c_2^2}\frac{\partial^2}{\partial t^2}\right)\right\}w(x_1 y_1 0) = 0. \qquad (1.17)$$

Introducing the symbol $O\left(\frac{1}{\lambda^2}\right)$ for the phrase "term of the order of" the above equation can be written as follows

$$O\left(\frac{1}{\lambda^6}\right) + O\left(\frac{1}{\lambda^4}\right) + O\left(\frac{\nu_\alpha^2}{\lambda^4}\right) + O\left(\frac{\nu_\alpha^2}{\lambda^2}\right) + O\left(\frac{\nu_\alpha^4}{\lambda^2}\right) + O(\nu_\alpha^2) + O(\nu_\alpha^4) = 0. \qquad (1.18)$$

In the elementary theory, terms 2 and 6 are retained; in the theory taking into account the influence of the rotatory inertia and shear displacement terms 2, 4, 6, and 7 appear; introducing terms 4 and 7 (from mechanical considerations referring to an element of height $2h$ treated as a whole) we eliminate the undesirable phenomenon of infinite velocity predicted by the elementary theory. However it is observed from (1.18) that this is not entirely consistent; in fact, the ratio of term 7 to 6 is $\nu_\alpha^2$; consequently if we retain term 4 we should also take into account term 5. If on the other hand we omit 5 we may also neglect 7. Retaining 2, 4, 6, and 7 we have

$$\left[\frac{2h^3}{3(1-\nu)}\nabla^4 - \frac{2h^3}{3}\frac{5-4\nu}{1-2\nu}\frac{1}{c_1^2}\frac{\partial^2}{\partial t^2}\nabla^2\right.$$

$$\left. + \frac{h}{c_2^2}\frac{\partial^2}{\partial t^2}\left(1 + \frac{h^2}{6}\frac{7-8\nu}{1-2\nu}\frac{1}{c_1^2}\frac{\partial^2}{\partial t^2}\right)\right]w(x_1 y, 0) = 0. \qquad (1.19)$$

This equation is very similar to that of the Timoshenko theory. However the latter contains the shear coefficient to be determined from additional considerations. Besides, the numerical values of the coeffi-

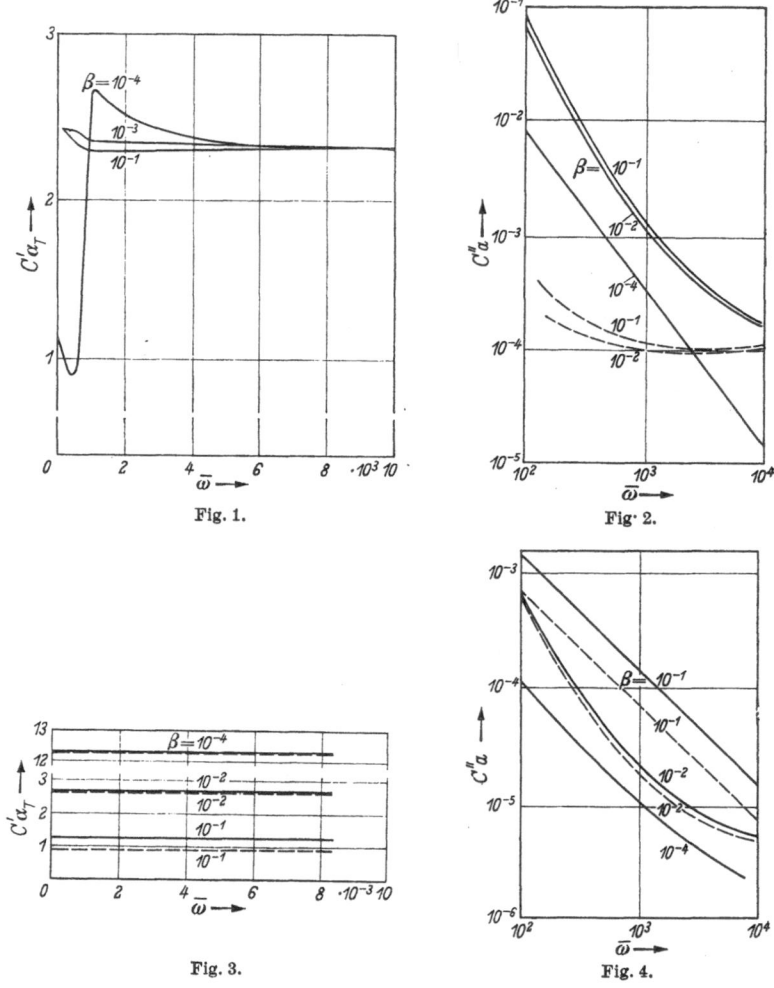

Fig. 1.

Fig. 2.

Fig. 3.

Fig. 4.

cients are also different. Setting in (1.19) $w = e^{i(\alpha'x + \beta'y - at)/\lambda}$, we obtain a biquadratic equation connecting the velocity of propagation $a$ with the wave length $\lambda$ (the vibrations are dispersive); this equation has two distinct roots:

$$a_{1,2}^2 = c_1^2 \frac{(1 - 2\nu)(5 - 4\nu)}{(1 - \nu)(1 - 8\nu)} \left(1 \pm \frac{1}{5 - 4\nu} \sqrt{11 - 24\nu + 16\nu^2}\right).$$

For $\nu = \dfrac{1}{4}$ the velocities are $a_1^2 = \dfrac{13}{15} c_1^2$, $a_2^2 = \dfrac{3}{15} c_1^2$. The corresponding velocities by the TIMOSHENKO theory are $\Big($the shear coefficient being equal to $\dfrac{2}{3}\Big)$ $\dfrac{8}{9} c_1^2$ and $\dfrac{2}{9} c_1^2$, respectively.

It was mentioned above that to be consistent we have to omit term 7; the resulting equation reads

$$\left( \frac{2h^3}{3(1-\nu)} \nabla^4 - \frac{2h^3}{3} \frac{5-4\nu}{1-2\nu} \frac{1}{c_1^2} \frac{\partial^2}{\partial t^2} \nabla^2 + \frac{h}{c_2^2} \frac{\partial^2}{\partial t^2} \right) w(x, y, 0) = 0.$$

This equation also yields a finite velocity of propagation of a disturbance; in this case however the equation for $a_{1,2}$ is quadratic and has one solution only, namely $a_1^2 = \dfrac{1-2\nu}{(1-\nu)(5-4\nu)} c_1^2$. For $\nu = \dfrac{1}{4}$ we have $a_1^2 = \dfrac{1}{6} c_1^2$.

In the case of anelastic bodies $c_1^2$, $c_2^2$ and $\nu$ depend on frequency $\omega$; Figs. 1—4 at the end of the paper show the results for the KELVIN-VOIGT material and the MAXWELL material, of solving Eq. (1.17) without the first term only, the solution being taken in the form of a plane wave. Various values of coefficient $\beta = \dfrac{\sigma h^2 \omega^2}{E}$ are taken, the independent variable being $\bar{\omega} = \dfrac{\omega \eta}{E}$, where $\eta$ is the viscosity coefficient. The influence of terms 3, 5, 7 is evident, on both the velocity of propagation and the attenuation.

## 2. The Stress-Strain Relations

We shall regard as characteristic displacements $u^a(x, y, 0)$, $w(x, y, 0)$ of the middle plane, and by the stress-strain relations we shall mean relations between these quantities and the total forces and moments defined in the usual way. The derivation is based on the linear law

$$\sigma_{ij} = \lambda \delta_{ij} \varepsilon_p^p + 2\mu \varepsilon_{ij},$$

where in general $\lambda$ and $\mu$ depend on frequency, and it is carried out in the spirit of the operational equations. Take first the stress $\sigma_{11}(x, y, z)$; we have for its image

$$\bar{\sigma}_{11} = -(2\mu \xi_1^2 + \lambda K_1^2)(A \cosh \gamma_1 z + B \sinh \gamma_1 z)$$

$$+ 2i\mu \xi_1 \gamma_2 (A_2 \sinh \gamma_2 z + B_2 \cosh \gamma_2 z).$$

Hence, substituting for $A, \ldots, B_2$ we obtain for $\tilde{M}_1(\xi_\alpha)$ the following result:

$$\tilde{M}_1 = \int_{-h}^{h} \sigma_{11} z \, dz = -\frac{\tilde{Q}_3}{M^{(2)}\gamma_1^2}\left[\left(K_1^2 - \frac{1}{2}K_2^2\right)\left(\frac{1}{2}K_2^2 - \xi_2^2\right)\cosh\gamma_1 h \cosh\gamma_2 h\right.$$

$$- \left(\xi_1^2 + \frac{1}{2}K_2^2 - K_1^2\right)\left(\xi^2 - \frac{1}{2}K_2^2\right)\cosh\gamma_2 h \frac{\sinh\gamma_1 h}{\gamma_1 h}$$

$$\left. + \xi_1^2 \gamma_1^2 \cosh\gamma_1 h \frac{\sinh\gamma_2 h}{\gamma_2 h}\right]. \tag{2.1}$$

Observe that this expression only contains the force $Q_3$ which is connected with $w(\xi_\alpha, 0)$ by means of the operational equation (1.8); thus, substituting we arrive at the required relation between $\tilde{M}_1(\xi_\alpha)$ and $\tilde{w}(\xi_\alpha, 0)$. Proceeding in the same way for the other moments, we obtain the invariant formulae

$$\mathcal{N}\tilde{M}_{\alpha\beta} = M_{\alpha\beta}\tilde{w}(\xi_\alpha, 0), \tag{2.2}$$

where

$$M_{\alpha\beta}(\xi_\alpha)$$

$$= \frac{8\mu h}{K_2^2}\left\{\left[-\left(K_1^2 - \frac{1}{2}K_2^2\right)\left(\xi^2 - \frac{1}{2}K_2^2\right)\cosh\gamma_2 h\left(\cosh\gamma_1 h - \frac{\sinh\gamma_1 h}{\gamma_1 h}\right)\right]g_{\alpha\beta}\right.$$

$$+ \left[\left(K_1^2 - \frac{1}{2}K_2^2\right)\cosh\gamma_1 h \cosh\gamma_2 h - \left(\xi^2 - \frac{1}{2}K_2^2\right)\cosh\gamma_2 h \frac{\sinh\gamma_1 h}{\gamma_1 h}\right.$$

$$\left.\left. + \gamma_1^2 \cosh\gamma_1 h \frac{\sinh\gamma_2 h}{\gamma_2 h}\right]\xi_\alpha\xi_\beta\right\}. \tag{2.3}$$

Similarly, for the shear force we have

$$\mathcal{K}^{(2)}R_\alpha = \mathcal{R}_\alpha\tilde{w}(\xi_\alpha, 0); \tag{2.4}$$

where

$$\mathcal{R}_\alpha(\xi_\alpha) = \frac{8\mu h}{K_2^2}(-i\xi_\alpha)\left(\xi^2 - \frac{1}{2}K_2^2\right)\left(\cosh\gamma_2 h \frac{\sinh\gamma_1 h}{\gamma_1 h} - \cosh\gamma_1 h \frac{\sinh\gamma_2 h}{\gamma_2 h}\right). \tag{2.5}$$

Retaining terms of order $\nu_{1,2}^6 \frac{\nu_{1,2}^4}{\lambda^2}, \frac{1}{\lambda^4}$ we obtain the expansion of the above operators

$$M_{\alpha\beta} = \frac{4\lambda}{3h}\left[(m^{(1)} + m^{(2)}\nabla^2 + m^{(3)}\nabla^4)g_{\alpha\beta} + (m^{(4)} + m^{(5)}\nabla^2)\nabla_\alpha\nabla_\beta\right],$$

$$\mathcal{N} = -(\nabla^2 + K_1^2)(n^{(1)} + n^{(2)}\nabla^2 + n^{(3)}\nabla^4), \tag{2.6}$$

$$\mathcal{R}_\alpha = -\frac{D}{h^2}(\nabla_\alpha)\left(\nabla^2 + \frac{1}{2}K_2^2\right)(r^{(1)} + r^{(2)}\nabla^2 + r^{(3)}\nabla^4), \quad D = \frac{E(2h)^3}{12(1-v^2)},$$

where

$$m^{(1)} = \frac{1}{2} v_1^4 \left[ 1 - \frac{v_2^2}{10} \left( 5 + \frac{v_1^2}{v_2^2} \right) \right],$$

$$m^{(2)} = v_1^2 h^2 \left[ \left( \frac{1}{2} + \frac{v_1^2}{v_2^2} \right) - \frac{v_2^2}{20} \left( 15 \frac{v_1^2}{v_2^2} + q \frac{v_1^4}{v_2^4} \right) \right],$$

$$m^{(3)} = h^4 \frac{v_1^2}{v_2^2} \left[ 1 + \frac{v_2^2}{120} \left( 51 - 82 \frac{v_1^2}{v_2^2} \right) \right],$$

$$m^{(4)} = \frac{\mu}{\lambda} h^2 \left[ v_1^2 + \frac{3}{10} v_1^2 v_2^2 \left( 1 - 3 \frac{v_1^2}{v_2^2} \right) \right],$$

$$m^{(5)} = \frac{\mu}{\lambda} h^4 \left[ 1 + \frac{v_2^2}{10} \left( 3 - 7 \frac{v_1^2}{v_2^2} - 8 \frac{v_1^4}{v_2^4} \right) \right],$$

$$n^{(1)} = \left( 1 - \frac{v_2^2}{2} + \frac{v_2^2}{24} \right),$$

$$n^{(2)} = -\frac{h^2}{4} \left[ \left( 3 - 2 \frac{v_1^2}{v_2^2} \right) \right] - \frac{v_2^2}{6} \left( 2 - \frac{v_1^4}{v_2^4} \right),$$

$$n^{(3)} = \frac{h^4}{48} \left[ \left( 5 - \frac{v_1^2}{v_2^2} \right) - \frac{v_2^2}{10} \left( 3 - 2 \frac{v_1^4}{v_2^4} \right) \right],$$

$$r^{(1)} = -\frac{1}{2} v_2^2 \left[ 1 - \frac{v_2^2}{10} \left( 1 + \frac{v_1^2}{v_2^2} \right) \right],$$

$$r^{(2)} = -h^2 \left[ 1 - \frac{v_1^2}{10} \left( 2 + \frac{v_1^2}{v_2^2} \right) \right],$$

$$r^{(3)} = \frac{h^4}{5} \left[ 1 - \frac{v_2^2}{21} \left( 3 + 2 \frac{v_1^2}{v_2^2} \right) \right]. \tag{2.7}$$

Observe that in general (for analytic operators) we have no explicit expression for the moment or force in terms of the displacements. In fact, the images $M_{\alpha\beta}$ and $R_\alpha$ are multiplied by operators and in general these relations are differential equations for the moments and forces.

Omitting higher order terms we have for Eqs. (2.2) and (2.4)

$$\left[ O\left(\frac{1}{\lambda^2}\right) + O(v_\alpha^2) \right] \left[ O(1) + O(v_\alpha^2) + O\left(\frac{1}{\lambda^2}\right) + O\left(\frac{v_\alpha^2}{\lambda^2}\right) + O\left(\frac{1}{\lambda^4}\right) + O\left(\frac{v_\alpha^2}{\lambda^4}\right) \right]$$

$$= h \left[ O(v_\alpha^4) + O(v_\alpha^6) + O\left(\frac{v_\alpha^2}{\lambda^2}\right) + O\left(\frac{v_\alpha^4}{\lambda^2}\right) + O\left(\frac{1}{\lambda^4}\right) + O\left(\frac{v_\alpha^2}{\lambda^4}\right) \right],$$

$$\left[ O(1) + O(v_\alpha^2) + O\left(\frac{1}{\lambda^2}\right) + O\left(\frac{v_\alpha^2}{\lambda^2}\right) + O\left(\frac{1}{\lambda^4}\right) + O\left(\frac{v_\alpha^2}{\lambda^4}\right) \right]$$

$$= \left[ O\left(\frac{1}{\lambda^2}\right) + O(v_\alpha^2) \right]$$

$$\times \left[ O(v_\alpha^2) + O(v_\alpha^4) + O\left(\frac{1}{\lambda^2}\right) + O\left(\frac{v_2^2}{\lambda^2}\right) + O\left(\frac{1}{\lambda^4}\right) + O\left(\frac{v_2^2}{\lambda^4}\right) \right]. \tag{2.8}$$

Neglecting terms of $O\left(\frac{1}{\lambda^2}\right)$ and $O(\nu_a^2)$ (and higher) as compared with $O(1)$ we have for the bending moments $M_{11} = M_1$ and $M_{22} = M_2$:

$$\left(\nabla^2 - \frac{1}{c_1^2}\frac{\partial^2}{\partial t^2}\right) M_\alpha(x, y, t) = -D\left(\nabla^2 - \frac{1}{c_1^2}\frac{\partial^2}{\partial t^2}\right)$$

$$\times \left[(1 - \nu)\frac{\partial^2}{\partial x^{\alpha 2}} + \nu\left(\nabla^2 - \frac{1-\nu}{1-2\nu}\frac{1}{c_1^2}\frac{\partial^2}{\partial t^2}\right)\right] w(x, y, 0, t). \qquad (2.9)$$

Introducing a normalization condition (e.g. $M_\alpha = 0$ when $w = 0$ if $M_\alpha = \frac{\partial}{\partial t} M_\alpha = 0$ for $t = 0$) we eliminate the operator $\left(\nabla^2 - \frac{1}{c_1^2}\frac{\partial^2}{\partial t^2}\right)$ from both sides of the equation

$$M_\alpha(x, y, t) = -D\left[(1 - \nu)\frac{\partial^2}{\partial x^{\alpha 2}} + \nu\left(\nabla^2 - \frac{1-\nu}{1-2\nu}\frac{1}{c_1^2}\frac{\partial^2}{\partial t^2}\right)\right] w(x, y, 0, t). \qquad (2.10)$$

Only assuming $\nu_1^2 \ll \frac{1}{\lambda^2}$, we arrive at the classical formula $M_2 = -D\left[(1 - \nu)\frac{\partial^2}{\partial x^{\alpha 2}} + \nu\nabla^2\right]w$. Since in general operator $M_{\alpha\beta}$ is not divisible by $(\xi^2 + K_1^2)$ we are faced with a wave equation for $M_{\alpha\beta}$ (velocity $c_1$). Observe now that the additional term in (2.10) implies that even when $w(x, y, t) = w(t)$ in a certain region of the plate, the moment does not vanish. Evidently this is due to the fact that $w$ is not constant along the thickness of the plate.

The relation corresponding to (2.10) for $M_{12}$ has the form

$$\left(\nabla^2 - \frac{1}{c_1^2}\frac{\partial^2}{\partial t^2}\right) M_{12} = D(1 - \nu)\left(\nabla^2 - \frac{1}{c_1^2}\frac{\partial^2}{\partial t^2}\right)\frac{\partial}{\partial x^1}\frac{\partial}{\partial x^2} w(x, y, 0, t). \qquad (2.11)$$

As before introducing a normalization condition we obtain the classical relation for $M_{12}$ (no additional term this time).

For the forces under the same assumptions we have

$$R_\lambda = -D\left(\nabla^2 - \frac{1}{2c_2^2}\frac{\partial^2}{\partial t^2}\right)\frac{\partial}{\partial x^\alpha} w(x, y, 0, t). \qquad (2.12)$$

Only for $\nu_2^2 \ll \lambda^{-2}$ we have the classical relation.

The results corresponding to (2.10), (2.11), and (2.12) for anelastic bodies are as follows (we assume for simplicity that POISSON's ratio is a constant):

for the KELVIN-VOIGT body

$$\left(1 + t^* \frac{\partial}{\partial t}\right) M_\alpha(x, y, t) = -\frac{(2h)^3}{6(1-\nu)} \mu \left(1 + t^* \frac{\partial}{\partial t}\right)$$

$$\times \left\{\left(1 + t^* \frac{\partial}{\partial t}\right)\left[(1 - \nu)\frac{\partial}{\partial x^{\alpha 2}} + \nu\nabla^2\right] - \frac{\nu\sigma}{2\mu}\frac{\partial^2}{\partial t^2}\right\} w(x, y, 0, t),$$

$$M_{12}(x, y, t) = \frac{2h^3}{6} \mu \left(1 + t^* \frac{\partial}{\partial t}\right) \frac{\partial^2}{\partial x^1 \partial x^2} w(x, y, 0, t), \qquad (2.13)$$

$$\left(1 + t^* \frac{\partial}{\partial t}\right) R_\alpha(x, y, t)$$

$$= -\frac{(2h)^3}{6(1-v)} \mu \left(1 + t^* \frac{\partial}{\partial t}\right)\left[\left(1 + t^* \frac{\partial}{\partial t}\right) - \frac{\sigma}{2\mu}\frac{\partial^2}{\partial t^2}\right]\frac{\partial}{\partial x^\alpha} w(x, y, 0, t);$$

for the MAXWELL body

$$\left(t_0^{-1} + \frac{\partial}{\partial t}\right) M_\alpha(x, y, t) = -\frac{(2h)^3}{6(1-v)} \mu$$

$$\times \left\{\frac{\partial}{\partial t}\left[(1 - \nu)\frac{\partial^2}{\partial x^{\alpha 2}} + \nu\nabla^2\right] - \frac{\nu\sigma}{2\mu}\left(t_0^{-1} + \frac{\partial}{\partial t}\right)\frac{\partial^2}{\partial t^2}\right\} w(x, y, 0, t),$$

$$\left(t_0^{-1} + \frac{\partial}{\partial t}\right) M_{12}(x, y, t) = \frac{(2h)^3}{6} \mu \frac{\partial^2}{\partial x^1 \partial x^2} w(x, y, 0, t), \qquad (2.14)$$

$$\left(t_0^{-1} + \frac{\partial}{\partial t}\right) R_\alpha(x, y, t)$$

$$= -\frac{(2h)^3}{6(1-v)} \mu \left[\frac{\partial}{\partial t}\nabla^2 - \frac{\sigma}{\mu}\left(t_0^{-1} + \frac{\partial}{\partial t}\right)\frac{\partial^2}{\partial t^2}\right]\frac{\partial}{\partial x^\alpha} w(x, y, 0, t).$$

Thus, the presence of the additional term changes substantially the form of the stress-strain relations, causing appearence of higher (second in the first approximation) time derivatives.

## 3. Initial and Boundary Conditions

Since the order of our equations depends on the degree of approximation, each approximation requires a different number of initial and boundary conditions. For definitness, we confine ourselves to system (3.16) and an elastic body.

7*

**Initial conditions.** Various combinations are possible. For instance we may prescribe

$$w(x, y, 0), \qquad \frac{\partial}{\partial t} w(x, y, 0),$$

$$M_{11}(x, y, 0) + M_{22}(x, y, 0),$$

$$\frac{\partial}{\partial t} [M_{11}(x, y, 0) + M_{22}(x, y, 0)]. \tag{3.1}$$

Since the moments contain $\frac{\partial^2 w}{\partial t^2}$ this is equivalent to prescribing $w(x, y, 0)$ and its first three derivatives at $t = 0$. The equations contain also fourth time derivatives of $u$ and $v$; consequently we have to prescribe

$$\frac{\partial^n}{\partial t^n} u^i(x, y, 0), \qquad\qquad n = 0, 1, 2, 3 \tag{3.2}$$

or equivalent quantities.

**Boundary conditions.** We confine ourselves to geometric boundary conditions. Since (3.16) contain sixth derivatives of $w$ and fourth derivatives of $u$ and $v$ we have to prescribe on the boundary

$$w(x, y, t), \qquad \frac{\partial}{\partial n} w(x, y, t), \qquad \frac{\partial^2}{\partial n^2} w(x, y, t),$$

$$u^i(x, y, t), \qquad \frac{\partial}{\partial n} u^i(x, y, t).$$

The above remarks outline only the problem of boundary and initial conditions. The new equations introduce new boundary value problems which require a further extensive investigation.

### References

[1] MIKUSINSKI, J.: Operational Calculus, London: Pergamon Press 1955.
[2] MINDLIN, R. D.: J. Appl. Mech. **73**, 31—38 (1951).
[3] NAGHDI, P. M.: Proceedings of the Symposium on the Theory of Thin Elastic Shells, Amsterdam: North-Holland Publ. Co. 1960, p. 301.
[4] NOVOZHILOV, V. V., and P. M. FINKELSTEYN: Prikladnaya Matematika i Mekhanika **7**, 331 (1943).
[5] MIKLOWITZ, J.: Appl. Mech. Reviews **13**, 865 (1960).
[6] KOLSKY, H.: Stress Waves in Solids, Oxford: Clarendon Press 1953.
[7] ZORSKI, H.: Warsaw: Biul. WAT 1957.
[8] HERRMANN, G., and I. MIRSKY: J. Appl. Mech. **23**, 563 (1956).
[9] LURIE, A. I.: Prikladnaya Matematika i Mekhanika **4**, 7 (1940).
[10] SNEDDON, I. N.: Fourier Transforms, New York/Toronto/London: McGraw-Hill 1951.

# Some Experimental and Theoretical Studies on the Propagation of Longitudinal Plastic Waves in a Strain-Rate-Dependent Material

By Giovanni Bianchi

Politecnico di Milano, Italy

## Summary

A series of tests has been performed on the propagation of longitudinal plastic waves in long prestressed specimens of annealed copper. The experimental results have been compared with those calculated with the strain-rate independent equation and with the linear strain-rate dependent equation. The latter is in good general agreement with the experimental findings, while the former shows marked discrepancies, especially with the higher values of the static preloading.

In the discussion of the results some general properties are demonstrated for the asymptotic values of a plastic wave. Some quantitative differences between calculated and experimental findings are then analyzed.

## 1. Introduction

This paper describes some of the results obtained in an investigation of the propagation of longitudinal plastic waves in prestressed bars.

The research was planned, at the Politecnico di Milano, with the aim of resuming that begun by the author at Cornell University in 1953 with a series of tests on the propagation of strain pulses in prestrained annealed copper ribbons conducted within the framework of an investigation of the dynamic properties of metals by RIPARBELLI [1a]. These tests [2a, 2b], in agreement with the results obtained by STERNGLASS and STUART [1b] and BELL [1c], had shown that while the amplitude of the wave attenuates, the wave front always propagates at a velocity equal to the elastic velocity $c_0 = \sqrt{E_0/\varrho}$ ($E_0$: YOUNG's modulus; $\varrho$: density of the material), no matter what the value of the static pre-loading.

As is known, this behavior cannot be adequately interpreted by a strain-rate independent theory. In order to test the validity of strain-rate dependent theories quantitatively, the new series of experiments, again on long specimens under tensile impact, were planned so that we could (a) obtain wave forms more regular at the impact end than those in the first series of tests, and such as would allow the study of the asymptotic behavior when the end conditions become constant; and (b) measure directly at the impact end those physical quantities which become the boundary conditions in the related analytical problem.

It was decided to perform most of the tests on specimens prestressed into the plastic range in the belief, that the strain-rate effects would be more evident, and numerical calculations more accurate, as a result of the smoother shape of the static stress-strain curve in the range of the strains produced by the impact.

## 2. The Experimental Apparatus

### (a) The Specimen

The specimens used were thirty-foot strips of annealed copper ribbon (1 in the schematic drawing, not in scale, of Fig. 2) with a rectangular cross section 0.39 in. × 0.079 in. The static stress-strain diagram

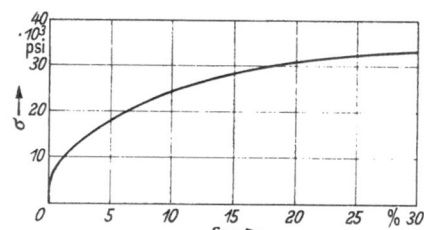

Fig. 1. Static stress-strain curve for the annealed copper specimen.

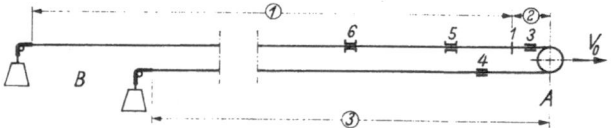

Fig. 2. Schematic of the specimen and the measuring devices.

obtained for these strips is shown in Fig. 1. One end of the ribbon was soldered (point 1) to a steel ribbon (2 and 3); the loads for the static prestressing were applied to the other end B. Resistance strain gages were attached on the short (two inch) upper strip (point 3) and on the lower strip (point 4) of the steel ribbon.

## (b) The Impact

The steel ribbon passed around a small aluminum cylinder at the impact end *A* and this cylinder was driven by a fly wheel impact machine (shown schematically in Fig. 3) with a prescribed velocity. Care

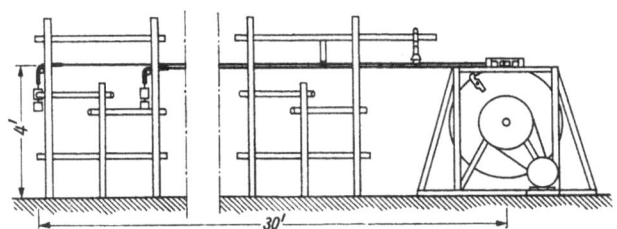

Fig. 3. Schematic of the experimental set-up.

was taken to keep the steel ribbon from slipping on the aluminum cylinder. The impact duration was always sufficiently short to prevent reflections from interfering with the original waves at all the measuring points. As is obvious, a new copper specimen had to be used for each test.

## (c) The Measuring Techniques

Particle velocities were measured on the specimen at points *5* and *6* by attaching thin copper wires to the ribbon and creating a magnetic field with permanent magnets. The plastic strains were not measured in this series of tests.

Since stresses in the steel ribbon were always below the limit of proportionality, we could deduce the force $\sigma(0, t)$ at the impact end of the specimen from the deformation at point *3*. From the strain wave at point *4* we deduced the velocity $v(0, t)$ of the impact end. The four signals were simultaneously recorded with a four channel cathode-ray oscillograph and a drum camera.

## (d) The Analysis of the Measuring Techniques

Details of the experimental apparatus and a critical analysis of the measuring techniques are given in references [*2c*] and [*2d*]. There, among other matters, the relation between the velocity measured at point *4* and the velocity at the impact end of the specimen, and the accuracy of the dynamometer are analyzed including a study of the elastic propagation within the dynamometer itself. The discontinuity due to the soldered joint on the copper strip is also taken into account. The criteria are given for determining the most convenient value of the cross section of the dynamometer. Eddy current effects on the particle velocity measurements at points 5 and 6 are defined and measured.

The accuracy of all the measuring methods was also checked with a series of elastic tests with a hard-drawn copper specimen, such as the one referred to in the oscillographic record of Fig. 4 (time scale: 1 in. $= 795 \times 10^{-6}$ sec). Velocity and stress at the impact end, particle velocities and strains at any section of the specimen were in this case represented by wave forms of the same shape. (Strain and velocity reflections from a fixed end are also shown in the oscillograms).

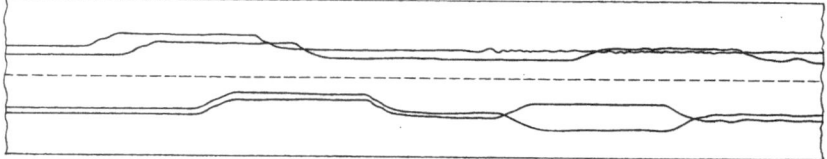

Fig. 4. Oscillographic record of an elastic test.

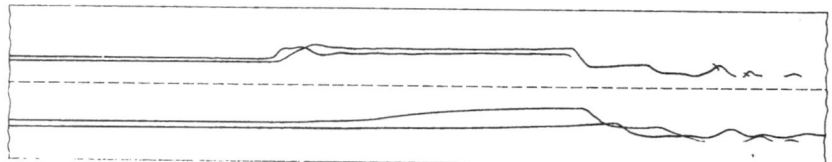

Fig. 5. Oscillographic record of a plastic test.

In the plastic tests, such as the one shown in Fig. 5, the measurements of end velocities and particle velocities still appear satisfactory. However, some oscillations, more evident in the tests with low preloading, appear in the signal from the dynamometer. This might be caused by some flexural vibrations at the impact end when, due to the sudden stress relaxation in the copper specimen, a difference arises between the tensions in the upper and lower strips. This difference is higher percentagewise when the static preloading is lower.

## (e) The Series of Tests

A series of tests was run with static preloading $\sigma_s$ ranging from $3 \times 10^3$ p.s.i. to $20 \times 10^3$ p.s.i.; steady-state values $V$ of the end velocity $v(0, t)$ at the impact end ranged from 13 ft/sec to 46 ft/sec; duration of the impact ranged from $2 \times 10^{-3}$ sec to $8 \times 10^{-3}$ sec; mean loading rate at the impact end — i.e., mean slope $\Delta V/\Delta t$ of the $v(0, t)$ at the beginning of the impact — ranged from $50 \times 10^3$ ft/sec² to $250 \times 10^3$ ft/sec². Particle velocities were measured at sections $x_1$ and $x_2$ distant 1.15 ft and 3.28 ft respectively from the impact end of the unstressed specimen.

### 3. The Experimental Results

The experimental results are summarized in the diagrams of Fig. 6 to 9, where the four experimental oscillographs of each test are plotted in the $(x, t)$ plane ($x$: distance of the section from the impact end on the unstressed specimen; $t$: time measured from the beginning of the impact). The solid lines show the experimental values of $v(x, t)$ at the

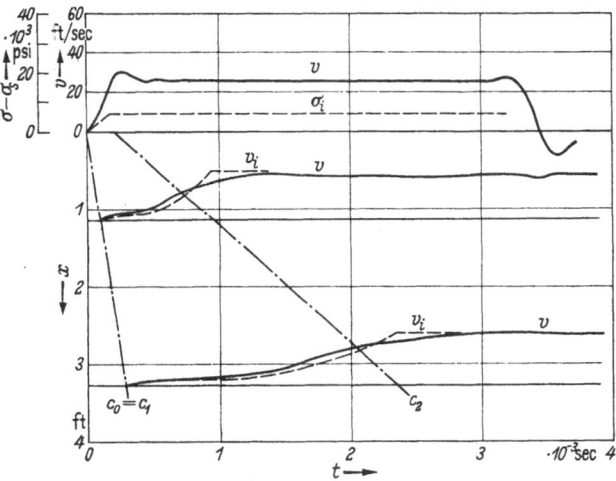

Fig. 6. The experimental waves and the waves calculated with the strain-rate independent theory (subindex $i$). Static preloading: 3.14 ksi.

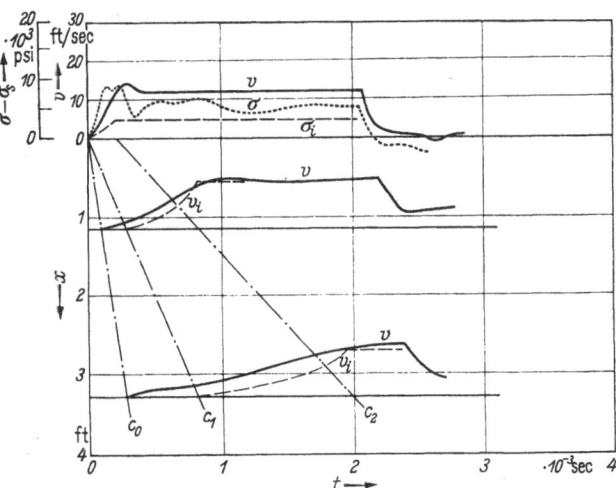

Fig. 7. Idem. Static preloading: 5.00 ksi.

impact end and at sections $x_1$ and $x_2$ of the specimen; the dotted line represents the experimental value of the stress $\sigma(0, t)$ at the impact end. The broken lines represent the values of the stress $\sigma_i(0, t)$ and of

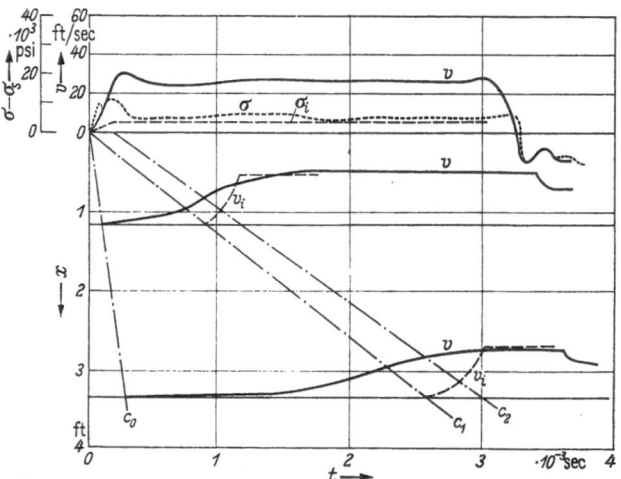

Fig. 8. Idem. Static preloading: 14 ksi.

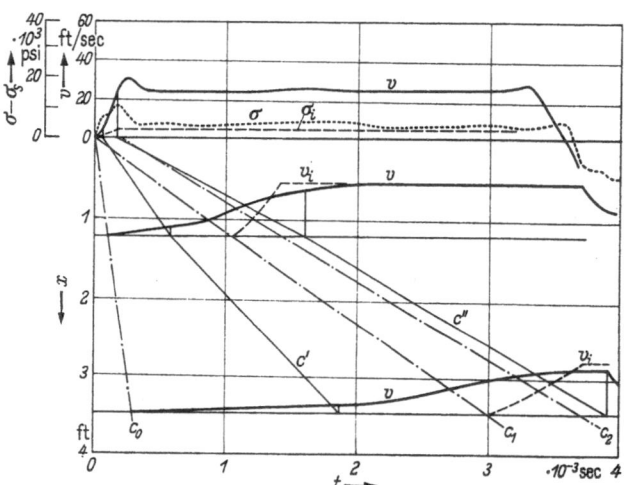

Fig. 9. Idem. Static preloading: 18.5 ksi.

particle velocities $v_i(x_1, t)$ and $v_i(x_2, t)$ calculated from the given experimental end velocity $v(0, t)$ according to the strain-rate independent theory (von KÁRMÁN's equation).

It appears that

(a) the velocity of propagation $c_0$ of the wave front is still independent of the value of the static preloading and always equal to $c_0 = 12{,}000\,\text{ft/sec}$;

(b) the waves of particle velocities at sections $x_1$ and $x_2$ are in fair agreement with the wave $v_i$ calculated with the strain-rate independent theory only when the static preloading $\sigma_s$ is below the limit of proportionality. The discrepancy with $v_i$ increases markedly with the increase of $\sigma_s$;

(c) the stress at the impact end $\sigma(0, t)$ is always greater than $\sigma_i$ and has a different shape, especially in the first instants of the impact;

(d) in the $(x, t)$ plane two zones take shape where the velocities with increasing $x$ tend to zero and to the steady state value $V$ respectively. This steady state wave is always in the region of $x < c_2 t$, where $c_2$ is the velocity of propagation of the steady state value of $v_i$;

(e) the velocities of propagation of smaller increments are higher than those calculated with the strain-rate independent equation and tend to decrease with the increase of $x$; the velocities of propagation of larger increments are lower and tend to increase with the increase of $x$ (see, for example, the lines $c'$ and $c''$ respectively in Fig. 9).

## 4. The Asymptotic Behavior Observed and the Asymptotic Behavior of a Generic Plastic Wave

The asymptotic behavior observed in the tests with higher static preloading agrees substantially with the asymptotic values given by RUBIN [3] for a linearized static stress-strain relation with a linear strain-rate function.

With the aid of some elementary physical considerations it can be shown that these asymptotic values — and the corresponding values in the case of lower static preloading, i.e., with a non linear static stress-strain curve — are to be expected with any form of the strain-rate function, so long as the velocity of propagation of the front $c_0$ remains a constant, and so long as the propagation has the general attenuation characteristics seen in the tests.

Let us now consider a plastic wave that propagates in a semi-infinite bar, initially at rest, the end of which is moved at a given velocity $v(0, t)$ rising in a finite time from zero to a constant value $V$ which is maintained.

At any section $x$ of the bar, the function $v(x, t)$ — given the attenuation characteristics evidenced in the experiments — will rise regularly with time from zero to a value not exceeding the maximum value $V$ reached at the impact end. Therefore, the plot of $v(x, t)$ in function of $c = x/t$ for any value of $x$ will always have the general shape indicated in

Fig. 10, that is, of a line which, although differing from section to section since $v(x, t)$ will not as a rule be a function of $c$, will always start at the same abscissa $c_0$ (constant velocity of propagation of the wave front) and rise regularly to a value not greater than $V$.

It appears, then, that when $t \to \infty$, finite increments of $v$ will occur in infinite intervals of time, i.e., $\partial v/\partial t \to 0$. We may thus reasonably assume that, asymptotically, the rate of loading of the material decreases to zero, so that the relation between stress and deformation will approach the static relation and the propagation will be independent of any strain-rate effect, as in VON KÁRMÁN's and TAYLOR's equation.

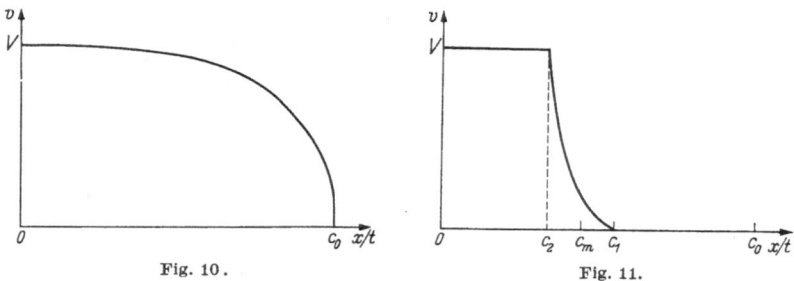

Fig. 10.              Fig. 11.

If we consider in particular the propagation in a bar prestressed to a value of $\sigma_s$, to which, according to VON KÁRMÁN's equation, the velocity of propagation $c_1$ corresponds, the asymptotic values of the wave (Fig. 11) will be zero for any $x > c_1 t$ and $V$ for any $x < c_2 t$, where $c_2$ is again calculated with VON KÁRMÁN's equation. If the range of the deformation due to the wave is so small that, given the smooth shape of the static $\sigma(\varepsilon)$ curve, the two values $c_1$ and $c_2$ become very close, it can easily be seen that the function $v(c)$ approaches a straight line between $c_1$ and $c_2$, with the result that the asymptotic value of the curve in correspondence with the mean value $c_m$ is $V/2$. These are, basically, the same results found by RUBIN for the linear case.

Furthermore if a steady state condition is reached at the impact end, in the sense that at a time $t_2$ the stress $\sigma(0, t)$ also becomes $\sigma(0, t) = \sigma_2 = $ constant, we may write, equating impulse and change of momentum,

$$\int_0^{t_2} \sigma(0, t)dt + \sigma_2(t - t_2) = \varrho \int_0^{c_0 t} v(x, t)dx = \varrho t \int_0^{c_0} v(x, t)d\left(\frac{x}{t}\right)$$

or

$$\varrho \int_0^{c_0} v(x, t)d\left(\frac{x}{t}\right) = \sigma_2 + \frac{1}{t}\int_0^{t_2}[\sigma(0, t) + \sigma_2]dt.$$

Since the integral on the right is bounded, the integral on the left becomes equal to $\sigma_2$ when $t$ approaches infinity. Therefore, the area under the curve $v(c)$, plotted for any large value of $t$, approaches a constant value.

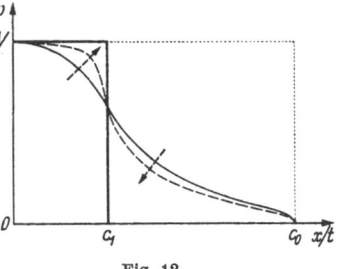

If we now consider the propagation in a prestressed bar with a linearized static relation $\sigma(\varepsilon) = E_1 \varepsilon$, we know that for large $t$, the diagram of $v(c)$ must approach the asymptotic form in such a way as to maintain the area under it practically constant (Fig. 12).

Fig. 12.

It seems, therefore, reasonable to expect, no matter what the form of the specific strain-rate function, that the velocity of propagation of smaller increments will in general decrease until it approaches the value $c_1 = \sqrt{E_1/\varrho}$, while the velocity of propagation of larger increments will in general increase until it approaches the same value[1].

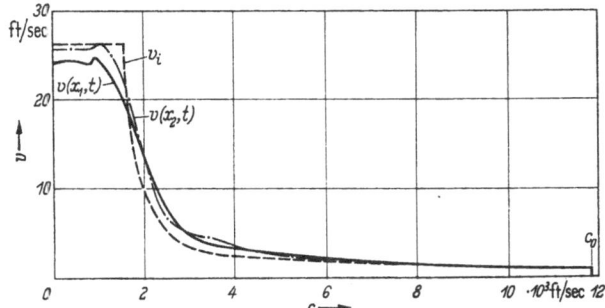

Fig. 13. The experimental velocity waves at $x_1$ and $x_2$ and the calculated wave $v$ as a function of $c = x/t$. Static preloading: 3.13 ksi.

In Figs. 13 and 14 we have plotted, together with $v_i(c)$ (broken line) calculated with von Kármán's equation, the experimental values of $v(x, t)$ at $x_1$ (solid line) and $x_2$ (dotted line) as a function of $c$ for the two

---

[1] It may be of interest to note that in the $(v, c)$ plane an elastic propagation of a step velocity impact would be represented by a rectangular wave (shown with the dotted line) of constant value $V$ for $0 \leq c \leq c_0$. Since the constancy of the velocity of propagation of the wave front $c_0$ is in itself a proof of the inital elastic behavior of the material (see for instance [2b]), we may deduce that the plastic propagation of a step velocity impact is represented in the $(v, c)$ plane by waves going with increasing time from the dotted line $(t \to 0)$ to the asymptotic line $(t \to \infty)$, while the stress at the impact end decreases from the elastic value $\sigma_0 = \varrho c_0 V$ to the plastic value $\sigma_2 = \varrho c_2 V$.

tests of Fig. 6 and Fig. 9. The two diagrams clearly show how in both cases the waves are approaching the form given by von KÁRMÁN's equation since, between $x_1$ and $x_2$, the main part of the wave, even with the higher value of the static preloading, is already traveling with von KÁRMÁN's velocities of propagation.

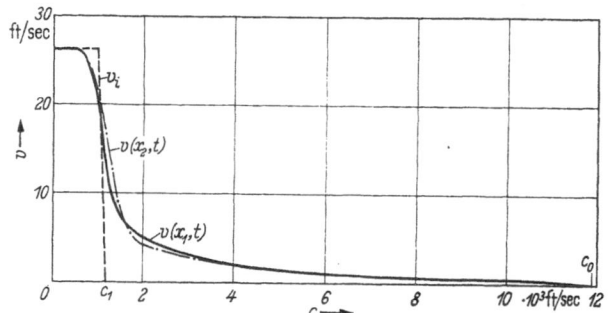

Fig. 14. Idem. Static preloading: 18.5 ksi.

## 5. The Comparison of the Experimental Results with the Results of the Linear Strain-Rate-Dependent Equation

### (a) The Differential Equations

Calculations were made with the linear strain-rate-dependent equation in order to compare the results of the tests with higher preloading $\sigma_s$ where the static stress-strain curve $\sigma(\varepsilon)$ may also be linearized, in the range of the strains produced by the impact, as $\sigma - \sigma_s = E_1(\varepsilon - \varepsilon_s)$. Given the high ratio of length to transverse dimensions of the bar, lateral inertia is neglected.

The partial differential equations are:

$$\frac{\partial \sigma}{\partial x} = \varrho \, \frac{\partial v}{\partial t},$$

$$\frac{\partial v}{\partial x} = \frac{\partial \varepsilon}{\partial t},$$

$$\frac{\partial \varepsilon}{\partial t} = \frac{1}{E_0} \, \frac{\partial \sigma}{\partial t} + \frac{k}{E_0} \, [(\sigma - \sigma_s) - E_1(\varepsilon - \varepsilon_s)],$$

where $x$ is the Lagrangian coordinate measured on the unstressed specimen; $\sigma$ and $\varepsilon$ are the nominal stress and strain referred to original area and length respectively; $E_0$ is YOUNG's modulus and $E_1$ the tangent modulus corresponding to $\sigma_s$. Given the test arrangement, the bar can be considered semi infinite in the calculations.

## (b) The Choice of the Parameter $k$

The order of magnitude of the parameter $k$ to be used in the equations was deduced from the tests with higher preloading by an approximate evaluation in finite terms of the mean derivative, between section $x = 0$ and $x_1$, of $v$ with respect to $x$ as a function of time ($\partial v/\partial x \simeq [v(x_1, t) - v(0, t)]/x_1$) and by calculation of the logarithmic decrement $\delta$ in an interval of time $\Delta t$ during which $\sigma(0, t)$ is already practically a constant. Since when $\partial \sigma/\partial t = 0$,

$$\frac{\partial v}{\partial x} = \frac{\partial \varepsilon}{\partial t} = A e^{-\frac{E_1}{E_0} kt},$$

where $A$ is an arbitrary constant, we have

$$k \simeq \frac{\delta}{\Delta t} \frac{E_0}{E_1}.$$

The range of values found was $k = 1.5 \div 1.9 \times 10^5$ sec$^{-1}$.

## (c) The Numerical Integration

The numerical integration was performed along the characteristic lines in the way indicated by MALVERN [4a] and followed by GLAUZ and LEE [4b] and PLASS and NENG-MING-WANG [4c]. The arithmetic mean of the strain-rate function was used in each integration step so as to obtain explicit resolving formulas of the finite difference system (with the usual symbols, which here indicate only the increments with respect to $\sigma_s$ and $\varepsilon_s$):

$$(\varepsilon_M - \varepsilon_E) - (\sigma_M - \sigma_B) = \frac{k}{E_0} \frac{(\sigma_M - E_1\varepsilon_M) + (\sigma_B - E_1\varepsilon_B)}{2} \Delta t,$$

$$(\sigma_M - \sigma_A) - \varrho c_0 (v_M - v_A) = - k \frac{(\sigma_M - E_1\varepsilon_M) + (\sigma_A - E_1\varepsilon_A)}{2} \cdot \frac{\Delta t}{2},$$

$$(\sigma_M - \sigma_C) + \varrho c_0 (v_M - v_C) = - k \frac{(\sigma_M - E_1\varepsilon_M) + (\sigma_C - E_1\varepsilon_C)}{2} \cdot \frac{\Delta t}{2}.$$

The choice of the integration step was decided by checking the accuracy of the calculation with a series development near the origin, of the kind used by GLAUZ and LEE, and by repeating the same calculation with smaller steps until no significant change was noticed in the results. A variable step was adopted, starting as $\Delta t = 0.14 \times 10^{-6}$ sec at the origin and increasing to a maximum value $\Delta t = 35 \times 10^{-6}$ sec.

The actual experimental end velocity $v(0, t)$ was used, with minor simplifications, as the end condition for the integration. The value of $E_1$ was measured on the static stress-strain curve (Fig. 1).

The computation was performed on the Remington Univac SS 90 computer.

### (d) Results of the Calculations

For a test like that shown in Fig. 9, the calculation was repeated with different values of $k$. The best agreement was found for $k = 1.8 \times 10^5 \text{ sec}^{-1}$.

In Fig. 15 the calculated stress at the impact end $\sigma_d$ and particle velocities $v_d$ at section $x_1$ and $x_2$ are shown (broken lines) together with

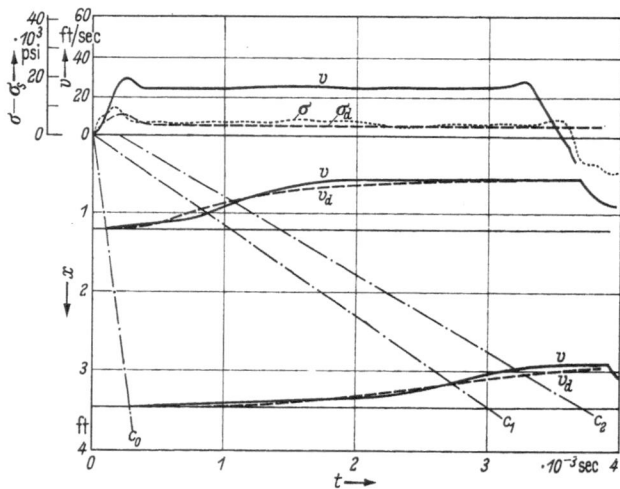

Fig. 15. The experimental waves and the waves calculated with the linear strain-rate-dependent theory (subindex $d$). Static preloading: 18.5 ksi.

the experimental values (solid lines). A very good agreement is found for the values of the velocities at $x_1$ and $x_2$. A good qualitative agreement is also found for the stresses at the impact end, although the calculated stress remains lower than the experimental stress.

In Fig. 16 the calculated values of $\sigma_d$ and $\varepsilon_d$ are plotted together with the calculated values of $v_d$ and the experimental value of $\sigma(0, t)$. The end condition actually used for the numerical calculations is plotted at $x = 0$. A mean curve has also been traced through the high frequency components of $\sigma(0, t)$. The scales are such that the asymptotic values of the three quantities are represented by the same ordinate. Since the static stress-strain relation is $(\sigma - \sigma_s) = E_1(\varepsilon - \varepsilon_s)$, the curve $v_d$ at the impact end also represents the values $\sigma_i$ and $\varepsilon_i$ calculated with the strain-rate independent theory (except for the small unloading at the peak of $v(0, t)$).

In appears very clearly that the stress $\sigma_d(0, t)$ calculated with the strain-rate-dependent equation is in much better agreement with the experimental value $\sigma(0, t)$ than the stress calculated with the strain-rate-independent equation.

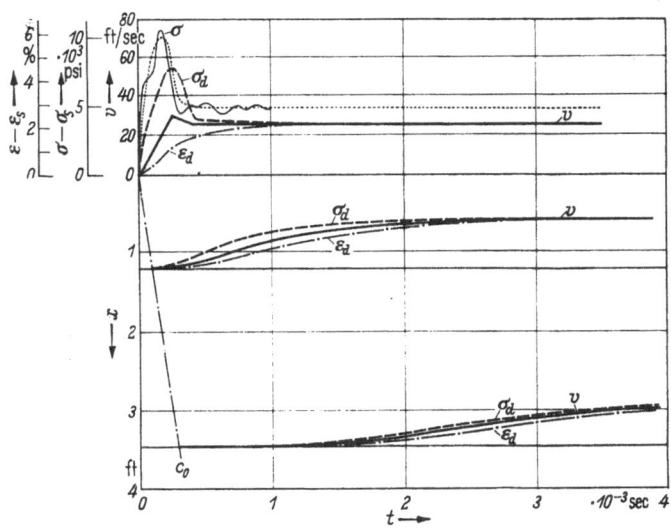

Fig. 16. The waves calculated with the linear strain-rate-dependent theory (subindex $d$) and the experimental stress (dotted line). Static preloading: 18.5 ksi.

## 6. Discussion of the Results

A precise analysis of the quantitative differences between the experimental results and those given by the linear strain-rate-dependent equation will necessarily entail the examination of a large number of tests with different values of static preloading and impact velocity. It must also be remembered that the results of the numerical calculation depend basically upon the evaluation of a tangent ($E_1 = d\sigma/d\varepsilon$) to a line of an experimental diagram (the static stress-strain curve).

Even at this stage, however, we should like to note the following points of interest which appear from the comparison of calculated and experimental values available to date:

(a) a smaller value of $k$ would bring the calculated stress at $x = 0$ into better agreement with the experimental stress which, as we have observed, is always the higher of the two. It would, on the other hand, lessen agreement between the particle velocities, in the sense that the calculated particle velocity wave at $x_1$ would be less attenuated than the experimental one. In other words, relative attenuations near the origin of $x$ and $t$ seem to be higher in the experimental waves;

(b) the mean percentage attenuation of the particle velocity wave between sections $x_1$ and $x_2$, when seen in the $(v, c)$ plane, as in Fig. 14, appears much smaller than in the calculated wave (see Fig. 17). This is indicative of a larger momentum in the experimental wave than in the calculated one and is in agreement with the difference between experimental and calculated stress;

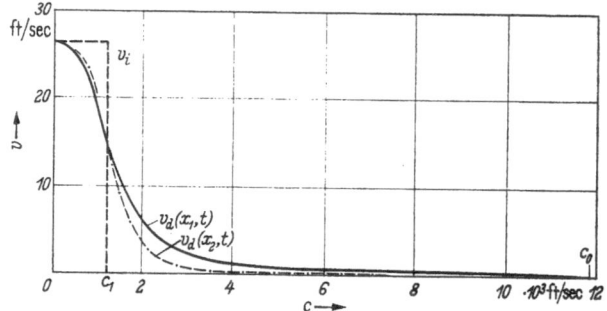

Fig. 17. The calculated waves $v$ at $x_1$ and $x_2$ and the asymptotic wave (broken line) as a function of $c = x/t$.

(c) in the comparison of tests with the same static preloading, but with different impact velocities, a fair proportionality appears to be maintained in the range of $V \leq 25$ ft/sec. With higher impact velocities, the stress at the impact end appears to be proportionally lower and the percentage attenuation of the velocity wave higher (Fig. 18).

All these points seem to indicate that a strain-rate function also including nonlinear terms should be used, that is

$$\frac{\partial \varepsilon}{\partial t} = \frac{1}{E_0} \frac{\partial \sigma}{\partial t} + \frac{k}{E_0} \left[ (\sigma - \sigma_{st}) + A (\sigma - \sigma_{st})^2 + \cdots \right].$$

Thus, for instance, an exponential one, might produce results in better agreement with the experiments since it would give, for the same $v(0, t)$, higher values of $\sigma(0, t)$ and higher relative attenuation of the velocity wave near the origin of $x$ and $t$ (see point a). It would give relatively smaller stresses and higher relative attenuation for higher impact velocity (see point c). For large $x$ and $t$ it would approach the asymptotic values at a lower rate than the linear function, in agreement with point (b) and with the higher value that the experimental $\sigma(0, t)$ maintains with respect to the calculated one. Furthermore, as we have shown in Section 4, it would still not contravene the general asymptotic values found in the tests.

A last point must be noted here, although it requires further experimental confirmation. In the first instants of the impact, the stress $\sigma(0, t)$

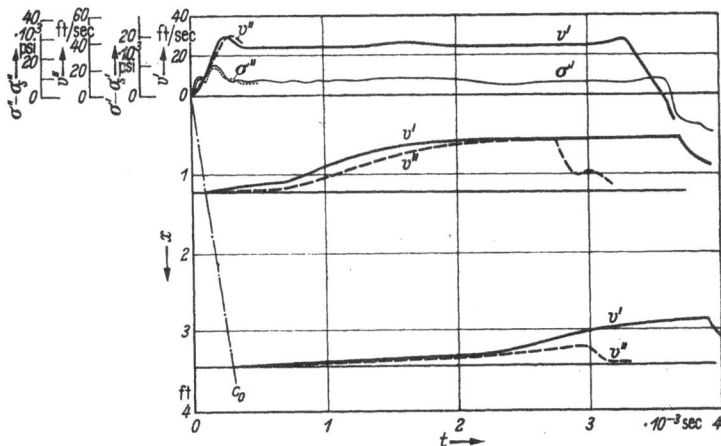

Fig. 18. Comparison between two tests with the same static preloading (18.5 ksi) and different impact velocities.

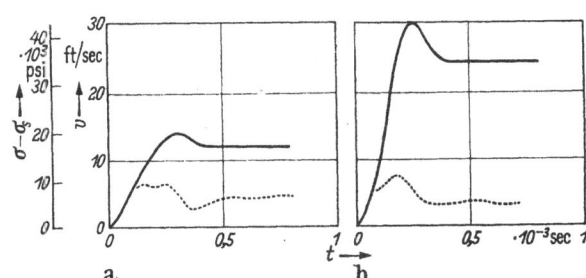

Fig. 19 a and b. The experimental velocity and stress at the impact end. Static preloading: a) 5.00 ksi; b) 18.5 ksi.

seems to maintain the elastic relation $\sigma(0, t) = \varrho c_0 v(0, t)$ for a finite interval of time, of the order of $\Delta t = 0.05 \div 0.1 \times 10^{-3}$ sec. This is shown in the diagram of Fig. 19, where the scale is such that the two lines coincide when $\sigma(0, t) = \varrho c_0 v(0, t)$, and of Fig. 20, where the experimental ratio $\sigma(0, t)/v(0, t)$ is plotted together with the corresponding calculated ratio (broken line) for the test with higher preloading.

8*

It is evident that were this result — which can, perhaps, be related to some of Turnbow and Ripperger's findings in [5] — generally true, the exact description of the phenomenon in the region near the origin of $x$ and $t$ would render the analytical problem much more complicated.

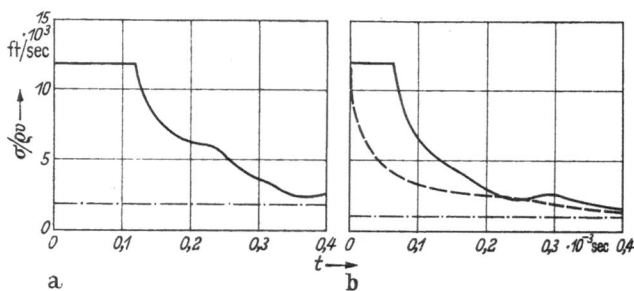

Fig. 20 a and b. The ratio between the experimental stress and velocity at the impact end. Static preloading: a) 5.00 ksi; b) 18.5 ksi.

## Conclusions

The comparison of the experimental results with the numerical calculations shows that, in the case of annealed copper and within the range of the impact velocities used in the tests, the strain-rate-dependent theory in the linear form, and possibly even more in a non linear form, gives a good description of the whole phenomenon for any value of time, distance from the impact end, and static preloading.

If, however, the impact velocity rises regularly to a steady state value, the asymptotic behaviour is given by the solution of the strain-rate-independent theory. Therefore the latter approximates the solution, at any section, for sufficiently large values of time and may be useful, in particular, if we are interested only in the permanent deformations. Furthermore, if there is no static preloading the approximation becomes good for any value of time at sections away from the origin. In this case, in fact, the strain-rate effects are percentagewise less noticeable since a portion of the wave is below the limit of proportionality and the bulk of the wave above that limit is already travelling with the velocities of propagation given by the strain-rate-independent theory.

### References

[1a] Riparbelli, C.: On the Time Lag of Plastic Deformation, Proceedings First Midwestern Conference on Solid Mechanics (1953).

[1b] Sternglass, E. J., and D. A. Stuart: An Experimental Study of the Propagation of Transient Longitudinal Deformation in Elasto-plastic Media, J. Appl. Mech. 20, 427 (1953).

[1c] BELL, J. F.: Propagation of Waves in Pre-Stressed Bars, Technical Report No. 5, Navy Contract N 6 — ONR 247, Johns Hopkins University 1951.

[2a] BIANCHI, G.: On the Propagation of Longitudinal Strain Pulses in a Bar Pre-stressed into Plastic Region, M.S. Thesis at Cornell University 1953.

[2b] —: La propagazione di onde d'urto in regime plastico, Zanichelli, Bologna 1957.

[2c] —: Il progetto di una apparecchiatura per prove d'urto, Ingegneria Meccanica 1960.

[2d] —: I moti transitori di un'asta pesante recante ad una estremità una massa concentrata soggetta a frenamento viscoso, Parti I, II, III, Rivista di Ingegneria 1960.

[3] RUBIN, R. J.: Propagation of Longitudinal Deformation Waves in a Pre-stressed Rod of Material Exhibiting a Strain-Rate Effect, J. Appl. Phys. 25, No. 4 (1954).

[4a] MALVERN, L. E.: The Propagation of Longitudinal Waves of Plastic Deformation in a Bar of Material Exhibiting a Strain Rate Effect, ONR Report N. A11—39/74 (1949).

[4b] GLAUZ, R. D., and E. LEE: Transient Wave Analysis in a Linear Time Dependent Material, J. Appl. Phys. 25, No. 8 (1954).

[4c] PLASS, H. J., and NENG-MING-WANG: Longitudinal Plastic Waves in Long Rods of Strain-Rate Dependent Material, Proceedings Fourth Midwestern Conference on Solid Mechanics 1959.

[5] TURNBOW, J. W., and E. A. RIPPERGER: Strain-Rate Effect on Stress-Strain Characteristics of Aluminum and Copper, Proceedings Fourth Midwestern Conference on Solid Mechanics 1959.

# Some Problems of the Mechanics of Extensible Strings

By N. Cristescu

Mathematical Institute of the Academy of the Rumanian People's Republic,
Bucharest, Rumania

## 1. Introduction

In the theory of plasticity, most of the constitutive laws used today for three-dimensional problems are well established only in the case of static simple (proportional) loading. Dynamic problems cannot, in general, be studied with the aid of these theories because dynamic boundary conditions rarely lead to simple loading. This is the reason why much of the present work dealing with dynamic problems is restricted to the one-dimensional case. Among the one-dimensional problems (involving only a single spatial coordinate and a single strain component) some that are interesting both from the mechanical and the mathematical points of view concern extensible strings, assumed to consist of elastic, visco-plastic materials. Several of these problems are discussed in the following; some are analyzed using various constitutive laws and pursuing especially the interaction between the two types of waves that can propagate in a string.

Bibliographical information on this problem up to 1960 is available in [1].

The equations of motion of an extensible string are

$$\frac{\partial}{\partial s_0}\left(\frac{T}{1+\varepsilon}\frac{\partial x}{\partial s_0}\right) + \varrho_0\frac{\partial^2 x}{\partial t^2} + (1+\varepsilon)X = 0, \tag{1}$$

and two similar equations for $y$ and $z$. The strain $\varepsilon$ is given by

$$\varepsilon = \frac{ds - ds_0}{ds_0} = \left[\left(\frac{\partial x}{\partial s_0}\right)^2 + \left(\frac{\partial y}{\partial s_0}\right)^2 + \left(\frac{\partial z}{\partial s_0}\right)^2\right]^{1/2} - 1, \tag{2}$$

$s_0$ being the Lagrangian coordinate along the string, $\varrho_0$ the initial linear density, and $X$, $Y$, $Z$ the components of the external force. The stress $T$ is connected to the strain $\varepsilon$ by a relation that characterizes the mechanical behavior of the material. From the mechanical point of view, it is important to know whether these constitutive relations explicitly contain the rate of strain, whether they are written in differential or finite form, and so on. Numerous cases must be separately discussed, because various materials or types of experiment require the use of different constitutive relations leading to mathematically different problems.

## 2. Finite Constitutive Equation: Basic Formulas

It will first be assumed that the constitutive relation may be written in finite form:

$$T = T(\varepsilon). \tag{3}$$

Here the function $T(\varepsilon)$ is monotonically increasing and, as a rule, nonlinear. It is known that two types of waves may propagate in a string of this kind; their velocities are

$$c_I = \left[\frac{T}{\varrho_0(1+\varepsilon)}\right]^{1/2}, \qquad c_{II} = \left[\frac{1}{\varrho_0}\frac{dT}{d\varepsilon}\right]^{1/2}. \tag{4}$$

The fronts of these waves, in the $s_0$, $t$ plane are the curves

$$\frac{ds_0}{dt} = \pm c_I(\varepsilon) \tag{5}$$

and

$$\frac{ds_0}{dt} = \pm c_{II}(\varepsilon). \tag{6}$$

If the relation (3) is nonlinear, neither of the families (5) or (6) is a family of straight lines. The first kind of wave is transverse, and the second longitudinal. On the fronts of transverse waves [the curves (5)] the relations

$$dx_t = \pm c_I(\varepsilon)dx_{s_0} + \frac{1}{\varrho_0}[(1+\varepsilon)X - x_{s_0}F]dt - \frac{x_{s_0}}{1+\varepsilon}(\pm c_I(\varepsilon)d\varepsilon - dv) \tag{7}$$

and four others for $y$ and $z$ are satisfied. On the fronts of longitudinal waves [the curves (6)] the relations

$$\pm c_{II}(\varepsilon)d\varepsilon - dv + \frac{1+\varepsilon}{\varrho_0}Fdt = 0 \tag{8}$$

are satisfied. In (7) and (8),

$$x_{s_0} = \partial x/\partial s_0, \qquad x_t = \partial x/\partial t, \qquad \text{etc.,}$$

while

$$dv = \frac{1}{1+\varepsilon}(x_{s_0}dx_t + y_{s_0}dy_t + z_{s_0}dt_t)$$

denotes the increment along the lines (5) or (6), of the component of the velocity of the typical point of the string, and

$$F = (X x_{s_0} + Y y_{s_0} + Z z_{s_0})/(1 + \varepsilon)$$

is the projection, on the tangent, of the components of external forces which act on a portion $ds$ of the string.

As is well known, the transverse waves modify only the shape of the string, and the longitudinal waves only its length (that is $\varepsilon$ and $v$). Of course, the two types of waves interact. This is mathematically expressed by the appearance of $\varepsilon$ and $v$ in (7) and that of $x_{s_0}$, $y_{s_0}$, $z_{s_0}$ in the expressions of $d\varepsilon$ and $dv$ that occur in (8). There is however no question here of coupling of waves as in two- and three-dimensional problems in the theory of plasticity.

Thus, the integration of the system (1)—(3) may be replaced by the integration of relations (7) and (8) along the curves (5) and (6), respectively. As a rule, this integration can only be performed approximately, resulting in the simultaneous construction of a family of curves (5) and (6) and the integration of the relations (7) and (8) (see [2]). The method is rather complicated, but it can be adapted to electronic computers. Some examples have been considered in [3, 4] where the relation (3) has been taken in a linear form to facilitate the computation. In this way one may study the simultaneous propagation of the longitudinal and transverse waves (direct and reflected) in the same portion of the string.

In certain cases, when the longitudinal and transverse waves propagate in different portions of the string, the integration of the system (7) to (8), or even that of the original system (1)—(3), is greatly facilitated. Many problems of this kind involving plane motion have been investigated by H. A. Rakhmatulin and his co-workers [1]. The sequence of propagation of different waves is essential in such problems. It is known that the sequence depends on the constitutive law (3) as well as on the initial and boundary conditions.

## 3. Finite Constitutive Equation: Initial and Boundary Conditions

In connection with the above methods of integration, we have to examine how the initial and boundary conditions must be formulated in the general, three-dimensional motion of an extensible string. We may consider both boundary conditions at fixed points of the string, and boundary conditions at points that move along the string. For example, a body that strikes the string may continue to make contact with it at the same material point or move along the string.

Similarly, the initial conditions, may take several forms. We may, for instance, start with a CAUCHY problem: at $t = 0$ and $0 \leq s_0 \leq l$ the functions

$$\varepsilon(s_0, 0), \quad v(s_0, 0), \quad x(s_0, 0), \quad \cdots, \quad x_{s_0}(s_0, 0), \quad \cdots, \quad x_t(s_0, 0), \quad \cdots \quad (9)$$

are known, $l$ being the initial length of the string. Thus the initial position (shape) and velocity of each point of the string are known as well as its deformation, while the ensuing motion and deformation of the string are to be determined. It is supposed that the functions (9) vary continuously along the string.

Naturally the functions (9) must satisfy certain compatibility conditions, because not all of them can be given arbitrarily. They may be given either analytically or numerically (at a certain number of points) on the segment $OB$ (Fig. 1). Using the numerical method mentioned above, we may then build the solution in a certain domain limited by this segment and two lines (6) that pass through the points $s_0 = 0$ and $s_0 = l$. More precisely, the lines $OA$ and $BA$ are those among the characteristics through $O$ and $B$, respectively, that have the smallest slopes. Usually, the curves (6) have a smaller slope than the curves (5) because $c_{II} > c_I$, but if the reverse inequality holds, the lines $OA$ and

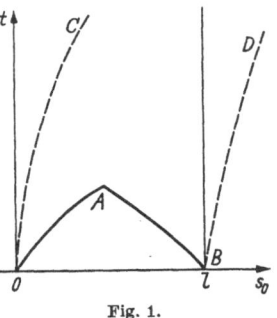

Fig. 1.

$BA$ are the two lines (5) that pass through $O$ and $B$, respectively. It is also, possible that a portion of the line $OA$ or $BA$ is a line (6), the remainder being a line (5); this happens if, at some point on $OA$ or $BA$, the slope of the lines (5) becomes smaller than the slope of the lines (6), depending on the initial shape or strain of the string or on its constitutive law. This case, however, occurs more rarely in practice. In the domain $OAB$ the motion of the string is influenced only by the initial conditions but not by the boundary conditions. Above the line $OAB$ the motion of the string also depends on the boundary conditions at the ends of the string. $OAB$ is therefore the domain of influence of the line segment $OB$.

If the boundary conditions involve at all times the same material points of the string, then they are given on the straight lines $s_0 = 0$ and $s_0 = l$ of the $s_0, t$ plane. If bodies that strike the string (or impart motion to its ends) move along it, then the boundary conditions must be given along curves such as $OC$ and $BD$ in Fig. 1. In both cases at each instant the positions and velocities of the end points of the string, that is, the functions

$$x(s_1, t), \quad y(s_1, t), \quad z(s_1, t), \quad x_t(s_1, t), \quad y_t(s_1, t), \quad z_t(s_1, t) \quad (10)$$
and
$$x(s_2, t), \quad y(s_2, t), \quad z(s_2, t), \quad x_t(s_2, t), \quad y_t(s_2, t), \quad z_t(s_2, t) \quad (11)$$

should be given numerically or analytically. In the latter case it is of course sufficient to give the equations of motion of these points. If the boundary conditions involve at all times the same points of the string, then $s_1 = 0$ and $s_2 = l$ in (10) and (11). If, on the other hand, the boundary conditions involve variable points, there exist relations of the form

$$s_1 = f_1(t), \qquad s_2 = f_2(t), \tag{12}$$

which represent the dotted lines in Fig. 1.

The curves (12) cannot be prescribed arbitrarily. Their shapes impose certain conditions upon the length of the string, because the horizontal distance between the two curves is related to the instantaneous length of the string. Because the strain at a certain point of the string cannot exceed a certain limit without rupture of the string, the lines (12) must be chosen in such a way that the resulting strain in the string will nowhere surpass this limiting strain. In general, the length of the string may vary. Only in some problems is this length constant. On the other hand it may be impossible to give the equations of the curves (12) in advance, because the shape of these curves only emerges during the process of integration.

For this method of integration to be applicable, the slope of the curves (12) must everywhere differ from that of the curves (5) or (6). If the former slope is greater than that of the curves (5) or (6), problems of mixed type must be solved in the regions between $OA$ and $OC$ or $BA$ and $BD$. This necessity arises when the bodies that strike the string move along it with a velocity smaller than the velocities of the transverse or longitudinal waves. If the slope of the curves (12) is smaller than that of the curves (5) or (6), a discontinuity of the slope of the tangent to the string or of the strain may develop. This discontinuity propagates along the curves $OC$ or $BD$ and this may cause the string to break or become entangled. Finally, if on certain portions, the curves (12) are tangent to the curves (5) or (6), a RIEMANN problem must be solved.

Only the first of these cases will be considered in the following. Along the curves $OA$ and $AB$, the values of $\varepsilon$, $v$, $x_t$, $y_t$, $z_t$, $x_{s_0}$, $y_{s_0}$, $z_{s_0}$ are known from the previous integrations or the initial conditions. Along the curves (12) the functions (10) and (11) must be given. The variation of $\varepsilon$ and $v$ along these curves as well as the variation of the slope of the tangent to the string are obtained by the appropriate differentiations [5]. In other words, from the described motions of the points of the string that are struck, the strain, the longitudinal velocity and the slope of the string in these points are obtained by calculus.

## 4. Examples

To illustrate the procedure, two examples have been worked out with the aid of the electronic computer CIFA 2: the semi-infinite string, and the finite string. We first explain the method of calculation.

We are constructing a network consisting of the lines (6). Let us consider a typical mesh of this network (the full lines in Fig. 2). At the vertices $A$, $B$ and $C$ the values of the functions $\varepsilon, v, x_{s_0}, y_{s_0}, z_{s_0}, x_t, y_t, z_t$ are known from the initial or boundary conditions, or from the previous computation. Using the relations (8), written as finite difference equations, in a first approximation we obtain the values of $\varepsilon$ and $v$ at the vertex $D$. With the help of the values of these functions, we compute $c_I(\varepsilon)$ at $D$ and draw the lines $DE$ and $DF$, which are segments of the characteristics (5). The values of $\varepsilon, v, x_t, y_t, z_t, x_{s_0}, y_{s_0}, z_{s_0}$ at $E$ and $F$ are calculated by interpolation, and the relations (7) are integrated along $ED$ and $FD$. The valuse of $x_t, y_t, z_t$ and $x_{s_0}, y_{s_0}, z_{s_0}$ at $D$ are obtained as a first approximation. Using these values, a second approximation of $\varepsilon$ and $v$ in $D$ may be calculated, and so on.

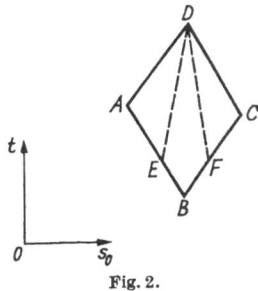

Fig. 2.

This approximate method has been discussed in detail in [2] and, with some modifications that slightly simplify the program for the electronic computer, in [3].

The evaluation of the unknown functions at the end points of the string is carried out by means of another rather complicated program, which will not be discussed here (see [5]).

1. In the case of the semi-infinite string it was supposed that initially the string coincides with the positive $x$ axis and is at rest. The motion of the string is supposed to be plane. The considered initial conditions are:

$$\left. \begin{array}{l} t = 0 \\ s_0 > 0 \end{array} \right\} \quad x_t = y_t = 0, \quad x_{s_0} = 1.04, \quad y_{s_0} = 0, \quad \varepsilon = 0.04\%, \quad v = 0. \quad (13)$$

Accordingly, the string has initially a given uniform strain. The boundary conditions are assumed to be as follows:

$$\left.\begin{array}{l} s_0 = 0 \\ t \geq 0 \end{array}\right\} \; x_t = 0, \quad y_t = a \cdot \sin bt, \qquad (14)$$

where $a$ and $b$ are constant: $a = -10^4 \, \text{cm sec}^{-1}$, $b = 200 \, \text{sec}^{-1}$. As the constitutive law (3), the relation

$$T = 70 \, (\varepsilon - 0.012) \qquad (15)$$

has been used, in which $T$ is given in grams/denier, and $\varepsilon$ in percent. To the constitutive law (15) corresponds $c_{II} = 2500 \, \text{m sec}^{-1}$ and $c_I(\varepsilon_0) = 400 \, \text{m sec}^{-1}$, while for $\varepsilon > \varepsilon_0$ we have $c_I(\varepsilon) > c_I(\varepsilon_0)$. Accordingly, the lines (6) are straight. For simplicity, the motion in the absence of external forces is discussed.

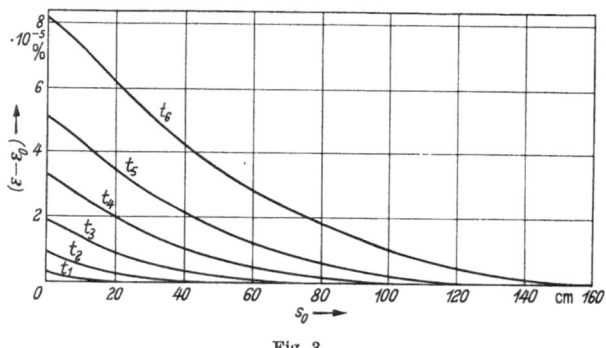

Fig. 3.

The relations (13), (14) and (15) have been chosen only for illustration. The shape, the deformation, and the initial velocities may be arbitrarily given for a three-dimensional motion. The boundary conditions may also be arbitrarily (even numerically) prescribed provided that no unloading occurs anywhere in the string. Finally, the function $T$ in (3) can be an arbitrary monotonically increasing function.

The results of the computations are printed in the form of tables for the values of the functions $x_{s_0}, y_{s_0}, x_t, y_t, \varepsilon, v$ and of the angle between the tangent to the string and the $x$ axis for different values of $t$ and $s_0$. From these tables graphs of these functions of $s_0$ may be obtained for different values of the time $t$ ($t_1 = 10 \cdot 10^{-5} \, \text{sec}$, $t_2 = 20 \cdot 10^{-5} \, \text{sec}$, $t_3 = 30 \cdot 10^{-5} \, \text{sec}$, $t_4 = 40 \cdot 10^{-5} \, \text{sec}$, $t_5 = 50 \cdot 10^{-5} \, \text{sec}$, $t_6 = 64 \cdot 10^{-6} \, \text{sec}$). For example, the variation of $\varepsilon$ is shown in Fig. 3. Similarly, the dotted lines in Fig. 4 show the variation of the component $x_t$ of the velocity.

Note that at any instant $t$, the component $x_t$ first increases and then decreases along the string. In the same figure, the variation of the longitudinal velocity $v$ is shown by the full lines for the same values of $t$.

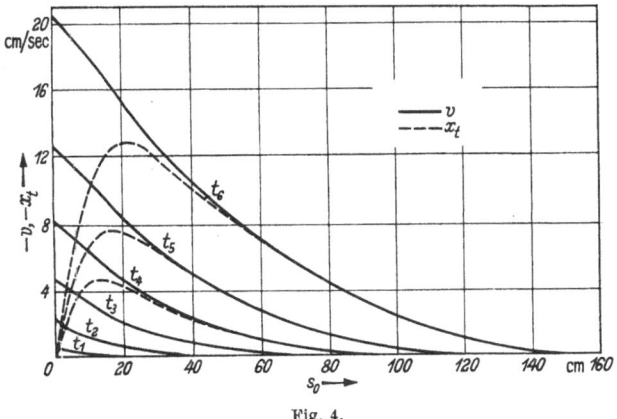

Fig. 4.

As is to be expected, beginning with a certain distance from the end of the string, one finds $v = x_t$ in the portion where the string is still rectilinear. The graphs for the variation of the component $y_t$ of the velocity, as well as of the slope of the tangent to the string, for different values of $t$, are not shown here. From the functions so obtained, one readily derives the shape of the string for a number of values of $t$.

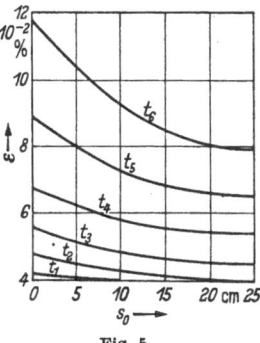

Fig. 5.

2. As a second example, the finite string will be considered, one of the ends being fixed, the other being given an accelerated motion. Initially the string is rectilinear and extends along the $x$ axis from $x = 0$ to $x = l$. The boundary conditions are as follows:

for $s_0 = 0$ and $t \geq 0$: $x_t = 0$, $y_t = -10^5 \sin (10^3 t)$,

for $s_0 = l$ and $t \geq 0$: $x_t = 0$, $y_t = 0$,        (16)

and the initial conditions (13) are used again, as is the constitutive equations (15). The calculation was performed on the electronic computer following the scheme outlined above. The network of the lines (6) was drawn with $\Delta t = 10^{-5}$ sec and $\Delta s_0 = 2{,}5$ cm between two neighboring vertices on the same characteristic. (For a more detailed discussion, see [4].)

The variation of $x_{s_0}$, $y_{s_0}$, $x_t$, $y_t$, $\varepsilon$ and $v$ was graphed for $t_t = 6 \cdot 10^{-5}$ sec, $t_2 = 12 \cdot 10^{-5}$ sec, $t_3 = 18 \cdot 10^{-5}$ sec, $t_4 = 24 \cdot 10^{-5}$ sec, $t_5 = 30 \cdot 10^{-5}$ sec, and $t_6 = 36 \cdot 10^{-5}$ sec. Fig. 5 shows the variation of $\varepsilon$, and Fig. 6 that

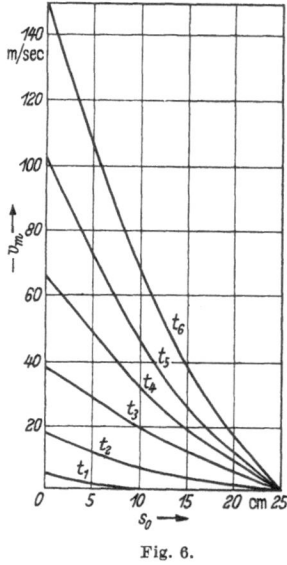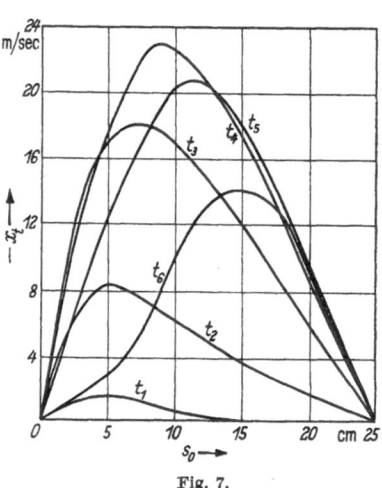

Fig. 6.                                    Fig. 7.

of $v$. We mention that due to the deformation, the initially homogeneous string becomes strongly non-homogeneous. The variation of $x_t$ (Fig. 7) shows the influence of the reflection of the longitudinal waves from the

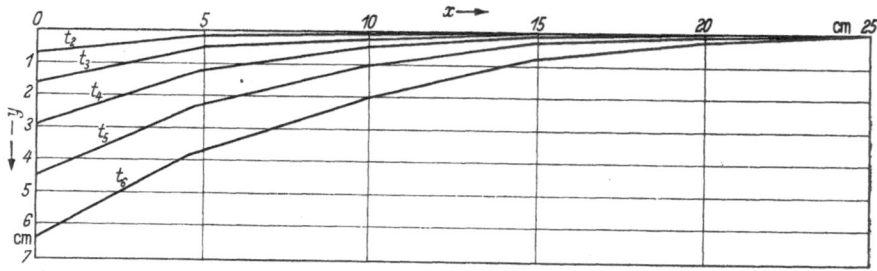

Fig. 8.

two ends of the string. The graph of the variation of $y_t$ is not shown here. Fig. 8 shows the string for different instants. The picture of different front waves is given in Fig. 9. $OA$ is the front of the first longitudinal wave which propagates in the string, and $AB$, $BC$, and $CD$ are successive reflections of this wave from the two ends of the string. $OM$ is the front of the first transverse wave which propagates in the string, and $QN$ and $BP$ are two further fronts of transverse waves. They were constructed approximately, starting from (5). It is interesting to follow the variation of the strain along the three fronts of transverse waves which are shown. This variation is given in Fig. 10 which indicates that along the fronts of transverse waves the strain varies strongly.

As these examples show, the deformation of strings can be obtained even when direct and reflected, longitudinal and transverse waves propagate simultaneously in the same portion of the string.

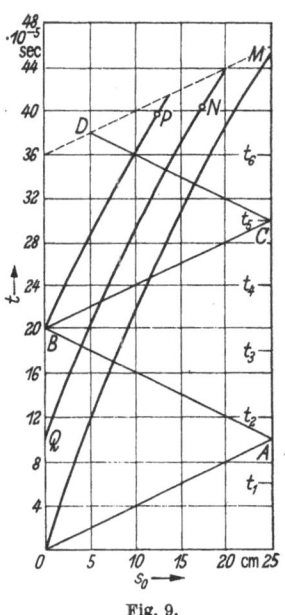

Fig. 9.

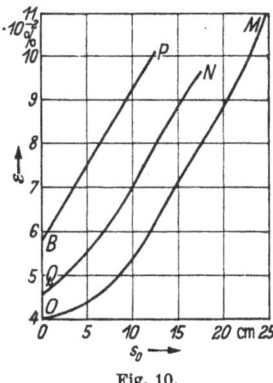

Fig. 10.

## 5. Differential Constitutive Equation

It has so far been assumed that the constitutive law can be written in the finite form (3). For some materials, however, the constitutive law must be written in a differential form, especially when dynamic problems are to be treated. In the following, some constitutive laws of this type will be considered.

*1.* We first consider constitutive laws of the general form

$$\frac{\partial T}{\partial t} = f\left(T, \varepsilon, \frac{\partial \varepsilon}{\partial t}\right) \tag{17}$$

relating the stress rate to the stress, the strain, and the strain rate. As particular cases we may consider the MAXWELL body, the standard linear body, and others. We shall suppose that the Eq. (17) is linear in the strain rate.

The following particular form of the relation (17), has often been used to take account of the influence of the strain rate:

$$\frac{\partial T}{\partial t} = E\,\frac{\partial \varepsilon}{\partial t} - g(T,\varepsilon); \tag{18}$$

here the function $g(T,\varepsilon)$ is frequently taken as

$$g(T,\varepsilon) = k[\sigma - f(\varepsilon)], \tag{19}$$

$k$ being a constant and $\sigma = f(\varepsilon)$ the quasistatic constitutive law of the material. If the material is perfectly plastic, the relation (18) has the form

$$\frac{\partial T}{\partial t} = E\,\frac{\partial \varepsilon}{\partial t} - k\mathfrak{F}[T - T_0(s_0)], \tag{20}$$

where $T_0(s_0)$ is the yield stress, the function $\mathfrak{F}(\xi)$ being defined as follows: $\mathfrak{F}(\xi) = 0$ if $\xi \leq 0$, and $\mathfrak{F}(\xi) = \xi$ if $\xi \geq 0$. This type of relation has been considered by V. V. Sokolovski, L. E. Malvern, V. N. Ku-kudjanov, L. V. Nikitin, and others.

The motion and deformation of a string, with a constitutive law of the form (18) or (20) can be studied as in Section 2 and the approximate method of integration can be used. In a certain sense these computations are even facilitated because for the constitutive laws (18) or (20) the fronts of longitudinal waves propagate with a constant velocity that equals the velocity of elastic longitudinal waves. The characteristic lines (6) are therefore straight.

If, for example, the constitutive law (20) is joined to the equations of motion (1) and to the relation (2), transverse and longitudinal waves are found to propagate with the velocities

$$c_I(\varepsilon) = \left[\frac{T}{\varrho_0(1+\varepsilon)}\right]^{1/2}, \qquad c_{II} = \left[\frac{E}{\varrho_0}\right]^{1/2} \tag{21}$$

respectively. The fronts of the longitudinal waves are the family of straight lines

$$\frac{ds_0}{dt} = \pm c_{II}. \tag{22}$$

Along them the differential

$$\mp \varrho_0 c_{II}\,dv \pm c_{II}(1+\varepsilon)^2 F\,dt + dT + k\mathfrak{F}[T - T_0(s_0)]dt = 0 \tag{23}$$

must be integrated. In these formulae the same notation as before has been used. The fronts of transverse waves are curves of variable slope, whose equations formally coincide with (5). Along these curves some differential relations, which also formally coincide with (7), must be integrated, but in all these formulae the stress $T$ must be obtained from the constitutive law (20) rather than from (3).

The scheme of integration as well as the formulation of the initial and boundary conditions remain in principle the same as above. There are still some slight differences. Because (23) contains both the stress and the strain, the relation (20) written in the form

$$dT = E\,d\varepsilon \pm \frac{\sqrt{1 + c_{II}^2}}{c_{II}}\,k\,\mathfrak{F}\,(T - T_0(s_0))\,dt \tag{24}$$

must be incorporated in the scheme of calculation[1]. With the help of this relation we find the strain at each vertex together with the stress.

There are also some modifications in the formulation of the initial conditions. When the constitutive law (20) is used, not only the initial position, velocity and deformation of the string must be given but also the initial stress for all $s_0$. In this case the boundary conditions are formulated in the same manner as before.

Similar remarks apply to the constitutive law (18), which is more general than the law (20).

A still more general constitutive law which is useful in the treatment of dynamic phenomena is given by

$$\frac{\partial T}{\partial t} = f(T, \varepsilon)\,\frac{\partial \varepsilon}{\partial t} + g(T, \varepsilon), \tag{25}$$

where the functions $f$ and $g$ may be functions of $s_0$ and $t$ and possibly of some characteristic constants of the material (yield stress, etc.). Relations of the type (25), with other forms of the functions $f$ and $g$ may also be used when the stress decreases that is for nonlinear unloading.

The rapid motions of strings with the constitutive law (25) also involve two types of waves. The fronts of the transverse waves are represented by (5); the differential relations satisfied on these characteristics formally coincide with (7). The fronts of longitudinal waves are given by

$$\frac{ds_0}{dt} = \pm c_{II}(T, \varepsilon) = \pm \sqrt{\frac{f(T, \varepsilon)}{\varrho_0}}; \tag{26}$$

along them the differential equations

$$\mp \varrho_0 c_{II}\,dv \pm c_{II}(1 + \varepsilon)^2 F\,dt + dT + g(T, \varepsilon)\,dt = 0 \tag{27}$$

must be integrated. We notice that of the two functions $f$ and $g$ in the constitutive law (25), the first appears explicitly in the velocity of propagation (26), and the second in the differential relations (27).

---

[1] This scheme will be examined in detail in another paper.

The method of integration for the constitutive law (25) remains the same as above, but in the program[1] we must also use the relation (25) in the form

$$dT = f(T, \varepsilon)d\varepsilon \pm \frac{\sqrt{1 + c_{II}^2}}{c_{II}} g(T, \varepsilon)dt. \tag{28}$$

The upper and lower signs in (26), (27), and (28) correspond to each other. The sole difficulty in this case is that both families of characteristics are in general curvilinear, with slopes that depend on the stress and strain in the point under consideration. As in the case of the constitutive law (20), the initial position, velocity and deformation in each point of the string must be given as well as the initial stress distribution.

2. A more difficult problem arises when the constitutive law has the form

$$T = T(\varepsilon, \dot{\varepsilon}), \tag{29}$$

that is, when the stress is a function, generally nonlinear, of strain and strain rate. A KELVIN-VOIGT material represents a special case of (29).

It is easy to show that the system of equations (1), (2) and (29) possesses only one family of characteristics, which formally coincides with (5), the differential relations satisfied along these characteristics also formally coinciding with (7). Therefore the same relations will govern the transverse movement of the string, independently of whether the constitutive law (3) or (29) is used.

The longitudinal motion has not yet been discussed in this case and we are as yet unable to integrate the equations of motion of the string which satisfies a constitutive law of the type (29).

## 6. Separable Constitutive Equation

Let us next consider a constitutive law of the form

$$T = \varphi(t)\psi(\varepsilon) \tag{30}$$

which is commonly used in the theory of creep, but whose value in the treatment of dynamic problems is as yet uncertain. Adding Eqs. (1) and (2) to the constitutive law (30), we readily show that two types of waves arise in dynamic problems of this kind. These waves propagate with the velocities

$$c_I = \left[\frac{1}{\varrho_0} \frac{\varphi(t)\psi(\varepsilon)}{1 + \varepsilon}\right]^{1/2} \quad \text{and} \quad c_{II} = \left[\frac{\varphi(t)\psi'(\varepsilon)}{\varrho_0}\right]^{1/2} \tag{31}$$

---

[1] This scheme will be examined in detail in another paper.

and the differential relations satisfied on the characteristics formally coincide with (7) and (8). Accordingly, the function $\varphi(t)$ in (30) only appears explicitly in the velocities of propagation. The scheme of integration remains in principle the same as above; we note only that, in general, both families of characteristics are curves in the plane $s_0, t$.

## 7. Initial Elastic Behavior

At the beginning of the deformation process most materials behave in a perfectly elastic manner according to the relation

$$T = E\varepsilon, \tag{32}$$

in which $E$ is a constant. The study of the simultaneous transverse and longitudinal motions of strings which satisfy (32) is carried out in manner similar to that employed in Section 2. A simplification arises from the fact that one of the families of characteristics [the curves (6)] now consists of straight lines, the velocity of propagation $c_{II} = \sqrt{\dfrac{E}{\varrho_0}}$ being constant. The differential relations satisfied on the characteristics coincide with (7) and (8). It must be mentioned that the equations of motion (1), (2), and (32) of the elastic string form a quasilinear system which can only be integrated numerically.

Similar conclusions are obtained when a linear relation between $T$ and $\varepsilon$ is used in the study of problems involving unloading. It is then more advantageous to write this relation in differential form.

Another much simpler particular case is that of the inextensible string. We shall consider both the case when the string remains inextensible at all times and the case when, after a certain uniform deformation, the string remains inextensible in certain domains of the $s_0, t$ plane. In these cases we have $T = $ const. in (1) and the equations are reduced to three quasilinear, or even linear, independent equations. The transverse motion is then studied easily.

## 8. Summary

Various constitutive laws and the influence of their form upon the rapid motions of strings have been reviewed above. We now present some conclusions.

If the constitutive law is in a finite or differential form, the order of the derivatives of $T$ and $\varepsilon$ being the same, then two types of waves can propagate in the string. These waves will in general interact. In these cases an approximate method of integration of the equations of motion can be constructed even when both longitudinal and transverse waves propagate simultaneously in the same portion of the string. We notice that formally, both the velocity of propagation of transverse waves and the differential relations satisfied on the fronts of these waves are the same for all the constitutive laws considered here [including the relation (29)]. Therefore in all these cases the transverse waves can propagate in the extensible or nonextensible strings.

The longitudinal waves, however, are to an important extent influenced by the form of the constitutive law as regards both the velocity of propagation and the differential relations satisfied on the fronts of the waves. In some cases these waves cease to propagate, in other cases they are not strictly speaking waves but perturbations.

Sometimes in the treatment of a problem the constitutive law for the string must be changed [for instance, after a certain limiting stress is reached, the law (32) may have to be replaced by a non-linear law (3)]. If, as a consequence of this change the velocity of propagation of the longitudinal waves decreases discontinuously, no difficulty arises. But, if this velocity increases discontinuously, shock waves will occur. We note that such a change of the constitutive law does not in general produce a discontinuous change of the velocity of propagation of transverse waves. The shock waves, longitudinal and transverse, and their interaction with the continuous waves, have not been considered in this paper. The unloading problem and the problem of the continuation of the transverse motion during the unloading, will be considered elsewhere.

### References

[1] LEE, E. H., and P. S. SYMONDS (editors): Plasticity, New York: Pergamon Press 1960, pp. 412—420.
[2] CRISTESCU, N.: J. Mech. Phys. Solids 9, 165 (1961).
[3] CRISTESCU, N.: Arch. Mech. Stos. 15, 47 (1963).
[4] CRISTESCU, N.: Inj. Journ. 1 (1964) (in press) (in Russian).
[5] CRISTESCU, N.: Anal. Univ. Bucuresti, Ser. Şt. Nat. Fasc. Mat. Fiz. 34 (1962) (in Rumanian).

# Mechanical Waves and Strain-Rate Effects in Metals

By Harry G. Hopkins

Royal Armament Research and Development Establishment,
Fort Halstead, Kent, England

## 1. Introduction

The dependence of the mechanical behaviour of certain metals, particularly mild steel, upon strain-rate has been well known and studied now for a long time. Investigations of strain-rate effects in metals have both interest and importance.

The direct evidence for the existence of the effect of strain-rate on the yield and post-yield behaviour of mild steel is quite conclusive (see, for example, [1, 2]). The yield and post-yield stresses have been found to increase by a factor of 2 or 3 at strain-rates of the order of $10^3$ sec$^{-1}$ and smaller increases have been observed at lower strain-rates. Mild steel shows a complex type of mechanical behaviour which is extremely sensitive to strain-rate, and this sensitivity begins to be exhibited as soon as there is any appreciable departure from 'quasi-static' strain-rates of the order of $10^{-4}$ sec$^{-1}$. In contrast, the evidence for the existence of strain-rate effects in certain other metals not exhibiting sharp yield-points (e.g. aluminium and its alloys) does not always appear to be entirely conclusive, quantitatively if not qualitatively. This is due to a number of factors linked sometimes with the lower or even negligible sensitivity of mechanical behaviour over extensive ranges of strain-rate. A particularly important matter is then the enhanced difficulty of interpreting experimental data at high strain-rates, when this interpretation cannot be achieved in the complete absence of a theory of plastic-elastic wave propagation which is sufficiently realistic in its physical basis and mathematical assumptions.

The present paper is concerned with a review and a commentary upon some aspects of the existing state of research on the topic of mechanical waves and strain-rate effects in metals. It is, however, difficult to undertake this task adequately without greatly exceeding the allowed limits. For this reason, the discussion given here is perforce restricted. In particular, attention is confined mainly to metals which do not have sharp yield-points and which have a face-centred cubic lattice structure (particularly, copper, aluminium and lead) and no very detailed account is included of mathematical aspects. Nevertheless, it is hoped that the resulting brevity will assist in focusing attention upon certain general conclusions that are suggested by the critical examination of the literature.

## 2. Preliminary Considerations

Solids have the well-known characteristic property of being able to transmit a wide variety of types of stress waves. In metals, these include, particularly, elastic, plastic and shock waves, which need be given no precise definition here (but see, for example, [3]). The reason for the existence of this wide variety of stress waves is due to the diverse features of mechanical behaviour exhibited over the possible ranges of variables, especially stress, strain and strain-rate, which features reflect the complex physical processes that can occur in a polycrystalline metal when this is transmitting mechanical energy, normally accompanied by some degree of dissipation into thermal energy. It is convenient to give first some account of this situation, from both macroscopic and microscopic viewpoints.

It is simple to envisage circumstances in which all the above three types of stress waves occur. Thus, suppose that a uniform compression force acts over the surface of a semi-infinite plate, this force being suddenly applied and subsequently increasing in magnitude so that a stress pulse of varying intensity is propagated into the plate. This type of situation has been discussed in detail, notably by MORLAND [4]. At stress intensities below the yield-point, propagation takes place at the elastic wave velocity. Above the yield-point, propagation takes place at the somewhat lower velocity of plastic waves, which varies with the amount of strain-hardening. However, at stress intensities greatly in excess of the yield-point, propagation now takes place at a varying velocity which may approach or even exceed the elastic (loading) wave velocity. This feature is due to the fact that the bulk modulus increases with increasing stress-intensity, and consequently it is possible for shock stress-wave fronts to be set up. There is interaction between the plastic and shock wave fronts, and the latter, after some degree of erosion, may overtake the leading elastic front. In addition to the loading waves, their mutual interactions cause unloading waves to be propagated backwards to the plate surface where these subsequently undergo reflection forwards. Thus, an intricate, changing stress wave system which involves elastic, plastic and shock waves is set up (see [4], Fig. 5). Any strain-rate sensitivity of mechanical behaviour will of course result in further complexities, not envisaged by MORLAND, however. It appears that the mode of transmission of mechanical energy in a solid is related in a complex manner to the levels of stress, strain, strain-rate, and temperature, and possibly other variables, involved. In any particular case, what matter most are the main modes of energy transmission and the manner in which mechanical energy is being dissipated.

Consider now the essential difference between elastic, plastic and shock waves from a molecular viewpoint. With an elastic wave, the atoms of the crystal lattice remain within their individual cells but undergo some relative translational motions, these being small. With a shock wave, the crystal lattice is appreciably distorted so that the atoms come closer together, excluding the possible occurrence of phase changes. With a plastic wave, although similarly as with an elastic wave a small elastic deformation of the lattice occurs, there are movements of dislocations (which may themselves become extensively multiplied in number) so that the lattice now undergoes permanent changes in overall deformation. In all three cases, there are, in general, additional vibrational motions of the atoms, which are recognized macroscopically as thermal changes. COTTRELL [5] has emphasized the fact that a solid undergoes abrupt changes of strain at a stress wave front which are associated with the operation of some particular molecular processes. At a stress wave front, the time-scales associated with the changes in strain and the molecular processes must be comparable, and under conditions of changing strain-rate it is to be expected that there will be changes in these processes. Thus, a given value of strain-rate may conveniently be termed *low* or *high* in relation to some molecular processes of deformation. Consequently, at a sharp elastic wave front, the strain-rate is so high that plastic flow processes cannot operate in the times available, and, in fact, elastic processes operate instead. On the other hand, at a plastic wave front, the strain-rate is sufficiently low for plastic flow processes to operate. It may be noted that at elastic and plastic wave fronts, although the stress intensities may be of the same order of magnitude, the strain-rates may be of very different orders of magnitude, say, for example, $10^{12}$ sec$^{-1}$ and $10^6$ sec$^{-1}$ respectively; and that at shock wave fronts, where the stress intensity is very high in comparison with the yield-point, the strain-rate may approach or even exceed the former value. Of course, it should be remembered that a distinction needs to be drawn carefully between the values of strain-rates measured at microscopic (as above) or macroscopic (as below) levels, which differ by several orders of magnitude. In this paper, attention will be given primarily to circumstances when the value of strain-rate is such that plastic flow processes can occur. Thus, in COTTRELL's sense, the range of values of strain-rate is then low in relation to such processes. The sensitiveness of plastic flow to stress and strain-rate (and also to temperature) is attributed to obstacles to slip of various kinds, including those due to strain-hardening, precipitation-hardening and polycrystalline effects and to lattice resistance. COTTRELL [5] has discussed this and indicated why there is much diversity and variability in the nature of plastic deformation of metals over ranges of strain-rate and temperature.

In proceeding, conventional or macroscopic values of strain-rates will be intended. Extremely low strain-rates of the order of $10^{-4}$ sec$^{-1}$ correspond to so-called static conditions, say, those obtaining in the ordinary tension test, and very high strain-rates of the order of $10^5$ sec$^{-1}$ correspond to severe dynamic loading conditions produced by means of the detonation of high explosives. These extremes perhaps delineate the range of strain-rate normally of interest in studies of plastic flow.

There are a number of different experimental techniques that have been developed for ordinary laboratory investigations of strain-rate effects in metals. Each technique is generally appropriate to fairly well defined ranges of strain and strain-rate. Here, however, mention will be made only of two main types of technique, namely those involving the use of either mechanical (or hydraulic) testing-machines or HOPKINSON pressure-bar apparatus (see, for example, [3, 6, 7]). The former technique can be employed only when it is possible first to treat the response of the test specimen as sufficiently uniform over its length and secondly to measure accurately the applied load and the overall strain. Accordingly, there are upper limits to the strain-rate at which a mechanical testing-machine can be used to provide experimental data that may be directly interpreted. In order to achieve a reasonably uniform distribution of strain, it is necessary for the time of strain-increments to permit the occurrence of several traverses of plastic waves, travelling forwards and backwards along the test specimen. Clearly, as the strain-rate increases, this condition is satisfied less well, particularly at large strains when the rate of strain-hardening and (hence) the plastic wave velocity become smaller. Ultimately, the condition will be definitely violated. Of course, this violation can be delayed by diminishing the test-specimen length, but that will eventually lead to the complication of significant departures from uni-axial conditions of stress and strain. The above general question has been discussed, for example, by LEE and WOLF [8], who gave a plastic-rigid stress-wave analysis by means of which estimates could be made of the limiting values of strain-rate when longitudinal inertia effects (leading to marked non-uniformities of strain distribution) can no longer be neglected. The second technique involving HOPKINSON pressure-bars has been extensively employed at high strain-rates, but, generally, the interpretation of the data obtained then requires analysis which takes explicit account of plastic wave propagation.

Quite generally, it is to be expected that at sufficiently high strain-rates it will not be possible to interpret experimental data without recourse to analysis requiring application of a theory of the propagation of stress waves. Only that procedure can lead to the elucidation of true material properties. However, there is now the immediate obstacle that the analysis of experimental data, undertaken in order to determine mechanical behaviour, requires the application of a theory of stress wave propagation which itself involves the as-yet-unknown mechanical behaviour expressed in the form of constitutive equations. Thus, there are obviously difficulties in the precise analysis of experimental data in order to obtain quantitative evidence on strain-rate effects. Specifically, these difficulties arise for the reasons that there is extremely little knowledge even of the general form of the constitutive laws (see [2]), so that inexact or hypothetical ones are used, and that recourse is made to simplified theories of stress wave propagation. Thus, inevitably, the analysis undertaken of experimental data involves uncertainties, so that the inferred strain-rate effects at best only approximate the true ones. The factors mentioned above have greatly impeded progress and often objectivity in the attempts made towards the resolution of the problems involved in studying strain-rate effects in metals, and a particular consequence has been the expression of apparently differing viewpoints in the literature.

In principle, the existence of stress wave propagation phenomena provides unique opportunities for extending the knowledge of the properties of metals under conditions that can be explored by no other methods, but non-uniformities of deformation make interpretation of data difficult. In the next section, a more detailed discussion will be given of this situation in respect of strain-rate effects in metals. Attention will be given primarily to the results obtained from theoretical and experimental studies of stress wave propagation in bars. For an account of experimental techniques, KOLSKY [9] should be consulted.

## 3. Stress-Wave Propagation in Bars

In the present discussion of stress-wave propagation in bars, it is necessary first to consider briefly elastic waves before considering plastic-elastic waves, the situation of main interest here. Reference throughout is principally to *longitudinal* (rather than torsional or flexural) stress waves in a *long cylindrical* bar, whose uniform cross-section is *circular* and material is homogeneous and isotropic, initially.

## (a) Elastic Waves

Arbitrary disturbances in infinite isotropic elastic media are propagated by dilatational and distortional waves at constant velocities $c_1 = [(k + 4\mu/3)/\varrho]^{1/2}$ and $c_2 = (\mu/\varrho)^{1/2}$, respectively (see, for example, [3, 10]). In the above, $k$ is bulk modulus, $\mu$ is shear modulus and $\varrho$ is density. In the case of a bounded medium, these types of waves each give rise to both types on reflection at the boundary and when there are continually repeated reflections this partial mode conversion leads to complex wave systems (see [11]). This is the situation for a bar, when the effect of the free surface is to produce geometrical dispersion so that the velocity of propagation of a disturbance is dependent upon the wavelength. The exact theory of stress wave propagation in infinite circular elastic bars was given by POCHHAMMER [12] and CHREE [13] (see also [3, 10, 11]) but, due to the complexity of this theory, development and applications are comparatively recent (for surveys of this work, see [14, 15, 16]). As DAVIES [14] has remarked, in analysing dispersion effects it is necessary first to derive dispersion curves (i.e. plots of phase velocity against frequency) from sinusoidal progressive-wave solutions of the governing equations under appropriate boundary conditions, and then secondly to synthesize (by means of FOURIER techniques) the solution of initial value problems of pulse propagation. In practice, this programme is rendered difficult by the boundary conditions, and exact procedures are seldom found to be possible. Thus, for an infinite bar, although the first stage of the analysis has been accomplished (see, in particular, [17, 18]), only approximate solutions of the pulse propagation problem have been found either by use of the exact dispersion curves and application of KELVIN's method of stationary phase or by use *ab initio* of approximate *engineering* theories, whose validity is gauged from the degree of accuracy of their predictions of dispersion curves. However, for *semi-infinite* or *finite* bars the development of exact theories (corresponding to the POCHHAMMER-CHREE theory for infinite bars) is more difficult still and only limited progress has been made (see [19, 20, 21]), and many investigations undertaken have been based *ab initio* upon approximate theories. A very brief account of the above work will now be given, and the references cited should be consulted for fuller details.

For longitudinal stress waves in infinite bars, the POCHHAMMER-CHREE theory involves attention to sinusoidal progressive-waves corresponding to radial and axial components of elastic displacement of the form

$$u_r = U(r) \exp [i(\gamma z + pt)], \qquad u_z = W(r) \exp [i(\gamma z + pt)], \qquad (1)$$

so that the wave frequency is $p/2\pi$, the wavelength $\Lambda$ is $2\pi/\gamma$ and the phase velocity $c_p$ is $p/\gamma$. Here, $r$ and $z$ denote radial and longitudinal co-ordinates and $t$ denotes time. The functions $U(r)$ and $W(r)$, which involve BESSEL functions, and the frequency equation are found by satisfying the elasticity equations under boundary conditions of zero stress at the bar surface. The frequency equation determines $c_p/c_0$ as a function of $a/\Lambda$ and POISSON's ratio $\nu$, where $c_0$ is the bar wave velocity $(E/\varrho)^{1/2}$, $E$ being YOUNG's modulus, and $a$ is the bar radius. This equation has multiple roots, and the dispersion curve therefore consists of an infinite number of branches corresponding to the fundamental and higher modes of vibration (see [17], Fig. 1 and [18], Fig. 13). It has been found that only the fundamental mode is usually excited in pressure-bar experiments. For this mode, $c_p$ has the values $c_0$ (bar wave velocity) and $c_s$ (RAYLEIGH wave velocity) for very large and very small wavelengths, respectively. Energy is propagated at the group velocity

$$c_g = c_p - \Lambda\, dc_p/d\Lambda,$$

which has the same limiting values but shows a pronounced minimum value at $a/\Lambda = 0.45$ approximately (see [18], Fig. 14).

DAVIES [18] has shown experimentally and theoretically (using the POCHHAMMER-CHREE theory and also the approximate RAYLEIGH-LOVE theory, discussed later) that a pulse, whose initial length is comparable with the bar radius, is distorted during propagation and the main pulse is followed by a train of high-frequency oscillations (cf. [19], Fig. 5 and [22], Fig. 2). Such results show clearly that the elementary theory of stress wave propagation (corresponding to constant phase velocity $c_0$) often does not provide an adequate description of the sequence of events for pulse propagation along a bar. It should be remembered that the POCHHAMMER-CHREE theory itself fails in the immediate neighbourhood of an end cross-section of a bar, although it is expected to apply at distant cross-sections.

For the reasons already mentioned, much emphasis has been placed upon the development of approximate theories of longitudinal elastic wave propagation in bars. Any approximate theory may be regarded as intermediate to the extremes of the elementary (see below) and exact theories, and it attempts to take some account of the three-dimensional nature of the bar but not (normally) of complications at end cross-sections. The status of these approximate theories has been discussed by GREEN [23]. Here, mention will be made first of the RAYLEIGH-LOVE [24, 10] and the MINDLIN and HERRMANN [25] theories. In the elementary theory (a limiting approximation to the POCHHAMMER-CHREE theory), plane sections of the bar remain plane and the only displacement con-

sidered is that parallel to the axis. Under these assumptions, the governing wave equation is

$$c_0^2 \frac{\partial^2 u_z}{\partial z^2} - \frac{\partial^2 u_z}{\partial t^2} = 0, \quad c_0 = (E/\varrho)^{1/2}, \tag{2}$$

where $u(z, t)$ is the axial displacement. The phase velocity is constant
and all stress pulses are propagated without change in form. In the
RAYLEIGH-LOVE theory (which may be derived as a direct approximation
to the POCHHAMMER-CHREE theory), the elementary theory is modified
by the inclusion of lateral inertia effects, and the governing wave equation
is now

$$c_0^2 \frac{\partial^2 u_z}{\partial z^2} - \frac{\partial^2 u_z}{\partial t^2} + \frac{1}{2} \nu^2 a^2 \frac{\partial^4 u_z}{\partial z^2 \partial t^2} = 0. \tag{3}$$

The phase and group velocities are given by

$$c_p/c_0 = 1 - \nu^2 \pi^2 (a/\Lambda)^2, \quad c_g/c_0 = 1 - 3\nu^2 \pi^2 (a/\Lambda)^2, \tag{4}$$

but clearly these results become invalid below some value of wavelength.
The MINDLIN and HERRMANN theory is more elaborate and takes account
of lateral inertia and shear effects, but it incorporates disposable constants and is not therefore a direct approximation to the POCHHAMMER-
CHREE theory. The frequency equation obtained is algebraically complicated, and of course there is dispersion. Finally, mention will be made of
the recent theory of HUNTER and JOHNSON [26], which appears to be
somewhat comparable to the MINDLIN and HERRMANN theory but, in
contrast, is a direct approximation to the exact POCHHAMMER-CHREE
theory. The accuracy of all these approximate theories may be gauged
from the comparison of the predicted dispersion curves with the exact
one (see [26], Figs. 2 and 4).

## (b) Plastic Waves

The investigation of longitudinal stress-wave propagation in bars
becomes more difficult under conditions of plastic strain, especially
when there is strain-rate dependent mechanical behaviour. J. HOPKINSON
[27] and subsequently B. HOPKINSON [28] studied the effect of strain-
rate upon the initial yield-stress for steel wires, and gave a theoretical
analysis of the situation which was based upon the elementary theory of
elastic wave propagation along bars (see also TAYLOR [6]). One of the
earliest systematic studies of strain-rate effects in metals was due to
LUDWIK [29] who found empirically a logarithmic dependence of yield
strength upon strain-rate. This result, for certain metals and for certain

ranges of strain-rate, has been found in many later studies sometimes to be a reasonable approximation. On the basis of investigations using mechanical testing machines, a constitutive equation appropriate to the conditions of the dynamic tensile test has been suggested in the form

$$\sigma = \varphi(\varepsilon_p, \dot{\varepsilon}_p) \qquad (5)$$

provided that the plastic strain distribution is sufficiently uniform, where $\sigma$ is nominal tensile stress, and $\varepsilon_p$, $\dot{\varepsilon}_p$ are nominal plastic strain and strain-rate. Thus, a logarithmic form for $\varphi$ was proposed by LUDWIK [29] and DEUTLER [30] on empirical grounds and by PRANDTL [31] as a prediction of a physical theory of plastic flow. MARSH and CAMPBELL [2] have discussed the physical basis of the type of constitutive eq. (5) (see also [32]) and their experimental determination of the relation between stress, strain and strain-rate for mild steel shows its complex nature. At sufficiently low strain-rates the form of the relation (5) can be directly determined, but at sufficiently high strain-rates the interpretation of experimental data necessitates the use of a theory of plastic wave propagation. This fact provides the starting-point for the following discussion.

The theoretical development of longitudinal plastic wave propagation in bars is due independently to several investigators, especially von KÁRMÁN and DUWEZ [33], WHITE and GRIFFIS [34], TAYLOR [35] and RAKHMATULIN [36], whose work was preceded by less complete work of DONNELL [37]. Their work essentially concerned the elaboration of the elementary theory from infinitesimal elastic strain to finite plastic strain. The theory neglects entirely lateral inertia effects and strain-rate sensitivity of the material and applies here only to concave stress-strain curves. Thus it is merely supposed that the static stress-strain relation $\sigma = f(\varepsilon)$ is known. The governing wave equation is

$$c^2 \frac{\partial^2 u_z}{\partial z^2} - \frac{\partial^2 u_z}{\partial t^2} = 0, \qquad c = (f'/\varrho_0)^{\frac{1}{2}}, \qquad (6)$$

where $z$ is the initial distance of a cross-section from the end of the bar and $\varrho_0$ is the initial density. A strain packet $\left(\varepsilon - \frac{1}{2}d\varepsilon, \ \varepsilon + \frac{1}{2}d\varepsilon\right)$ is therefore propagated at the velocity $c(\varepsilon)$, so that an incremental strain $d\varepsilon$ is propagated at a velocity which varies with the mean level of strain $\varepsilon$. Here, $c$ decreases monotonically with $\varepsilon$ (or $\sigma$), and accordingly a stress pulse in excess of the elastic limit tends to be progressively smoothed out, higher stresses being propagated more slowly than lower ones, an amplitude-dispersion effect. These are general results, and the basic problem first solved in detail by von KÁRMÁN concerns the situation when the end of a semi-infinite uniform wire is suddenly constrained to move with uniform velocity $V_1$, extension of the wire occurring. A similarity type of solution applies, which essentially expresses relevant quantities functionally in terms of a single independent variable $\xi = z/t$. The solution predicts that the front of the disturbance travels at the velocity of longitudinal elastic waves, and immediately at this elastic wave front there is a discontinuity corresponding to that for initial yield, say $\varepsilon_y$. In the rear of this front, there is a plastic disturbance in which the velocity of wave propagation diminishes with the strain over the range $\varepsilon_y$ to $\varepsilon_1$, the latter value obtaining at the plastic wave front and being determined by $V_1$ and $f$. Thereafter, right up to the end of the wire, the strain has the uniform value $\varepsilon_1$. In other words, the largest plastic strain and corresponding largest stress proceed down the wire most slowly and leave behind a region of uniform plastic strain extending up to the end. If the strain $\varepsilon$ is plotted as a function of $\xi$ then the curve obtained is the same for all time (see [33], Fig. 1). A most remarkable feature of the solution is the presence of a flat plateau of strain extending over a region that steadily grows outwards from the end of the wire, the occurrence of fracture being excluded. Experimental investigations undertaken at the California Institute of Technology ([33, 38]), although substantiating some of the theoretical predictions, showed that there were significant and systematic discrepancies mainly between measured and predicted strain variations. These discrepancies were ascribed to strain-rate sensitivity, and VON KÁRMÁN and DUWEZ [33] remarked that it was not logical to attempt to provide basic strain-rate data from the analysis of results of experiments involving plastic wave propagation in the complete absence of a physical theory.

The limitations of the VON KÁRMÁN-TAYLOR-RAKHMATULIN strain-rate independent theory of longitudinal plastic wave propagation were early appreciated, the conclusion being made that the strain-rate sensitivity in the mechanical behaviour of certain metals was not always a negligible effect. Since that time much effort has been made to develop more general plastic wave propagation theories and to elucidate more accurately the separate nature of and the interdependence between mechanical waves and strain-rate effects in metals. However, in spite of this effort the understanding of these matters remains very incomplete. Some of the main investigations made will now be described.

SOKOLOVSKY [39, 40] and MALVERN [41, 42] developed theories of plastic wave propagation in bars, taking account of the dependence of mechanical behaviour upon strain-rate in a phenomenological manner. MALVERN expressed the constitutive equation in the form

$$\sigma = f(\varepsilon) + a \ln (1 + b \dot{\varepsilon}_p), \qquad (7)$$

where $\sigma = f(\varepsilon)$ is the static stress-strain relation and $a$, $b$ are physical parameters. Eq. (7) may be rewritten as

$$b\dot{\varepsilon}_p = \exp\left[(\sigma - f)/a\right] - 1, \tag{8}$$

so that the plastic strain-rate therefore depends upon the difference between the stresses at the same strain under dynamic and static conditions. MALVERN proposed a wave propagation theory based on the more general constitutive equation

$$E\dot{\varepsilon} = \dot{\sigma} + g(\sigma, \varepsilon), \tag{9}$$

where $g$ is an arbitrary function expressing the strain-rate dependency. In this theory, elastic strain at a given stress is assumed independent of strain-rate and no account is taken of lateral inertia effects. Once the form of $g$ and the conditions of the problem are assigned, the governing equations of MALVERN's theory may be integrated numerically by the use of characteristics techniques.

MALVERN's constitutive equation carries the implication that there is an apparent increase in the value of the initial yield-stress at high strain-rates, because plastic flow does not occur instantaneously but takes time in which to become appreciable, although the choice of time-scale here has to be made somewhat empirically. Furthermore, small increments of strain superimposed upon a static, plastic strain distri-bution are propagated at the elastic wave velocity and not at the appropriate plastic wave velocity as predicted by the VON KÁRMÁN-TAYLOR-RAKHMATULIN theory, which neglects strain-rate effects. Thus, the gross result of strain-rate sensitivity in MALVERN's theory is to decay, rather than to delay, the stress wave. These predictions agree broadly with certain experimental results, and the second one will be referred to later. For purposes of illustration and for simplicity, MALVERN took the quasi-linear form $g = k(\sigma - f)$ where $k$ is a constant specifying the magnitude of the strain-rate effect, particularizing values of quantities by reference to experimental data. On this basis, MALVERN derived a strain-rate dependent solution of the basic problem discussed earlier by VON KÁRMÁN and DUWEZ [33]. It was now found that there was no uniform strain plateau near the end and that the maximum strain was increased (see [42], Fig. 6). In MALVERN's view, these results were in opposition to improved agreement between experiment and theory because there was failure to predict the uniform strain plateau apparent-ly observed. However, not all the experimental results had definitely predicted a plateau, and MALVERN himself said that one was not found in certain tests made on lead bars. LEE [43] also later observed that more precise analysis of the experimental data obtained for plastic waves in copper wires (see [33]) also did not reveal a plateau.

MALVERN's work has generally been regarded as achieving an improvement in the theory of plastic wave propagation. It stimulated renewed interest in the subject and since that time many further experimental studies involving pressure-bar techniques have been made, some of which will now be mentioned.

BELL [44] in tests made on steel bars found that incremental impact loads superimposed upon a static load in excess of the elastic limit were propagated at the elastic wave velocity. STERNGLASS and STUART [45] investigated the same situation as that in BELL's work, and their tests made on copper strips showed a similar result. Physically these results are to be expected, agreeing and disagreeing with MALVERN's and VON KÁRMÁN-TAYLOR-RAKHMATULIN's theories respectively. ALTER and CURTIS [46] considered a related type of experiment in which a lead bar was subjected to two loadings, applied consecutively so that the first produced plastic preloading, and it was found that the front of the second disturbance propagated at the elastic wave velocity. However, to attach a wider significance to this result may be imprudent in view of the anomalous mechanical behaviour of lead. These studies and also others indicate that the fronts of low-amplitude disturbances travel at the elastic wave velocity and that there is amplitude dispersion during propagation (see also [26]). These results show of course a qualitative difference from the predictions of the strain-rate independent theory. RUBIN [47] has given a theoretical discussion of this situation based upon MALVERN's [42] theory and has provided a qualitative description of the effects described (see also [48]). The viewpoint of BELL and STEIN [49] is different and they attribute the observed effect (in annealed aluminium) to serrations in stress-strain curves due to discontinuous mechanical behaviour and not to strain-rate effects.

Other, more general work has also involved the use of pressure-bars, and the analysis of data has been based upon either VON KÁRMÁN-TAYLOR-RAKHMATULIN's or MALVERN's theories or modifications of these theories. BELL and his collaborators (see [50]) have undertaken a comprehensive series of studies of constant velocity impact between identical, annealed aluminium or copper bars in free flight, and have concluded that there is substantial agreement between experimental data and the predictions of the strain-rate independent theory. The maximum impact velocities and strain-rates considered were of the order of 250 ft/sec and $3 \times 10^4$ sec$^{-1}$, respectively. It should be noted that the VON KÁRMÁN-TAYLOR-RAKHMATULIN theory was assumed valid (and was employed in analysing experimental data) here for some definite stress-strain relation, not *necessarily* the static one, under given conditions. BELL has also given attention to an examination of three-dimensional conditions near the interface and to the effect of temperature. KOLSKY and DOUCH [51] have obtained dynamic stress-strain curves for annealed copper, aluminium and aluminium alloy from experiments in which short bars of these materials were fired from a compressed-air gun at a steel pressure-bar. It was found that for copper and aluminium the dynamic curves lay appreciably above the static ones, whereas for the aluminium alloy there appeared to be no appreciable strain-rate effect (cf. [52]). These *dynamic* stress-strain curves were then used to obtain predictions from the VON KÁRMÁN-TAYLOR-RAKH-MATULIN theory. It was found that the theory gave a reliable relation between the permanent strain produced at the end of a bar and its velocity of impact. The theoretical predictions of strain distribution were less satisfactory and there was some indication, in the results for copper at low velocities of impact, that the simple theory needed modification in the manner suggested by MALVERN. The maximum impact velocities considered were of the order of 150 ft/sec but the magnitude of the strain-rate was not too well defined, although the loading time was of the order of 10 μsec. At the University of Texas, RIPPERGER, PLASS and their co-workers have undertaken extensive investigations of the stress-strain characteristics of materials at high strain-rates. In particular, extensions have been made of MALVERN's theory to include lateral inertia and shear effects and some comparison has been made between experimental data for copper and theoretical predictions (see [53]). This work appears to indicate that MALVERN's strain-rate dependent theory (or modifications of it) can be applied except perhaps in the vicinity of faces of impact.

The limited extent of investigations and the considerable variations in test conditions and the metals chosen for experimentation preclude, at the present time, the derivation of general, precise quantitative conclusions concerning strain-rate effects in metals from experimental data provided by pressure-bar tests. Moreover, for some metals (aluminium and copper), apparently differing conclusions have been reached by different investigators (see [49, 50, 51]) and therefore careful attention to their precise test conditions and precise analyses must be given in any comparative discussion, which will not be attempted here.

Although some investigations seem to indicate that MALVERN's strain-rate dependent theory is in broad agreement with certain trends, this theory may not be sufficiently realistic in its simple form and without due account of certain three-dimensional effects, such as those arising from lateral inertia and shear. Furthermore, it must be remembered, as VON KÁRMÁN and DUWEZ [33] have pertinently observed, that it is not logical to attempt to provide basic strain-rate data from the analysis of results of experiments involving plastic wave propagation in the complete absence of a physical theory. HAUSER, SIMMONS and

DORN [54] have emphasized this viewpoint, concluding from their work that the MALVERN strain-rate dependent theory fails to take proper account of crystal structure in determining strain-rate effects and that a description of such effects can be given in terms of dislocation theory.

## 4. Closure

The detailed appraisal of reported investigations of strain-rate effects would obviously be lengthy, and for this reason it has not been attempted in the present paper. However, it is possible to draw certain conclusions concerning that part of the subject of mechanical waves and strain-rate effects in metals with which the discussion has been particularly concerned.

For many reasons, the pressure-bar test is complicated. The bar itself is a mechanical wave guide whose surface confines the propagation of an intricate, changing system of plastic-elastic waves originating from the impact end, these waves undergoing reflections at the surface and mutual interactions. It is apparent that the situation near the impact end is at least initially three-dimensional, it being sensitive to the precise experimental conditions (e.g. the degree of alignment) and perhaps involving stress intensities considerably in excess of the yield-point, but it may be sensibly one-dimensional at distant cross-sections. The interpretation of experimental data for dynamic states of non-uniform stress and strain necessarily involves recourse to a theory of wave propagation, and the correctness or otherwise of this interpretation is inevitably linked to the physical and mathematical limitations of the theory. If these limitations are not properly appreciated and assessed, then the data cannot be interpreted correctly in terms of intrinsic mechanical properties of the material experimented with, but they depend to an unknown extent upon the conditions of the experiment. In spite of the difficulties, some already apparent from the theory of elastic wave propagation in bars, research needs to be directed towards radical improvements of the existing strain-rate independent theories of plastic wave propagation to take direct account of three-dimensional effects (see [26]). If sufficient progress could be made, it would be possible to determine unequivocally, for prescribed conditions, whether the mechanical behaviour of a given metal was strain-rate independent or dependent. If there is strain-rate dependence (which, from considerations of crystal physics, must be expected at sufficiently high strain-rates provided that plastic flow processes still occur), then the determination of basic strain-rate data requires physically-based theories, currently lacking (see [55]). The alternative procedure is to assume some type of strain-rate dependence and to determine experimentally the values of the disposable parameters involved, but this would seem to be merely a pragmatic expedient of not certain value.

## References

[1] CAMPBELL, J. D., and J. DUBY: Proc. Roy. Soc. London A **236**, 24 (1956).
[2] MARSH, K. J., and J. D. CAMPBELL: J. Mech. Phys. Solids **11**, 49 (1963).
[3] KOLSKY, H.: Stress Waves in Solids, Oxford: Clarendon Press 1953.
[4] MORLAND, L. W.: Phil. Trans. Roy. Soc. London A **251**, 341 (1959).
[5] COTTRELL, A. H.: Proceedings of the Conference on the Properties of Materials at High Rates of Strain, London: Institution of Mechanical Engineers 1957, p. 1.
[6] TAYLOR, G. I.: J. Instn. Civ. Engrs. **26**, 486 (1946).
[7] HARDING, J., E. O. WOOD and J. D. CAMPBELL: J. Mech. Engng. Sci. **2**, 88 (1960).
[8] LEE, E. H., and H. WOLF: J. Appl. Mech. **18**, 379 (1951).
[9] KOLSKY, H.: Structural Mechanics (edited by J. N. GOODIER and N. J. HOFF), Oxford, etc.: Pergamon Press 1960, p. 233.
[10] LOVE, A. E. H.: The Mathematical (Theory of Elasticity, fourth Edition, Cambridge: University Press 1927.
[11] REDWOOD, M.: Mechanical Waveguides, Oxford, etc.: Pergamon Press 1960.
[12] POCHHAMMER, L.: J. reine angew. Math. **81**, 324 (1876).
[13] CHREE, C.: Trans. Camb. Phil. Soc. **14**, 250 (1889).
[14] DAVIES, R. M.: Surveys in Mechanics (edited by G. K. BATCHELOR and R. M. DAVIES), Cambridge: University Press 1956, p. 64.
[15] ABRAMSON, H. N., H. J. PLASS and E. A. RIPPERGER: Advances in Applied Mechanics, Vol. 5 (edited by H. L. DRYDEN and TH. VON KÁRMÁN, New York: Academic Press 1958, p. 111.
[16] MIKLOWITZ, J.: Appl. Mech. Rev. **13**, 865 (1960).
[17] BANCROFT, D.: Phys. Rev. (2) **59**, 588 (1941).
[18] DAVIES, R. M.: Phil. Trans. Roy. Soc. London A **240**, 375 (1948).
[19] SKALAK, R.: J. Appl. Mech. **24**, 59 (1957).
[20] FOLK, R., G. FOX, C. A. SHOOK and C. W. CURTIS: J. Acoust. Soc. Amer. **30**, 552 (1958).
[21] McNIVEN, H. D., and D. C. PERRY: J. Acoust. Soc. Amer. **34**, 433 (1962).
[22] HSIEH, D. Y., and H. KOLSKY: Proc. Phys. Soc. **71**, 608 (1958).
[23] GREEN, W. A.: Progress in Solid Mechanics, Vol. I (edited by I. N. SNEDDON and R. HILL), Amsterdam: North-Holland Publ. Co. 1960, p. 225.
[24] RAYLEIGH: The Theory of Sound, Vol. I, second Edition, London: MacMillan 1894, p. 251.
[25] MINDLIN, R. D., and G. HERRMANN: Proceedings of the First U.S. National Congress of Applied Mechanics, New York: American Society of Mechanical Engineers 1952, p. 187.
[26] HUNTER, S. C., and I. A. JOHNSON: These Proceedings, p. 149.
[27] HOPKINSON, J.: Proc. Manchester Lit. Phil. Soc. **11**, 40, 119 (1872).
[28] HOPKINSON, B.: Proc. Roy. Soc. London A **74**, 498 (1904).
[29] LUDWIK, P.: Phys. Z. **10**, 411 (1909).
[30] DEUTLER, H.: Phys. Z. **33**, 247 (1932).
[31] PRANDTL, L.: Z. angew. Math. Mech. **8**, 85 (1928).
[32] HUNTER, S. C.: unpublished U.K. War Office report 1962.

[33] VON KÁRMÁN, TH., and P. DUWEZ: J. Appl. Phys. **21**, 987 (1950).
[34] WHITE, M. P., and L. GRIFFIS: J. Appl. Mech. **14**, p. A-337 (1947).
[35] TAYLOR, G. I.: The Scientific Papers of G. I. Taylor, Vol. I, Mechanics of Solids (edited by G. K. BATCHELOR), Cambridge: University Press 1958, pp. 456, 467.
[36] RAKHMATULIN, KH. A.: Prikl. Mat. Mekh. **9**, 91 (1945).
[37] DONNELL, L. H.: Trans. Amer. Soc. Mech. Engrs. **52** (I), APM, p. 153 (1930).
[38] DUWEZ, P. E., and D. S. CLARK: Proc. Amer. Soc. Test. Mat. **47**, 502 (1947).
[39] SOKOLOVSKY, V. V.: Prikl. Mat. Mekh. **12**, 261 (1948).
[40] SOKOLOVSKY, V. V.: Doklady Akad. Nauk. SSSR **60**, 775 (1948).
[41] MALVERN, L. E.: Quart. Appl. Math. **8**, 405 (1950).
[42] MALVERN, L. E.: J. Appl. Mech. **18**, 203 (1951).
[43] LEE, E. H.: Verformung und Fließen des Festkörpers (edited by R. GRAMMEL), Berlin/Göttingen/Heidelberg: Springer-Verlag 1956, p. 129.
[44] BELL, J. F.: Technical Report No. 5, Baltimore: Department of Mechanical Engineering, The Johns Hopkins University 1951.
[45] STERNGLASS, E. J., and D. A. STUART: J. Appl. Mech. **20**, 427 (1953).
[46] ALTER, B. E. K., and C. W. CURTIS: J. Appl. Phys. **27**, 1079 (1956).
[47] RUBIN, R. J.: J. Appl. Phys. **25**, 528 (1954).
[48] RIPARBELLI, C.: Proceedings of the First Midwestern Conference on Solid Mechanics, Urbana: University of Illinois 1953, p. 148.
[49] BELL, J. F., and A. STEIN: J. Mécan. **1**, 395 (1962).
[50] BELL, J. F.: J. Appl. Phys. **34**, 134 (1963).
[51] KOLSKY, H., and L. S. DOUCH: J. Mech. Phys. Solids **10**, 195 (1962).
[52] JOHNSON, J. E., D. S. WOOD, and D. S. CLARK: J. Appl. Mech. **20**, 523 (1953).
[53] PLASS JR., H. J., and E. A. RIPPERGER: Plasticity (edited by E. H. LEE and P. S. SYMONDS), Oxford, etc.: Pergamon Press 1960, pp. 453, 475.
[54] HAUSER, F. E., J. A. SIMMONS, and J. E. DORN: Response of Metals to High Velocity Deformation (edited by P. G. SHEWMON and V. F. ZACKAY), New York and London: Interscience Publishers 1961, p. 93.
[55] KRÖNER, E.: Appl. Mech. Rev. **15**, 599 (1962).

# The Propagation of Small Amplitude Elastic-Plastic Waves in Pre-Stressed Cylindrical Bars[1]

By S. C. Hunter and I. A. Johnson[2]

Royal Armament Research and Development Establishment
Fort Halstead, Kent, England

## Summary

The effects of geometrical dispersion on the propagation of axi-symmetric incremental stress waves in pre-loaded elastic-plastic cylin-drical bars are analysed using an approximation in which radial and axial displacements are expanded as power series in the radial coordinate. The approximation is shown to lead to a tolerable description of the propagation characteristics of elastic waves, for which an exact solution is known. For the elastic-plastic case it is shown that incremental pulses may travel at group velocities of the order of one half the elastic velocity, even when the plastic wave velocity vanishes.

## 1. Introduction

The propagation of elastic-plastic waves along one-dimensional bars has been a subject of experimental and theoretical investigation since the original papers of VON KÁRMÁN and DUWEZ [1], TAYLOR [2] and RAKHMATULIN [3]. These papers treat the problem of elastic-plastic wave propagation in a filament approximation, in which the only signi-ficant dependent variables are nominal axial stress $\Sigma_z(z, t)$ and axial displacement $u_z(z, t)$ both assumed uniform over the cross section of the bar.

Denoting the nominal stress — nominal strain curve by -

$$\Sigma = f(e) = f(\partial u_z/\partial z),$$

---

[1] Published with the permission of the Controller of Her Britannic Majesty's Stationary Office. British Crown Copyright reserved.
[2] Now at the University of Manchester.

we see from the axial equation of motion that

$$\varrho^{-1} f'(e) \partial^2 u_z / \partial z^2 = \partial^2 u_z / \partial t^2, \tag{1.1}$$

where the prime denotes differentiation with respect to the argument.

Eq. (1.1) is the basic equation of the filament approximation and, as is well known, may be integrated by the methods of characteristics to yield the solution of a variety of transient impact problems for both finite and semi-infinite bars. For the particular case of constant velocity impact of a semi-infinite bar, an elementary solution of (1.1) is possible (see KOLSKY [4]) and this solution has been extensively used in the interpretation of experiments on bar impact[1]. The most recent work along these lines is that of BELL [5, 6] and KOLSKY and DOUCH [7].

While the results of these investigations confirm the general validity of the filament approximation, there is one implication of the theory concerning the propagation of small amplitude incremental waves, which is not experimentally realised. Suppose a bar of material, statically pre-stressed to nominal stress and strain levels denoted respectively by $\Sigma_1$ and $e_1$, to be subjected to a further small dynamic load $\Sigma - \Sigma_1$, which results in a corresponding incremental displacement $u_z(z, t)$. If $z$ now denotes the axial coordinate with respect to the pre-stressed bar, the increment in nominal strain is

$$\eta = (1 + e_1) \partial u_z / \partial z,$$

whence

$$\begin{aligned} \Sigma - \Sigma_1 &= f(e_1 + \eta) - f(e_1) \\ &\approx \eta f'(e_1) \\ &= (1 + e_1) f'(e_1) \partial u_z / \partial z. \end{aligned}$$

Combining this result with the equation of motion leads to the one-dimensional wave equation

$$\partial^2 u_z / \partial t^2 - c_p^2 \partial^2 u_z / \partial z^2 = 0$$

where the propagation velocity $c_p$ is given by

$$c_p^2 = \varrho^{-1} (1 + e_1) f'(e_1) \tag{1.2}$$

and depends on the slope of the stress-strain curve at the pre-stressed point $\Sigma_1$, $e_1$.

Of the few attempts to investigate the corresponding experimental situation, the most important is that of STERNGLASS and STUART [8]

---

[1] To accomodate strain-rate effects, it is usual to take $f(e)$ as a dynamic stress-strain curve which nevertheless is assumed independent of strain rate over the range of strain rates present in the experimental situation.

who induced incremental pulses of duration of order $10^{-3}$ secs. into pre-loaded copper strips of cross sectional dimensions $1/2'' \times 1/8''$. The results of these experiments showed that incremental pulses travelled at velocities well in excess of those predicted by (1.2) and that the pulses suffered frequency dependent but amplitude independent dispersion[1]. In particular the wave fronts of the pulses were observed to travel at the elastic velocity $c_0 = (E/\varrho)^{1/2}$ while the overall pulse velocities were slightly smaller.

The above experimental results are commonly believed to be a consequence of strain-rate-dependent mechanical properties; the simplest argument along these lines presupposes the existence of a static and dynamic curve (Fig. 1), and the initial loading is assumed to bring the mechanical conditions of the specimen to a point $\Sigma_1$, $e_1$ located on the static curve. Subsequent transient loading entails a jump to the dynamic curve and this is assumed to occur along an elastic loading line as in Fig. 1, thus leading to the observed propagation velocity. A somewhat more sophisticated model may be based on the constitutive equation postulated by MAL-VERN [9] for strain-rate sensitive solids:

$$E\dot{e} - \dot{\Sigma} = q(\Sigma, e)$$

where $q$ is a function such that $q = 0$ specifies the static curve. If the stress and strain rates in the pulse are sufficiently large for the MALVERN equantio to be dominated by the time derivative terms, the pulse velocity is again given by the elastic value $c_0$.

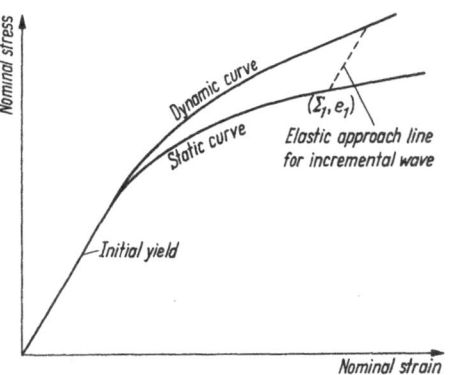

Fig. 1. Schematic diagram for explanation of elastic propagation velocity of incremental pulse.

Neither of the above explanations is entirely adequate since the mechanism used to explain the magnitude of the propagation velocity is essentially elastic, while the initial strain amplitudes in the STERN-GLASS-STUART experiments are predominantly plastic. While strain-rate effects are undoubtedly of major importance in determining the propagation characteristics of incremental pulses in most metals, an equally important (and neglected) factor is that of geometric dispersion, associated with the finite lateral dimensions of the specimen.

---

[1] The amplitude independence is important in indicating that irrespective of the detailed mechanism responsible, the propagation of incremental pulses may be adequately treated in a linear approximation.

The purpose of the present paper is to study the phenomenon of geometrical dispersion for the simplest case of a preloaded bar of an elastic-plastic solid whose mechanical behaviour is insensitive to strain-rate effects. It transpires that for sufficiently small values of $c_p/c_0$, such as obtain in real materials, geometric dispersion has a profound effect on the propagation characteristics of incremental pulses, and leads in particular to pulse velocities much larger than $c_p$ and comparable with $c_0/2$. However since velocities as large as $c_0$ are not obtained, it is clear that geometric dispersion does not account for the STERNGLASS-STUART results and that strain-rate effects are an important factor in these experiments.

The phenomenon of geometric dispersion arises from the failure of two assumptions implicit in the filament approximation. These assumptions are (a) neglect of inertially induced stresses and (b) initially plane sections of the bar remain plane during a transient disturbance. A quantitative criterion for the limit of validity of the assumptions is provided by the exact solutions of the axisymmetric problem of the propagation of elastic waves in bars of circular cross section (POCHHAMMER [10], CHREE [11] — for a readily accessible account see KOLSKY [4].) The criterion is

$$a\omega/c_0 \ll 1, \qquad (1.3)$$

where $a$ is the bar radius and $\omega$ is the circular frequency of the highest frequency components to contribute significantly to the FOURIER integral

$$u_z(t - z/c_0) = \int_{-\infty}^{\infty} b(\omega) \exp[i\omega(t - z/c_0)]d\omega$$

defining the displacement profile.

Although (1.3) is a mathematical deduction from the POCHHAMMER-CHREE theory for circular bars, physical arguments may be adduced to show that the inequality is valid for bars of arbitrary cross section, provided that the parameter $a$ is taken as a typical linear dimension of the cross section.

In the inequality (1.3), $c_0$ is a physical parameter defining the propagation velocity of elastic waves in the filament approximation. Clearly the corresponding quantity for elastic-plastic incremental waves is $c_p$ and we infer tentatively that for these waves the inequality

$$a\omega/c_p \ll 1 \qquad (1.4)$$

defines the limits of validity of the filament approximation; this result is subsequently verified. The important difference between (1.3) and (1.4) lies in the relative magnitudes of $c_0$ and $c_p$. Typically for metals strained initially by a few percent, $c_p/c_0 \sim 10^{-1}$ to $10^{-2}$, so that the inequality

(1.4) is violated for values of $\omega$ one or two orders of magnitude smaller than in the case of (1.3). The effects of geometric dispersion therefore are of correspondingly greater importance for elastic-plastic waves.

A numerical estimate of the circumstances in which geometric dispersion is likely to be important may be obtained as follows. For a reasonably smooth pulse the quantity $\omega$ may be identified with $\pi/T$ where $T$ is the pulse duration. From the experiments of DAVIES [12], geometric dispersion in elastic steel bars of $1''$ diameter is of importance for pulse durations of the order 30 microseconds or less; for elastic-plastic pulses in bars with comparable cross sectional dimensions, the critical duration lies in the millisecond region.

The remainder of this paper comprises a detailed theoretical investigation of the geometric dispersion problem for axisymmetric incremental elastic-plastic waves in circular bars. It is not possible to give an exact solution of this problem (corresponding to the POCHHAMMER-CHREE theory of elastic waves) and we have recourse to an approximate method based on expanding displacement and stress fields in (finite) power series in the radial coordinate. The method is more powerful than many of the previous approximations employed for the elastic problem, in that most of these latter methods devolve on variational principles which have no validity outside the theory of elasticity (for a recent survey article see GREEN [13]). We have applied the present method to the elastic problem and compared the results obtained with those of the exact theory; the agreement is tolerable and lends confidence to the use of the approximation for the elastic-plastic case.

The treatment of the elastic problem is given in Section 2; the elastic-plastic problem is discussed in Section 3.

## 2. Elastic Problem

The propagation of axisymmetric disturbances along an elastic cylindrical rod is governed by the equations of motion

$$\frac{\partial \sigma_r}{\partial r} + \frac{\sigma_r - \sigma_\theta}{r} + \frac{\partial \tau_{rz}}{\partial z} = \varrho \frac{\partial^2 u_r}{\partial t^2}, \tag{2.1}$$

$$\frac{\partial \sigma_z}{\partial z} + \frac{\partial \tau_{rz}}{\partial r} + \frac{\tau_{rz}}{r} = \varrho \frac{\partial^2 u_r}{\partial t^2}, \tag{2.2}$$

together with the stress-strain relations

$$\varepsilon_r \equiv \partial u_r/\partial r = E^{-1}\{\sigma_r - \nu(\sigma_\theta + \sigma_z)\}, \tag{2.3}$$

$$\varepsilon_\theta \equiv u_r/r = E^{-1}\{\sigma_\theta - \nu(\sigma_z + \sigma_r)\}, \tag{2.4}$$

$$\varepsilon_z \equiv \partial u_z/\partial z = E^{-1}\{\sigma_z - \nu(\sigma_r + \sigma_\theta)\}, \tag{2.5}$$

$$\gamma_{rz} \equiv 1/2\{\partial u_r/\partial z + \partial u_z/\partial r\} = E^{-1}(1 + \nu)\tau_{rz}, \tag{2.6}$$

and the boundary conditions

$$\sigma_r = \tau_{rz} = 0 \qquad \text{for} \qquad r = a. \tag{2.7}$$

The exact POCHHAMMER-CHREE solutions of these equations are steady-state eigenfunctions of the form

$$u_z = Q_1(r, \omega) \exp\left[i\omega(t - z/c(\omega))\right],$$

$$u_r = Q_2(r, \omega) \exp\left[i\omega(t - z/c(\omega))\right],$$

where the frequency dependent phase velocity $c(\omega)$ is given by a transcendental equation involving BESSEL functions. Of the various branches of the frequency equation, the lowest mode of vibration (i.e., for given $\omega$, the smallest value of $c$ which solves the equation) is the branch of greatest physical significance. (In the experiments of DAVIES [12] only this lowest vibration mode was detected.) Fig. 2 shows a plot of the lowest mode as a function of circular frequency for $\nu = 0.29$. This plot is taken from DAVIES [12].

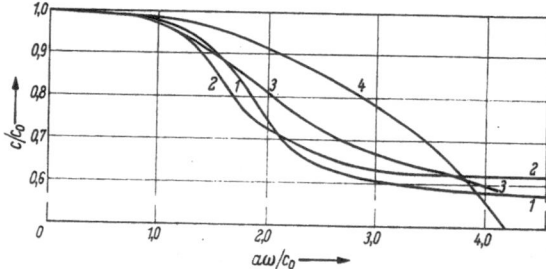

Fig. 2. Dispersion curves for elastic waves.

Curve 1 Exact POCHHAMMER-CHREE curve;          Curve 3 MINDLIN and HERRMANN (1951);
  "   2 Present approximation;                      "   4 RAYLEIGH approximation.

The present method of approximation proceeds as follows. We seek solutions of Eqs. (2.1) through (2.7) in the form

$$\begin{aligned}
u_z &= u_0(z, t) + r^2 u_2(z, t) + \cdots, \\
u_r &= r\omega_1(z, t) + r^3\omega_3(z, t) + \cdots, \\
\sigma_z &= h_0(z, t) + r^2 h_2(z, t) + \cdots, \\
\sigma_r &= f_0(z, t) + r^2 f_2(z, t) + \cdots, \\
\sigma_\theta &= f_0(z, t) + r^2 g_2(z, t) + \cdots, \\
\tau_{rz} &= r k_1(z, t) + r^3 k_3(z, t) + \cdots.
\end{aligned} \tag{2.8}$$

In writing down these expansions we have made use of the obvious symmetry requirements $\sigma_r = \sigma_\theta$ and $\tau_{rz} = 0$ for $r = 0$.

Substituting from (2.8) into the equations of motion and equating coefficients of $r^n$ ($n = 0, 1, 2$), we obtain three equations; to terms of the same order in $r$; the stress-strain equations give a further six equations. Finally two equations derive from the boundary conditions, which we impose on the assumption that the first two terms dominate the expansions for $\sigma_r$ and $\tau_{rz}$. In this manner the following eleven homogeneous linear differential equations may be derived for the eleven functions $u_0$, $u_2$, $\omega_1$, etc.:

$$E\omega_1 = (1 - \nu)f_0 - \nu h_0,$$

$$3E\omega_3 = f_2 - \nu(h_2 + g_2),$$

$$E\omega_3 = g_2 - \nu(f_2 + h_2),$$

$$Eu_0' = h_0 - 2\nu f_0,$$

$$Eu_2' = h_2 - \nu(f_2 + g_2),$$

$$E[u_2 + (1/2)\omega_1'] = (1 + \nu)k_1, \tag{2.9}$$

$$3f_2 - g_2 - k_1' = \varrho\ddot{\omega}_1,$$

$$h_0' + 2k_1 = \varrho\ddot{u}_0,$$

$$h_2' + 4k_3 = \varrho\ddot{u}_2,$$

$$f_0 + a^2 f_2 = 0,$$

$$k_1 + a^2 k_3 = 0.$$

In these equations the prime and dot denote respectively differentiation with respect to the axial coordinate $z$ and time $t$.

Eqs. (2.9) comprise the basis for the lowest order non trivial treatment of the wave propagation problem using the power series expansion method. The method is approximate since terms of order $r^3$ and higher have been ignored in the equations of motion and stress-strain relations, while additionally the boundary conditions have been forced on the first two terms of the expansions for $\sigma_r$ and $\tau_{rz}$. By taking account of higher order terms, a more elaborate treatment is possible but the resulting algebra would be very involved. In fact, the present approximation predicts a dispersion curve $c(\omega)$ in fair agreement with the exact POCH-HAMMER-CHREE equation and further improvements obtained from taking account of terms in $r^3$ and higher would be marginal.

After some algebra, Eqs. (2.9) lead to the following sixth order equation for any one of the functions $u_0, u_2, u_1, \ldots$, etc.

$$D^2 - c_0^{-2}\delta^2 - a^2(AD^4 - Bc_0^{-2}D^2\delta^2 + Cc_0^{-4}\delta^4)$$
$$+ a^4(FD^6 - Gc_0^{-2}D^4\delta^2 + Hc_0^{-4}D^2\delta^4 - Ic_0^{-6}\delta^6) = 0. \tag{2.10}$$

Here $D \equiv \partial/\partial z$, $\delta \equiv \partial/\partial t$, while the coefficients $A$, $B$, ... are constants determined by Poisson's ratio:

$$A = 1/4, \qquad B = \frac{7(1 - \nu) - 2\nu^2}{8(1 - \nu)}, \qquad C = \frac{(1 + \nu)(5 - 10\nu + 4\nu^2)}{8(1 - \nu)},$$

$$F = \frac{3 + \nu}{64(1 + \nu)}, \qquad G = \frac{6 - 7\nu - \nu^2}{32(1 - \nu)}, \qquad H = \frac{(1 + \nu)(15 - 28\nu + 4\nu^2)}{64(1 - \nu)},$$

$$I = \frac{(3 - 2\nu)(1 - 2\nu)(1 + \nu)^2}{32(1 - \nu)}.$$

The lowest order terms in (2.10) reproduce the wave equation, whose solutions in terms of arbitrary functions of $t \pm z/c_0$ correspond to the propagation of undispersed wave in the negative and positive $z$ directions.

An improvement on the zero order approximation, essentially due to Lord Rayleigh, incorporates a mixed 4th order term in $D^2\delta^2$ and may be written

$$D^2 - c_0^{-2}\delta^2 + (1/2)\nu^2 a^2 c_0^{-2} D^2 \delta^2 = 0. \tag{2.11}$$

This equation, derivable from variational principles (see [14], p. 428) has been used extensively to describe the propagation of pulses whose dominant circular frequencies $\omega$ satisfy the inequality

$$a\omega/c_0 < 0.5 \tag{2.12}$$

(as in Davies [12]).

At first, it seems odd that the 4th order terms in (2.10) and (2.11) are not identical. However, for values of $\omega$ conforming to (2.12) the predictions of (2.10) and (2.11) are identical, and this resolves the apparent contradiction.

We assume eigenfunction solutions of (2.10) of the form

$$\exp\left[i\omega\left(t - z/c(\omega)\right)\right], \tag{2.13}$$

from which more general solutions can be constructed by Fourier analysis. Substitution of (2.13) into (2.10) yields the algebraic equation

$$1 - v^2 + x^2(A v^{-2} - B + C v^2) + x^4(F v^{-4} - G v^{-2} + H - I v^2) = 0, \tag{2.14}$$

where $v$ and $x$ are the dimensionless phase velocity and frequency defined by $v = c/c_0$ and $x = a\omega/c_0$, respectively. For small $x$, Eq. (2.14) gives

$$v^2 = 1 - v^2 x^2/2 - [v^2(1 - 2v^2)(3 - 2v^2)/16(1 - v^2)]x^4 + \cdots, \tag{2.15}$$

while the Rayleigh equation (2.11) yields

$$v^2 = 1 - v^2 x^2/2. \tag{2.16}$$

On the other hand the exact POCHHAMMER-CHREE transcendental equation furnishes for the lowest mode

$$v^2 = 1 - v^2 x^2/2 - v^2[(2 - 3v)(1 + v)/12 - 1/48(1 - v^2)]x^4 + \cdots.$$

To terms of order $x^2$ the exact relation agrees both with the present approximation and the RAYLEIGH equation. In view of this result the fact that the fourth order terms of (2.10) and (2.11) are not identical is of no significance; in this connection it is of interest to note GREEN's comment [13] that the agreement between the predictions of the RAYLEIGH approximation and the exact result for circular bars is to some extent fortuitous and results from the cancellation of two (neglected) compensating errors (the corresponding errors for bars of noncircular cross section are not self-compensating).

Eq. (2.14) is a cubic in $v^2$ with one real positive branch for $0 < x < \infty$. For $x \to 0$ we have $v \to 1$, and the behavior for small values of $x$ is given by (2.15); for $x \to \infty$, $v$ tends to one of the three real roots of

$$Fv^{-4} - Gv^{-2} + H - Iv^2 = 0.$$

These roots are

$$v_1^2 = \frac{1}{2(1 + v)}, \qquad v_2^2 = \frac{1 - v}{(1 + v)(1 - 2v)}, \qquad v_3^2 = \frac{3 + v}{(3 - 2v)(1 + v)}. \tag{2.17}$$

The first of these roots is the relevant assymptotic limit corresponding to a limiting high frequency phase velocity

$$c \to c_0/[2(1 + v)]^{1/2} = (\mu/\varrho)^{1/2},$$

where $\mu$ is the shear modulus. The exact asymptotic limit is known to be the velocity of RAYLEIGH waves

$$c \to c_r = \beta(\mu/\varrho)^{1/2}$$

where $\beta$ is a numerical parameter slightly dependent on $v$ (see GOLDSMITH [15]):

$$\beta = 0.955 \qquad v = 0.5$$
$$0.926 \qquad 0.29$$
$$0.919 \qquad 0.25$$

For intermediate values of $v$ in the range $1 > v > v_1$, Eq. (2.14) has been solved numerically and the results are displayed in Fig. 2 together with the lowest mode of the POCHHAMMER-CHREE equation and the best previous approximation due to MINDLIN and HERMANN [16]. In the latter treatment, which is based on variational principles, the dispersion equation contains a free parameter ultimately chosen to give agreement with the exact result at the high frequency limit. A full account of the various approximation methods used to describe wave propagation in elastic bars is given by GREEN [13].

It is apparent that the present approximation leads to results in reasonable agreement with the exact solution and provides a feasible method for solving the more complex problem of the dispersion of elastic-plastic waves. This analysis is presented in the next section.

### 3. The Elastic-Plastic Problem

It proves convenient to work in terms of logarithmic (or "natural") strain $\varepsilon = \log(1 + e)$ and normal stress $\sigma$ rather than with the nominal values.

Consider a bar pre-stressed uniformly to a mechanical state specified by $\sigma_1$, $\varepsilon_1$ the corresponding bar radius being given by $a$. We suppose the bar subject to further incremental stresses $\sigma_z$, $\sigma_r$, $\sigma_\theta$, $\tau_{rz}$ which result in incremental axial and radial displacements $u_z(z, t)$ and $u_r(r, t)$ where $z$ is the axial coordinate in the pre-stressed bar. (For convenience we have adopted a slightly different notation to that of Section 2; in particular the total axial stress is now given by $(\sigma_1 + \sigma_z)$).

We assume the incremental stresses and displacements to be expanded in the form of Eqs. (2.8) and make the same assumptions regarding the neglegibility of terms of order $r^3$ and higher. Of the required eleven equations for $u_0$, $u_2$, $\omega_1$, etc. the last five Eqs. (2.9), which derive from the boundary conditions and equations of motion, remain valid. Of the remaining required equations, two are provided by the assumption of plastic incompressibility, two by the yield criterion, and two by the plasticity equations of REUSS.

From Eqs. (2.8), the components of incremental plastic strain may be written as follows:

$$\varepsilon_r^p = \partial u_r/\partial r - E^{-1}\{\sigma_r - \nu(\sigma_\theta + \sigma_z)\}$$
$$= \omega_1 - E^{-1}\{(1-\nu)f_0 - \nu h_0\} + r^2\{3\omega_3 - E^{-1}[f_2 - \nu(g_2 + h_2)]\} + \cdots,$$
$$\varepsilon_\theta^p = \omega_1 - E^{-1}\{(1-\nu)f_0 - \nu h_0\} + r^2\{\omega_3 - E^{-1}[g_2 - \nu(f_2 + h_2)]\} + \cdots,$$
$$\varepsilon_z^p = u_0' - E^{-1}\{h_0 - 2\nu f_0\} + r^2\{u_2' - E^{-1}[h_2 - \nu(f_2 + g_2)]\} + \cdots,$$
$$\gamma_{rz}^p = r\{u_2 + (1/2)\omega_1' - E^{-1}(1+\nu)k_1\} + \cdots.$$

The assumption of incompressible plastic behaviour,

$$\varepsilon_r^p + \varepsilon_\theta^p + \varepsilon_z^p = 0,$$

leads immediately to two equations

$$2\omega_1 + u_0' = E^{-1}(1 - 2\nu)(2f_0 + h_0), \tag{3.1}$$

$$4\omega_3 + u_2' = E^{-1}(1 - 2\nu)(f_2 + g_2 + h_2). \tag{3.2}$$

Also, with $\omega_1$ and $\omega_3$ from (3.1) and (3.2), the plastic strain equations may be re-written as follows:

$$\varepsilon_r^p = -u_0'/2 + \theta_1, \tag{3.3}$$

$$\varepsilon_\theta^p = -u_0'/2 + \theta_2, \tag{3.4}$$

$$\varepsilon_z^p = u_0' + \theta_3, \tag{3.5}$$

$$\gamma_{rz}^p = \theta_4, \tag{3.6}$$

where the $\theta_i$ are given by

$$E\theta_1 = h_0/2 - \nu f_0 + r^2\{-3Eu_2'/4 - (1 + 6\nu)f_2/4 \\ + (3 - 2\nu)(g_2 + h_2)/4\}, \tag{3.7}$$

$$E\theta_2 = h_0/2 - \nu f_0 + r^2\{-Eu_2'/4 - (3 + 2\nu)g_2/4 \\ + (1 + 2\nu)(f_2 + h_2)/4\}, \tag{3.8}$$

$$E\theta_3 = -h_0 + 2\nu f_0 + r^2\{Eu_2' - h_2 + \nu(f_2 + g_2)\}, \tag{3.9}$$

$$E\theta_4 = r\{E(u_2 + (1/2)\omega_1') - (1 + \nu)k_1\}. \tag{3.10}$$

Following HILL ([17], p. 30) we assume the criterion for continued plastic yielding to be given by

$$(\sigma_1 + \sigma_z - \sigma_r)^2 + (\sigma_1 + \sigma_z - \sigma_\theta)^2 + (\sigma_r - \sigma_\theta)^2 + 6\tau_{rz}^2 = 2\varphi^2(\bar{\varepsilon}) \tag{3.11}$$

where $\bar{\varepsilon}$ is the following scalar measure of total plastic deformation

$$\bar{\varepsilon} = \varepsilon^* + \int_0^t \{(2/3)\,\dot{\varepsilon}_{ij}^p(t')\,\dot{\varepsilon}_{ij}^p(t')\}^{1/2}dt', \tag{3.12}$$

and where $\varepsilon^* = \varepsilon_1 - \sigma_1/E$ is the plastic strain of the pre-stressed bar prior to transient impact. According to this model the static stress versus plastic strain curve is given by

$$\sigma = \varphi(\varepsilon_z^p),$$

and the filament approximation leads to

$$\varrho c_p^2 = \varphi'(\varepsilon^*)/\{1 + \varphi'(\varepsilon^*)/E\} \tag{3.13}$$

for the plastic wave velocity $c_p$.

From (3.3) through (3.6) we derive

$$\bar{\varepsilon} = \varepsilon^* + \int_0^t \{(\dot{u}_0' + \dot{\theta}_3)^2 + (1/3)(\dot{\theta}_1 - \dot{\theta}_2)^2 + (2/3)\dot{\theta}_4^2\}^{1/2} dt',$$

or on neglecting $(\dot{\theta}_1 - \dot{\theta}_2)^2$ and $\dot{\theta}_4^2$ in comparison with $(\dot{u}_0' + \dot{\theta}_3)^2$

$$\bar{\varepsilon} = \varepsilon^* + u_0' + \theta_3. \tag{3.14}$$

(This will be justified later.) Thus in the present approximation $\bar{\varepsilon}$ is given by the axial plastic strain as under static conditions.

From (3.11) and (3.14) and the relation $\sigma_1 = \varphi(\varepsilon^*)$, we find on neglect of squares of small order quantities that

$$2\sigma_z - \sigma_r - \sigma_\theta = 2(u_0' + \theta_3)\varphi'(\varepsilon^*),$$

or on substituting for $\sigma_z$, $\sigma_r$, $\sigma_\theta$, $\theta_3$ from (2.8) and (3.9) and equating coefficients of terms in $r^n$ ($n = 0, 2$):

$$h_0 - f_0 = \{u_0' - E^{-1}(h_0 - 2\nu f_0)\}\varphi'(\varepsilon^*), \tag{3.15}$$

$$2h_2 - g_2 - f_2 = \{u_0' - h_2/E + \nu(f_2 + g_2)/E\}\varphi'(\varepsilon^*). \tag{3.16}$$

The REUSS equations may be written in the form:

$$\frac{\dot{\varepsilon}_r^p}{\dot{\varepsilon}_z^p} = \frac{2\sigma_r - \sigma_\theta - \sigma_z - \sigma_1}{2\sigma_1 + 2\sigma_z - \sigma_r - \sigma_\theta}, \qquad \frac{\dot{\varepsilon}_\theta^p}{\dot{\varepsilon}_z^p} = \frac{2\sigma_\theta - \sigma_r - \sigma_z - \sigma_1}{2\sigma_1 + 2\sigma_z - \sigma_r - \sigma_\theta},$$

$$\frac{\dot{\gamma}_{rz}^p}{\dot{\varepsilon}_z^p} = \frac{6\tau_{rz}}{2\sigma_1 + 2\sigma_z - \sigma_r - \sigma_\theta}$$

Substituting for strain rates from (3.3) through (3.6) and again neglecting terms in the squares of the incremental stresses, we find

$$2\dot{\theta}_1 + \dot{\theta}_3 = (3/2)\frac{(\sigma_r - \sigma_\theta)}{\sigma_1}(\dot{u}_0' + \dot{\theta}_3),$$

$$2\dot{\theta}_2 + \dot{\theta}_3 = -(3/2)\frac{(\sigma_r - \sigma_\theta)}{\sigma_1}(\dot{u}_0' + \dot{\theta}_3),$$

$$\dot{\theta}_4 = \frac{3\tau_{rz}}{\sigma_1}(\dot{u}_0' + \dot{\theta}_3).$$

However, by virtue of the identity $\theta_1 + \theta_2 + \theta_3 = 0$, there are only two independent equations, which we take in the form

$$\dot{\theta}_1 - \dot{\theta}_2 = (3/2)\frac{\sigma_r - \sigma_\theta}{\sigma_1}(\dot{u}_0' + \dot{\theta}_3), \tag{3.17}$$

$$\dot{\theta}_4 = \frac{3\tau_{rz}}{\sigma_1}(\dot{u}_0' + \dot{\theta}_3). \tag{3.18}$$

Since we are assuming that the incremental stresses are small compared with $\sigma_1$, Eqs. (3.17) and (3.18) show that $\dot{\theta}_1 - \dot{\theta}_2$ and $\dot{\theta}_4$ are small compared with $\dot{u}_0' + \dot{\theta}_3$, which justifies the approximation leading to (3.14). In fact the stresses $\sigma_r$, $\sigma_\theta$, and $\tau_{rz}$ are entirely inertial in origin, and in the spirit of the present approximation are to be assumed small compared to $\sigma_z$. Thus $\sigma_r$, $\sigma_\theta$ and $\gamma_{rz}$ are of second order compared with $\sigma_1$ and we take as our final pair of equations

$$\theta_1 - \theta_2 = 0, \qquad \theta_4 = 0. \tag{3.19}$$

Physically, Eqs. (3.19) specify equal radial and tangential components of plastic strain and zero plastic shear strain.

The final set of equations governing the propagation of incremental elastic-plastic waves are given by (3.1), (3.2), (3.15), (3.16), (3.19), and the last five of the set (2.9). Thus

$$2\omega_1 + u_0' = E^{-1}(1 - 2\nu)(2f_0 + h_0),$$

$$4\omega_3 + u_2' = E^{-1}(1 - 2\nu)(f_2 + h_2 + g_2),$$

$$h_0 - f_0 = \{u_0' - E^{-1}(h_0 - 2\nu f_0)\}\varphi'(\varepsilon^*),$$

$$2h_2 - g_2 - f_2 = \{u_2' - E^{-1}[h_2 + \nu(f_2 + g_2)]\}\varphi'(\varepsilon^*),$$

$$(1/2)\omega_1' + u_2 = E^{-1}(1 + \nu)k_1,$$

$$u_2' = E^{-1}\{(1 - 2\nu)h_2 + 3g_1 - (1 + 4\nu)f_2\},$$

$$f_0 + a^2 f_2 = 0, \tag{3.20}$$

$$k_1 + a^2 k_3 = 0,$$

$$3f_2 - g_2 + k_1 = \varrho\ddot{\omega}_1,$$

$$h_0' + 2k_1 = \varrho\ddot{u}_0,$$

$$h_2' + 4k_3 = \varrho\ddot{u}_2.$$

In the limit $\Phi'(\varepsilon^*) \to \infty$, we have $c_p \to c_0$, and the set of Eqs. (3.20) are equivalent to the elastic set (2.9).

The remainder of the analysis parallels that of the elastic case. After some algebra, it may be shown that each of the variables $u_0$, $u_2$, etc. satisfy the following sixth-order equation

$$D^2 - c_p^{-2}\delta^2 - a^2 J^{-1}(A D^4 - B c_0^{-2} D^2 \delta^2 + C c_0^{-4} \delta^4)$$

$$+ a^4 J^{-1}(F D^6 - G c_0^{-2} D^4 \delta^2 + H c_0^{-4} D^2 \delta^4 - I c_0^{-6} \delta^6) = 0, \tag{3.21}$$

where the constants $A, B, \ldots, J$ are given by

$$A = 4(1 - \nu^2) - (1 + \nu)\gamma,$$

$$B = 2(1 - \nu)(7 - 7\nu - 2\nu^2) + 6(3 - 2\nu - 2\nu^2)\gamma + 3(1 - 2\nu)\gamma^2/2,$$

$$C = 2(1 + \nu)^2(5 - 10\nu + 4\nu^2) + (25 - 19\nu - 32\nu^2 + 12\nu^3)\gamma$$
$$\quad + (31 - 46\nu + 4\nu^2)\gamma^2/2,$$

$$F = (2\nu + 6 + \gamma)(2 - 2\nu + \gamma)/16,$$

$$G = (1 + \nu)(6 - 7\nu - \nu^2)/2 - (7 - 4\nu - 2\nu^2)\gamma/2 + 3(2 - \nu)\gamma^2/8,$$

$$H = (1 + \nu)^2(15 - 28\nu + 4\nu^2)/4 + (1 + \nu)(7 - 11\nu)\gamma$$
$$\quad + (41 - 44\nu - 4\nu^2)\gamma^2/16,$$

$$I = (1 + \nu)(1 - 2\nu)(2 + 2\nu + 3\gamma)[6 + 2\nu - 4\nu^2 + (7 - 2\nu)\gamma]/8,$$

$$J = 4\{4(1 - \nu^2) + (5 - 4\nu)\gamma\},$$

in which

$$\gamma = E/\varphi'(\varepsilon^*) = \chi^2 - 1 \quad \text{and} \quad \chi = c_0/c_p. \tag{3.22}$$

The general solution of (3.21) may be synthesised from basic periodic eigenfunctions of the form (2.13), where $c(\omega)$ satisfies the sixth order equation

$$1 - v^2 + x^2 J^{-1}(A v^{-2} - B\chi^2 + C\chi^4 v^2)$$
$$\quad + x^4 J^{-1}(F v^{-4} - G\chi^2 v^{-2} + H\chi^4 - I\chi^6 v^2) = 0 \tag{3.23}$$

and where

$$v = c(\omega)/c_p, \quad x = a\omega/c_p.$$

For small values of $x$, Eq. (3.23) admits the expansion[1]

$$v^2 = 1 - (1/8)\left[\frac{2\nu + \chi^2 - 1}{\chi^2}\right]^2 x^2 + \cdots$$

i.e.

$$c^2(\omega) = c_p^2 - (1/8)\left[\frac{\chi^2 + 2\nu - 1}{\chi^2}\right]^2 a^2\omega^2 + \cdots. \tag{3.24}$$

It is evident that this reduces to the elastic result for $\chi = 1$. Moreover, for $\nu = 1/2$ we have

$$c^2 = c_p^2 - a^2\omega^2/8 + \cdots, \tag{3.25}$$

which, for the limiting cases in question, can be deduced from somewhat simpler arguments along the lines indicated by Kolsky [18].

---

[1] This expansion fails for $\chi \to \infty$.

Since the coefficient of the term in $a^2\omega^2$ necessarily lies between $1/8$ and $\nu^2/2$ it is evident that the validity of the filament approximation (i.e. $c(\omega) \approx c_p$) is restricted to small values of $a\omega/c_p$ thus validating the inituitive physical arguments of Section 1. However, as is shown below, for realistic values of $\chi$ ($\chi \gg 1$), the second order approximation (3.24) is also of very limited validity, in contrast to the corresponding approximation for the elastic problem.

Eq. (3.23) possesses one physically significant branch $v(x)$ which is continuous over the entire range of $x$ and tends to unity as $x \to 0$. This function is displayed graphically for $\nu = 0.29$ and various values of $\chi$ [$\chi = 3, 5, 7, 10, \infty$] in Fig. 3. For convenience we have plotted $v/\chi \equiv c(\omega)/c_0$ as a function of $x/\chi \equiv a\omega/c_0$.

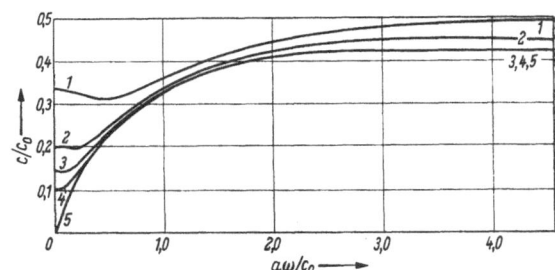

Fig. 3. Dispersion curves for elastic-plastic waves.

Curve _1_ $c_0/c_p = 3$;
  „  _2_ $c_0/c_p = 5$;
  „  _3_ $c_0/c_p = 7$;
  „  _4_ $c_0/c_p = 10$;
  „  _5_ $c_0/c_p = \infty$.

For all values of $\chi$ the resulting dispersion curves are of quite different character from those of the elastic case (Fig. 2). The initial parabolic behaviour specified by the approximation (3.24) is confined to a very limited region for $\chi = 3, 5$, and virtually disappears for the larger values of $\chi$. Beyond the parabolic region the dispersion curves are monotonically increasing functions, in contrast to the elastic curves of Fig. 2.

A curiosity emerges in the asymptotic behaviour of these curves. It is evident that the asymptotic value of $v$ is given by one of the three roots of the coefficient of the $x^4$ term in (3.23). These roots are

$$\frac{v_1^2}{\chi^2} = \frac{1}{2(1+\nu)}, \quad \frac{v_2^2}{\chi^2} = \frac{\chi^2+1-2\nu}{(1-2\nu)(3\chi^2-1+2\nu)}, \quad \frac{v_3^2}{\chi^2} = \frac{5+2\nu+\chi^2}{(7-2\nu)\chi^2-(1-2\nu)}$$

and correspond respectively to the three roots (2.17). Of these three roots for $(c/c_0)^2$, the first is independent of the plastic properties and provides the relevant asymptotic limit for the elastic case $\chi = 1$. When these observations are coupled with the intuitive physical concept that a very high frequency wave should be primarily elastic in character, it seems probable that the first root should also provide the relevant asymptotic limit for the elastic-plastic case. In fact for the values of $\chi$ chosen here, the relevant root is the third one. It is not clear whether this is a physically significant result or whether, in contrast to the elastic case, the approximations employed are grossly inadequate at the high frequency limit.

Figs. 4 and 5 show plots of the group velocity as a function of frequency. The group velocity

$$c_g(\omega) = c(\omega)\left[1 - \frac{d(\log c)}{d(\log \omega)}\right]^{-1}$$

is of direct physical significance since in the context of the stationary phase approximation it defines the physical propagation velocity of a

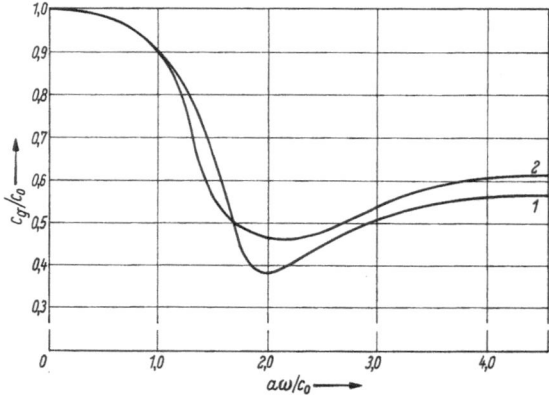

Fig. 4. Group velocity of elastic waves.
Curve *1* Exact Pochhammer-Chree curve;
  „   *2* Present approximation.

pulse dominated by Fourier components with circular frequencies clustered around the value $\omega$. Again the elastic-plastic curves are of entirely different character from the elastic case; in particular, there is virtually no elastic-plastic counterpart of the initial region of the elastic curve in which $c_g$ is a rapidly decreasing function of $\omega$. Thus in contrast to the case of elastic waves we do not anticipate the developments of high frequency tails in transient elastic-plastic pulses.

Since Fig. 5 fails to predict transient pulse velocities as large as $c_0$, it is clear that geometric dispersion does not provide an explanation of the observed pulse velocities in the Sternglass-Stuart experiments. However, it is believed that geometric dispersion is likely to be an important factor in such experiments and that detailed interpretation of experimental observations must allow for the effect.

In conclusion it should be noted that the eigenfunction solutions of Eq. (3.21) are only of physical significance in the context of a Fourier synthesis for a continuously loading pulse; the periodic nature of the individual eigenfunctions entails alternate loading and unloading and such functions are unacceptable solutions of the equation.

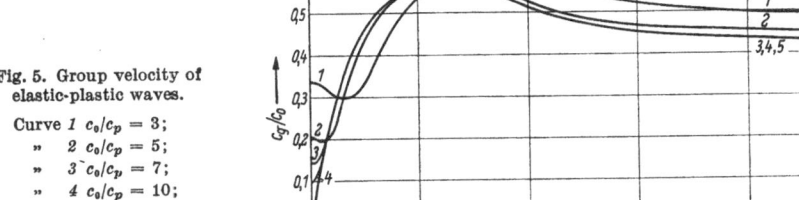

Fig. 5. Group velocity of elastic-plastic waves.

Curve 1 $c_0/c_p = 3$;
" 2 $c_0/c_p = 5$;
" 3 $c_0/c_p = 7$;
" 4 $c_0/c_p = 10$;
" 5 $c_0/c_p = \infty$.

## References

[1] VON KÁRMÁN, TH., and P. DUWEZ: J. Appl. Phys. 21, 987 (1950).
[2] TAYLOR, G. I.: Scientific Papers, Vol. 1, Cambridge University Press, 1958, p. 467.
[3] RAKHMATULIN, KH. A.: Prikl. Mat. Mekh. 9, 91 (1948).
[4] KOLSKY, H.: Stress waves in solids, Oxford: Clarendon Press 1953.
[5] BELL, J. F.: J. Appl. Phys. 31, 277, 2188 (1960).
[6] BELL, J. F.: J. Mech. Phys. Solids 9, 1, 261 (1961).
[7] KOLSKY, H., and L. S. DOUCH: J. Mech. Phys. Sol. 10, 195 (1962).
[8] STERNGLASS, E. J., and D. A. STUART: J. Appl. Mech. 20, 427 (1953).
[9] MALVERN, L. E.: J. Appl. Mech. 18, 203 (1951).
[10] POCHHAMMER, L.: J. Reine u. Angew. Math. (Crelle) 81, 324 (1876).
[11] CHREE, C. F.: Proc. Cambridge Philos. Soc. 14, 250 (1889).
[12] DAVIES, R. M.: Phil. Trans. Roy. Soc. A 821, 47 (1948).
[13] GREEN, W. A.: Contribution to Progress in Solid Mechanics, vol. 1, Amsterdam: North-Holland Publ. Co. 1960.
[14] LOVE, A. E. H.: Mathematical theory of elasticity, 4th ed., Cambridge: University Press 1944.
[15] GOLDSMITH, W.: Impact, London: Edward Arnold Ltd. 1960.
[16] MINDLIN, R. D., and G. HERMANN: Appl. Mech. Reviews 5 (1951) Rev. 1308.
[17] HILL, R.: Mathematical theory of plasticity, Oxford: Clarendon Press 1950.
[18] KOLSKY, H.: Proc. Phys. Soc. B 62, 676 (1949).

# The Initiation of Finite Amplitude Waves in Annealed Metals

By James F. Bell

The Johns Hopkins University, Baltimore, Maryland, U.S.A.

## Abstract

Plastic wave initiation in the immediate vicinity of impacting surfaces for high purity large-grained multicrystals of fully annealed aluminum is shown to be consistent with first diameter behavior which has
been obtained in fully annealed commercial purity aluminum, copper,
and pure lead. These new data introduce considerable simplicity into
the description of the process of plastic wave initiation since for the high
purity polycrystal, the quasi-static stress-strain curve of the material
and the parabolic law governing the finite amplitude wave propagation
are identical. Thus for high purity aluminum, not only does the strain-
rate-independent finite amplitude wave theory apply, but the governing
stress-strain curve is the quasi-static stress-strain curve of the material.

## 1. Introduction

An experimental program designed to compare large amplitude plastic
wave propagation with the one-dimensional finite amplitude wave theory
must give consideration to the manner of wave initiation at the impact
face. The available solution of the differential equations of the finite
amplitude wave theory is for the physical situation of a constant velocity
impact. In such an impact, as for example in the axial collision of two
identical cylindrical specimens, it is implicitly assumed that a total
plastic instantaneous wave front occurs at the instant of impact and
then propagates away from the impact face as a dispersive plastic front.
Even a casual reflection upon the situation existing in the bar in the
immediate vicinity of the impact face during the first few microseconds
following initiation of impact raises a question as to whether a constant
velocity impact would immediately produce the simple situation of a
one-dimensional behavior.

To compare the finite amplitude wave theory with experiment, then,
one must examine propagation detail for at least three positions, all at

a considerable distance from the impact face. One must examine the detail for at least three positions inasmuch as one would expect that the wave shape arriving at the first position would be distorted in traversing the three-dimensional first diameter, and also since one is not given *a priori* the governing stress-strain relation. By comparing wave speeds between positions one and two with those between positions two and three, their constancy as prescribed by the theory may be determined. The maximum strain at all positions should be the same and determinable, as prescribed by the theory, from the measured wave speeds, provided they are constant as predicted. If the above conditions are found to be applicable, the governing stress-strain curve then follows as a consequence and is independent of impact velocity.

In an extensive series of measurements outside the first diameter in several fully annealed polycrystalline metals [1—5], BELL has found from the *direct* measurement of strain and surface angle (using the diffraction grating technique) that the one-dimensional finite amplitude wave theory for which stress is a single-valued function of strain is in excellent agreement with experiment. And, it must be stated once more that the strain-rate-independent finite-amplitude wave theory does *not* presume that the governing stress-strain curve is necessarily the static stress-strain curve of the material, although many investigators persist in presuming that it does. In fact, the governing stress-strain curve for the face-centered cubic metals for which accurate strain-time experimental data has been obtained, has been found to be parabolic, with coefficients linear functions of the absolute temperature [3—6]. These parabolas are, in turn, determinable through the summation processes of the aggregate theories of TAYLOR [7] and of BISHOP and HILL [8] from quasi-static measurements in single crystals. Furthermore, BELL has found experimentally that aluminium, copper, gold, silver, and lead all have the same parabolic stress-strain law at absolute zero [5, 9]. Hence, a single stress-strain law applicable to the dynamic deformation of any of these solids is provided solely from a knowledge of the melting point of the material.

Measurements of strain near the impact face inside the first diameter exhibit departures from the above behaviour. Obviously, then, the behaviour in the first diameter and the manner of wave initiation is a separate problem. It is a problem involving the consideration of the behaviour in a three-dimensional domain in which reflections from the side walls of the cylinder in the vicinity of the impact face play a prominent role [10, 11, 12, 6, 3]. The manner of this development in the first diameter would be expected to be dependent upon the geometry of the bar as well as upon the nature of the three-dimensional elastic fronts which must reach the side walls of the cylinder from the impact face to bring about the growth of the plastic wave front which emanates from the first diameter.

No theory has as yet been developed which includes the non-linear complications of large deformation wave fronts containing both elastic and plastic components reflecting from cylindrical surfaces. The only description available at the present time is that given by experiment. BELL [11, 13, 3, 6, 12] has carried out several series of experiments with the purpose of empirically describing this behavior in the first diameter. These experiments have included the measurement of strain and surface angle by means of diffraction gratings at numerous positions in the first diameter up to within .020 in. from the impact face. They have included simultaneous measurement of stress at the impact face using both thin piezo crystals and elastic load bars. They have included optical displacement measurements inside the first diameter, as well as time of contact measurements in symmetrical free-flight impact which were shown to be related to the wave initiation behavior. These experiments of BELL relating to this first diameter behavior have been at moderate impact velocities; FILBEY [14] has extended these results to relatively higher impact velocities; and BELL and SUCKLING [12] have studied the intermediate region of transition. An additional method of investigating the behavior in the first diameter has involved the study of the distortion detail of the strain-time front which is emitted from the first diameter [11, 3, 4].

From all these experimental studies, the following wave initiation behavior has been established in fully annealed metals. Immediately following impact, a fully elastic wave front is inaugurated with extremely high measured rates of stress and strain. At any material point on the surface, the duration of this elastic front after the maximum has been reached is of the order of $10^{-7}$ sec., although 2 or 3 $\mu$ secs. may be involved in the initial stage of its collapse once this collapse has begun. The existence of this elastic front has been established from the study of arrival time at the side wall of the cylinder in the immediate vicinity of the impact face, using .001 in. to .005 in. long diffraction gratings close to the impacting surfaces, and from piezo crystal measurements at the impact face. Employing HUYGENS' principle, and considering the fact that these elastic stress levels are one to two orders of magnitude higher than the elastic limit of the material, large plastic deformation has been seen to commence at the side wall of the cylinder during the reflection process [3, 6].

The experimental data have shown that by 1/2 diameter, however, this initial elastic front has produced a one-dimensional stepped shock front such as that which would have been developed had the elastic limit of the bar exceeded the maximum elastic stress. Plastic deformation is produced immediately behind this elastic shock front in two sections. First, the deviatoric component of the elastic step collapses into plastic deformation with extreme rapidity. Then, additional plastic deformation develops at a much slower rate from the remaining dilatational or hydrostatic component of the elastic front. The total stress during the latter process is invariably 3/2 the plastic stress of the maximum of the first stage. It is not unreasonable to presume that this delay in the development of plastic deformation from the dilatational portion of the initial elastic front is because, unlike the equivoluminal component, the dilatational component is reflected from the side wall of the cylinder as it would have been in a total elastic situation. In any event, the plastic strain level, developed first from the deviatoric portion of the elastic step front, has a strain level corresponding to 1/2 the energy of deformation per unit volume of the maximum strain obtained experimentally outside the first diameter and given theoretically from a substitution of the governing parabolic stress-strain law into the appropriate energy integral of the deformation.

In a recent paper by the present writer [6], this development of the plastic wave in the first diameter has been described in full, and has been shown empirically to be related throughout the process to theoretical wave speeds given by TRUESDELL [15] in his general theory of wave propagation in non-linear elasticity. There is no point in repeating all the detail summarized in that paper [6] and contained in a number of earlier papers. But, it is relevant to the present paper to consider two experimental aspects of this process. The first is that the strain-time data at one or more diameters when extrapolated to the impact face show that the strain up to $\bar{\varepsilon}$, or the strain of the mean energy, does rise as nearly an infinite step. The second is that the maximum strain of this first wave, as determined from the downstream distorted wave front, is such that there is an equi-partition of energy between the two sections of the wave.

Since in the diffraction grating technique surface angle measurements are obtained simultaneously with strain measurements, one may examine both the strain and the surface angle distortion produced in the first diameter from measurements outside the first diameter. One of the pronounced effects in these experimental data, as has been pointed out earlier [11, 3, 4], is that this surface angle undergoes a maximum at an intermediate level of strain, and that this strain at the surface angle maximum is invariably, in all materials investigated, the strain $\bar{\varepsilon}$ at 1/2 the maximum energy of deformation per unit volume.

The experimental values shown in Table 1 are averaged data obtained from diffraction grating measurements of the strain at surface angle maxima between one and six diameters from the impact face. They are compared with the calculated strain at this mean energy $\bar{\varepsilon}$, from the parabolic law for the stated impact velocity.

Table 1

| Impact velocity in/sec. | Average experimental $\bar{s}$ at surface angle Maximum % | Calculated strain at mean energy $\bar{s} = .63s_{max}$ % |
|---|---|---|
| 472 | .660 | .682 |
| 528 | .793 | .786 |
| 680 | 1.080 | 1.112 |
| 800 | 1.385 | 1.380 |
| 960 | 1.710 | 1.760 |
| 1053 | 2.005 | 2.000 |
| 1250 | 2.510 | 2.505 |

In all materials, for different impact velocities, the initial maximum stress obtained from load bar measurements are experimentally found to be $\sigma_D = 3/2\bar{\sigma}$ where $\bar{\sigma}$ is the stress of the quasi-static stress-strain curve corresponding to the strain $\bar{\varepsilon}$. Despite the fact that $\bar{\sigma}$ at $\bar{\varepsilon}$, from which the dynamic overstress $\sigma_D$ is determined, is given by the quasi-static stress-strain curve, $\bar{\varepsilon}$ is obtained from integrating under the parabolic stress-strain law governing the wave speeds. These two stress strain curves are not necessarily the same.

This fact—that two stress-strain curves are involved in the determination of the dynamic overstress—has been the cause of considerable difficulty in understanding the wave initiation behavior. This phenomenon of wave initiation would be much more easily understood if the dynamic and quasi-static stress-strain curves were identical and of a simple analytical form. Such a situation has been found experimentally.

In a recent examination of the quasi-static deformation of large-grained high purity polycrystals, it was found [5, 9] that unlike the commercial purity, fine-grained materials, the quasi-static stress-strain curves were parabolic and were the same curves that were predicted for the dynamic deformation. This coincidence of static and dynamic curves thus offers the opportunity for the further examination of this matter of wave initiation under much simpler conditions than those described earlier [5, 9]. The new experimental data in the present paper are concerned with the wave initiation for such high purity polycrystals.

## 2. Wave Initiation in High Purity Columnar Multicrystals

When 99.99% pure molten aluminum is poured into a chilled cylindrical mold, one obtains a large-grained polycrystal for which the individual single crystal components are roughly arranged in a pie-shaped configuration, shown in idealized form in Fig. 1. In the actual specimens, of course, the physical faces of the individual crystals do not lie in the plane nor do they have the regular geometry indicated in Fig. 1. The columnar multicrystal specimens considered in this paper have approximately fifty individual single crystals in the one inch diameter crosssection. The specimens are one inch in diameter and ten inches in length.

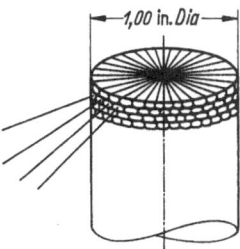

Originally this writer investigated the columnar multicrystal to examine the deformation behavior of a polycrystalline specimen for which each grain had at least one free surface in order to study the role of the normal stress, or hydrostatic component, in large deformation. These studies are discussed in considerable detail in a recent paper and hence will not be repeated here [9]. Of greatest importance for the present consideration of wave initiation is that the quasistatic stress-strain curve of the high purity co-

Fig. 1. A highly idealized version of the columnar multicrystal specimen for which each grain has at least one free surface.

lumnar multicrystal is parabolic, and the same curve governs the finite amplitude wave propagation. It also is of interest that this parabola differs significantly *in a predictable fashion* from the parabola governing the finite amplitude wave propagation in fine-grained commercial purity polycrystals.

Fig. 2 shows the quasi-static compression stress-strain curve of a one inch diameter, 3 in. long, high purity columnar multicrystal. Also shown in Fig. 2 are the experimental data of a quasi-static compression test in a one inch diameter, 3 in. long, 99.99% pure aluminum single crystal. From x-ray diffraction measurements, the SCHMID factor has been determined, from which the resolved shear stress, resolved shear strain curve of the single crystal has been computed and is shown as the lowest solid curve in Fig. 2. It will be noted that the static compression stress-strain curve of the columnar multicrystal, as well as that of the single crystal, are parabolic.

The stress and strain ratios of the TAYLOR [7] and BISHOP and HILL [8] theory of the aggregate may be used to determine the parabolic stress-strain law of the multicrystal

$$\frac{\sigma}{\tau} = \bar{m} = \frac{\gamma}{\varepsilon}, \quad \text{where} \quad \bar{m} = 3.06. \quad (1)$$

Introducing the numerical parabola coefficient of the resolved single
crystal parabola shown in Fig. 2, one obtains for the quasi-static or
dynamic stress-strain curve of the multicrystal,

$$\sigma = 2.4 \times 10^4 \varepsilon^{1/2} \, psi, \qquad\qquad (2)$$

shown as the solid line through the multicrystal data of Fig. 2. That the
parabola of Eq. (2) is also a governing stress-strain curve of the dynamic
deformation will be shown below.

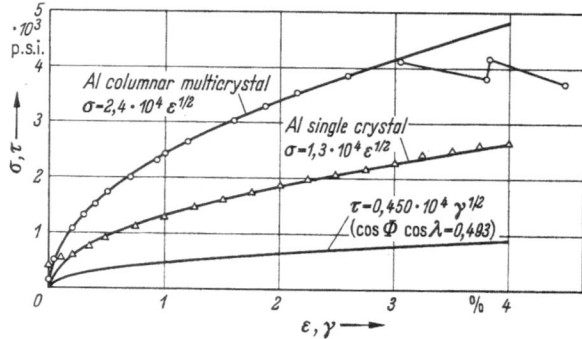

Fig. 2. Experimental data (triangles) from uniaxial compression in one inch diameter, 99.99% purity
aluminum single crystal are compared with the uniaxial compression of a one inch diameter, 99.99%
purity columnar aluminum multicrystal. The lowest solid line is the calculated resolved shear stress,
resolved shear strain parabola of the single crystal. Buckling began in the high purity multicrystal
at approximately 3% deformation.

To study wave initiation or wave propagation in *large-grained* polycrystalline specimens by means of the diffraction grating technique requires the averaging of a very large amount of data, due to the fact
that the gauge length of the diffraction grating, which is from .001 in.
to .005 in., is so short that each measurement is made on an individual
single crystal in the aggregate. The effect of variation of orientation in
the individual crystal elements thus requires the obtaining of large
amounts of data to average out the SCHMID factor.

Two other methods of investigation may be used to explore the
dynamic behavior of the large-grained polycrystals, which do not require the experimental averaging of very large amounts of data. The first
of these is the study of the stress-time behavior at the impact face,
using either the load bar or the piezo crystal technique; and the second
is the optical study of the time of contact and longitudinal displacement,
using the optical displacement technique developed earlier by the present
writer and discussed in recent papers [16, 13, 12]. Solely from a determination of the time of contact, by either of these techniques the parabola
coefficient of the dynamic stress-strain curve may be determined in a

way that has been described earlier [*16, 13, 12, 3, 4*]. Fig. 3 shows several experimental situations employing either load bars or piezo crystals, from which the stress-time behavior at the ends of cylindrical specimens may be inferred.

The measurements discussed in the present paper are of Type I, in which the 99.99% purity columnar aluminum multicrystal is in axial collision with a 6 ft. long, hard aluminum load bar. This bar is sufficiently long so that the time of contact of the impact is less than the time

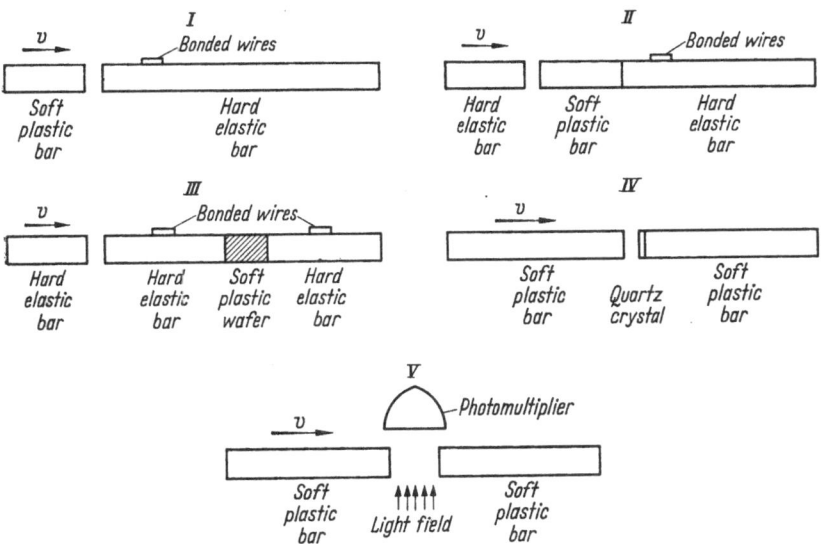

Fig. 3. Methods for the measurement of dynamic overstress.

required for the reflected wave in the hard elastic load bar to return from the free end. The load bar is instrumented with wire resistance strain gauges from whose strain-time data, given YOUNG's modulus of the material, the stress-time behavior at the impact face is inferred.

In earlier papers this writer has shown [*11, 13, 3, 4*] that the dynamic overstress obtained from load bar measurements in commercial purity polycrystals may be determined from stress and energy considerations of the quasi-static stress-strain curve of the material of interest. This calculation of the dynamic overstress was carried out by two different methods, each giving approximately the same result.

In the first method, it was observed from the surface-angle vs. time behavior in diffraction grating measurements that surface angle maxima in hundreds of impact tests in many different materials invariably coincided with a strain, $\bar{\varepsilon}$, which is the strain of the mean energy of deformation per unit volume for the impact velocity under consideration. It was further shown that if one determined the stress, $\bar{\sigma}$, corresponding to the stress of the quasi-static stress-strain curve at the strain, $\bar{\varepsilon}$, then the dynamic overstress, $\sigma_D$, was invariably $3/2\bar{\sigma}$. This behavior arises, as was pointed out above, from the fact that the first stage of plastic deformation develops from the deviatoric component of the initial elastic shock front.

Thus we have for the situation in which the quasi-static stress-strain curve is parabolic and coincides with the dynamic stress-strain curve,

$$\sigma_D = \frac{3}{2}\,\beta\,\bar{\varepsilon}^{1/2}. \tag{3}$$

The energy integral of Eq. (4),

$$U = \int_0^{\varepsilon_{\max}} \sigma\,d\varepsilon, \tag{4}$$

gives for the stress-strain law of Eq. (2):

$$U = \frac{2}{3}\,\beta\,\varepsilon_{\max}^{3/2}. \tag{5}$$

Since $\bar{\varepsilon}$ is the strain of the mean energy of deformation, one may write:

$$\frac{1}{2}\cdot\frac{2}{3}\,\beta\,\varepsilon_{\max}^{3/2} = \frac{2}{3}\,\beta\,\bar{\varepsilon}^{3/2}, \tag{6}$$

which furnishes $\bar{\varepsilon} = .63\,\varepsilon_{\max}$ for a parabolic stress-strain law. Substituting in (3), we have

$$\sigma_D = \frac{3}{2}\,\beta\,(.63)^{1/2}\,\varepsilon_{\max}^{1/2}. \tag{7}$$

But, from a substitution of the parabolic law of Eq. (2) into the integral (8) of the finite amplitude wave theory one obtains Eq. (9):

$$v_0 = \int_0^{\varepsilon_{\max}} c_p\,d\varepsilon, \qquad \text{where} \qquad \varrho c_p^2 = d\sigma/d\varepsilon, \tag{8}$$

$$v_0^2 = \frac{8}{9}\,\frac{\beta}{\varrho}\,\varepsilon_{\max}^{3/2}, \tag{9}$$

which furnishes

$$\varepsilon_{\max}^{1/2} = \left(\frac{9}{8}\,\frac{\varrho}{\beta}\right)^{1/3} v_0^{2/3}. \tag{10}$$

Substituting this value of $\varepsilon_{\max}^{1/2}$ into (7), one obtains the relation

$$\sigma_D = \frac{3}{2}\,(.63)^{1/2}\varrho^{1/3}\beta^{2/3}\,v_0^{2/3}, \tag{11}$$

which gives the dynamic overstress in terms of the impact velocity.

The second empirical method of computing the dynamic overstress was from shock wave considerations in which one writes (12) in the form

$$\frac{1}{2}\,\sigma_D\varepsilon_0 = T_0 = \frac{1}{2}\,\varrho v_0^2 = \int\limits_0^{\varepsilon_0}\sigma d\varepsilon. \tag{12}$$

The integral in this equation is for the quasi-static stress-strain curve.

In the commercial purity aluminum it has been shown by a graphical determination of the integral in (12) that, given the impact velocity $v_0$, the strain $\varepsilon_0$ is determined, from which, in turn $\sigma_D$ is found.

For the columnar multicrystal one has

$$\int\limits_0^{\varepsilon_0}\sigma d\varepsilon = \frac{2}{3}\,\beta\varepsilon_0^{3/2}. \tag{13}$$

Thus, from equation (12) we find

$$\sigma_D = \varrho v_0^2/\varepsilon_0; \tag{14}$$

on the other hand, substitution of the integral (13) into Eq. (12) yields

$$\varepsilon_0 = \left(\frac{3}{4}\,\frac{\varrho}{\beta}\right)^{2/3} v_0^{4/3}. \tag{15}$$

Substituting (15) into (14), one obtains the following equation for the dynamic overstress

$$\sigma_D = \left(\frac{4}{3}\right)^{2/3} \varrho^{1/3}\beta^{2/3}v_0^{2/3}. \tag{16}$$

For Eqs. (11) and (16) to be equivalent, $k$ in Eq. (17) should be unity, whereas

$$\frac{(4/3)^{2/3}}{3/2\,(.63)^{1/3}} = k = 1.02\,. \tag{17}$$

The difference of approximately 2% shown in Eq. (17) between the two methods of calculating the dynamic overstress is less than the experimental error in measuring $\sigma_D$. The strain $\bar\varepsilon$ is experimentally identified from surface angle measurements at all impact velocities, whereas the somewhat larger $\varepsilon_0$ has not been thus found experimentally. A similar correspondence between dynamic overstress calculations is obtained in commercial purity materials.

A proper understanding of the significance of the experimental fact that the measured dynamic overstress may be determined by either of the methods outlined above, must await the development of a three-dimensional theory which will include the reflection behavior of finite amplitude waves.

Fig. 4 shows three load bar measurements in 99.99% purity columnar multicrystals for which the average hitter velocity was 832 in./sec. In the impact collision of soft and hard specimens, the hitter velocity $v_h$, is equal to the sum of the maximum particle velocity $v_e$, in the elastic load bar and the maximum particle velocity $v_0$, in the plastically deformed specimen:

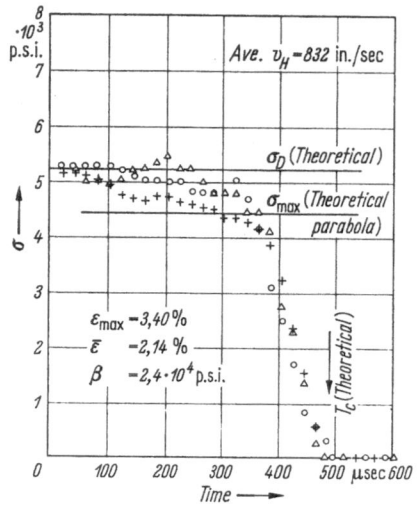

$$v_h = v_e + v_0 \qquad (18)$$

or since $\sigma_s = \varrho c_0 v_e$, where $c_0 = (E/\varrho)^{1/2}$ is the elastic bar velocity, and the maximum particle velocity $v_0$ in the plastically deformed bar is given by the integral in (8), Eq. (18) becomes

$$v_h = \frac{\sigma}{\varrho c_0} + \int_0^{\varepsilon_{max}} c_p \, d\varepsilon. \qquad (19)$$

For the parabolic law of Eq. (2) this equation takes the form

$$v_h = \frac{\sigma}{\varrho c_0} + \left[ \frac{8}{9 \varrho \beta^2} \right]^{1/2} \sigma^{3/2}. \qquad (20)$$

Fig. 4. Experimental measurements of the stress-time behavior at the impact face from load bar tests, compared with calculated value of the dynamic overstress $\sigma_D$ and the time of contact, $T_C$.

Thus, given the hitter velocity $v_h$, the maximum stress $\sigma$ of the finite amplitude wave theory may be computed, from which, in turn, the maximum particle velocity of the plastically deformed specimen is given. From this value of $v_0$ upon substitution in Eq. (11) the dynamic overstress may be determined. For the average $v_h$ of Fig. 4, the calculated $\sigma_D$ is shown as the theoretical value. The maximum stress of the parabolic stress-strain law, which is obtained after the first diameter (and which is also given in Fig. 4) is of the order of magnitude of the stress levels to which the stress at the impact face has fallen prior to the beginning of unloading. The manner in which the stress at the impact face decreases has been studied in detail by BELL and SUCKLING [12].

The predicted value and the experimental value of the dynamic overstress shown in Fig. 4 are in agreement. The calculated value of $\sigma_D$ for these high purity columnar multicrystals is 5,240 psi for the experimental hitter velocity $v_h$ of 832 in./sec. For this hitter velocity the dynamic overstress for fully annealed commercial purity aluminum would be $\sigma_D = 8,750$ psi which, indeed, is the value obtained experimentally in that material. For commercial purity aluminum the parabola coefficient is $5.60 \times 10^4$ psi instead of the $2.4 \times 10^4$ psi of the high purity multicrystal, and its quasi-static stress-strain curve is non-parabolic.

An agreement between predicted dynamic overstress and experimental values similar to those shown here has been obtained in commercial purity fully annealed and half-hard aluminum, fully annealed and hard copper, and pure lead. In each instance, calculations are made from the appropriate quasi-static stress-strain curve and parabolic laws of the material. In a recent paper [6], it was also shown that this dynamic overstress in fully annealed commercial purity aluminum may be determined entirely from strain measurements inside the first diameter, empirically using the theoretical wave speed relations in nonlinear elasticity given by TRUESDELL [15].

That the stress-strain curve governing the dynamic plastic behavior is the same as the quasi-static stress-strain curve for these high purity columnar multicrystals may be seen from an inspection of the time of contact given experimentally in Fig. 4. As has been shown in detail in earlier papers [16, 13, 3], the time of contact is given by the time required on the Lagrangian diagram for the characteristic of 1/2 the maximum stress to propagate to the free end of the specimen and return to the impact face with the elastic velocity.

For a parabolic stress-strain law equation, such as (2), the wave speeds $\varrho c_p^2 = d\sigma/d\varepsilon$ become:

$$\varrho c_p^2 = \frac{\beta}{2\varepsilon^{1/2}} = \frac{\sigma}{2\varepsilon}. \tag{21}$$

For the characteristic of 1/2 the maximum stress, the wave speed $c_p^*$ is given by

$$\varrho c_p^{*2} = \frac{\sigma_{\max}}{\varepsilon_{\max}}, \tag{22}$$

and the time of contact is

$$T_c = \frac{L}{c_p^*} + \frac{L}{c_0}, \tag{23}$$

where $L$ is the length of the bar, and $c_0$ the elastic bar velocity. The value for the time of contact for the indicated hitter velocity $v_h$ for the columnar multicrystal whose parabola coefficient is $2.4 \times 10^4$ psi is shown by the arrow in Fig. 4. The close agreement between the time of contact given by these load bar measurements and the predicted value demonstrates both the applicability of the finite amplitude wave theory in the columnar multicrystal and the fact that the governing stress-strain curve is the predicted parabolic law.

It should again be emphasized that this parabola coefficient of $2.4 \times 10^4$ psi for the columnar multicrystal differs considerably from the parabola coefficient of $5.60 \times 10^4$ psi for the commercial purity, fine-grained polycrystal, and yet, in each instance, the dynamic overstress and times of contact are experimentally in agreement for the respective parabolas, thus providing further evidence for the consistency of the relations presented here between the dynamic overstress and the quasi-static stress-strain curve.

## 3. Further Discussion of Wave Initiation in Commercial Purity Polycrystals

For the commercial purity fine-grained materials the quasi-static stress-strain curve from which the dynamic overstress is determined is not the same as the parabolic law governing the wave speeds from which $\bar{\varepsilon}$ is determined. One may, however, formalize the earlier graphical procedures [11] of this determination of the dynamic overstress by fitting an empirical equation to the static stress-strain curve of the commercial purity materials from which the dynamic overstress $\sigma_D$ may be approximated.

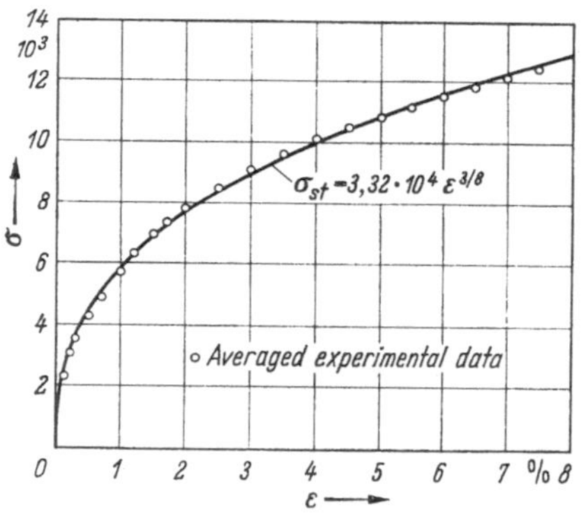

In Fig. 5 the average of 14 quasi-static stress-strain curves in fully annealed fine-grained commercial purity polycrystalline aluminum at room temperature is found to fit approximately a 3/8 power law:

$$\sigma = 3.32 \times 10^4 \varepsilon^{3/8}. \quad (24)$$

The parabolic law for wave propagation in this material is given by

$$\sigma = 5.60 \times 10^4 \varepsilon^{1/2}. \quad (25)$$

Fig. 5. The averaged experimental data (circles) from 14 uniaxial compression tests in one inch diameter, fully annealed, fine-grained commercial purity aluminum. The solid line is an empirical fit to the data.

For a given impact velocity $v_0$, $\varepsilon_{\max}$ may be obtained, as before, from Eq. (9), using the parabola coefficient (25). Independent of the coefficient from the integral in (5), $\bar{\varepsilon} = .63\, \varepsilon_{\max}$. A substitution of this value into (9) gives

$$\bar{\varepsilon} = \left( \frac{9}{16} \frac{\varrho}{\beta} \right)^{2/3} v_0^{4/3}. \quad (26)$$

Since the dynamic overstress is 3/2 the stress of the quasi-static stress-strain curve (24) at the strain $\bar{\varepsilon}$, we find the dynamic overstress

$$\sigma_D = \frac{3}{2} \times 3.32 \times 10^4 \bar{\varepsilon}^{3/2}. \tag{27}$$

Upon substitution of the value of $\bar{\varepsilon}$ from (26), we obtain

$$\sigma_D = 343 v_0^{1/2}. \tag{28}$$

Eq. (28) has been used in an earlier paper [12] for the comparison of the experimental dynamic overstress with the calculated value.

Averaged experimental results of the present writer and of JOHNSON, WOOD, and CLARK [17] in fully annealed commercial purity aluminum in which the dynamic overstress is determined from load bar measurements, are compared with the values calculated from Eq. (28).

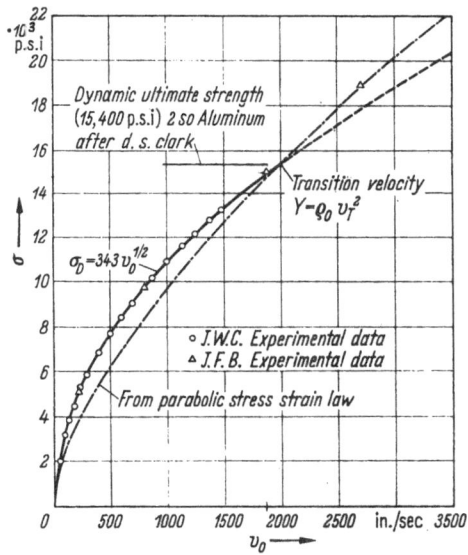

Fig. 6. The dynamic overstress experimental data compared with prediction.

Eq. (28) has been shown by BELL and SUCKLING [12] to apply only up to a transition velocity, given by Eq. (29), in which $Y$ is the original elastic limit of the material

$$\varrho v_t^2 = Y, \tag{29}$$

$\varrho$ being the mass density, and $v_t$ the impact velocity, which has a value of approximately 2000 in/sec in fully annealed aluminium and approximately 1000 in/sec in fully annealed copper. It is of great interest, and clearly significant, that the value of the dynamic overstress at the transition velocity, which coincides with the stress of the parabolic law, is the same as the dynamic ultimate strength obtained earlier in dynamic tension testing by CLARK [18] in this material.

The experimental points, shown as triangles in Fig. 6, were obtained from averaging the data both from the load bar experiment (Type I, Fig. 3) and from quartz crystal measurements (Type IV, Fig. 3). The latter technique alone was used in obtaining the values at 1900 in/sec and at 2800 in/sec. The value at an impact velocity of 2800 in/sec was obtained by FILBEY [14] who has shown that even at this high impact velocity the wave propagation outside the first diameter is still that of the finite amplitude wave theory governed by the parabolic law. Above the transition velocity the wave initiation effects described in the present paper are no longer found and the wave initiation, which is accompanied by large mushrooming, has been shown by FILBEY [14] to be entirely different. It is of more than a little interest that the experimental measurements of the dynamic overstress made by KOLSKY and DOUCH [19] in half-hard commercial purity aluminum also are given by Eq. (28), as may be seen in Table 2.

Table 2

| Impact velocity in/sec | KOLSKY and DOUCH Experimental $\sigma_D$ psi | $\sigma_D$ Calculated from Eq. (28) psi |
|:---:|:---:|:---:|
| 466 | 7,900 | 7,420 |
| 1010 | 11,300 | 10,900 |
| 1596 | 13,600 | 13,700 |
| 1745 | 14,500 | 14,350 |

The present writer has carried out dynamic overstress determinations by all five of the methods shown in Fig. 3. A particularly interesting one is the wafer experiment of Type III for fully annealed commercial purity aluminum. Simultaneous diffraction grating measurements at the center of the wafer have exhibited the importance both of propagation detail and of the wave initiation behavior described in the present paper. These measurements of instantaneous plastic strain differ in a calculable manner from those obtained by gross averaging procedures of particle velocities in the hard elastic bars. *Apparent* strain rate effects obtained from averaged elastic particle velocity data for small wafer specimens whose length is of the order of one diameter, which have been interpreted by HAUSER, SIMMONS, and DORN [20] as evidence of the non-applicability of the finite amplitude wave theory, arise from their neglect of wave propagation detail.

## 4. Summary and Conclusion

In the present paper a series of measurements of the stress-time behavior at the impact face in high purity multicrystals have been presented. For high purity multicrystals the quasi-static stress-strain law is parabolic and coincides with the dynamic stress-strain curve governing the finite amplitude wave propagation, thus permitting a direct determination of the dynamic overstress.

Agreement is obtained between calculated values of the dynamic overstress and experiment, even though the parabola coefficient for the high purity columnar multicrystal is far below that for the fully annealed commercial purity aluminium.

It is further shown that introducing an equation constituting an empirical fit to the quasi-static stress-strain curve of the commercial purity material permits the calculation of the dynamic overstress without the necessity for graphic analysis.

The empirical description of the wave initiation behavior in the first diameter for which plastic deformation arises in two stages behind a short-lived unstable high-stressed initial elastic wave front, has been further generalized to include high purity, large-grained multicrystals. Initial overstress following collision impact clearly does not depend upon strain rate, but arises from considerations of the reflection and instability characteristics of the initiation of high elastic fronts, the stability of weak plastic shock fronts, the geometry of the bar, and the axiality of the impact.

## Acknowledgments

The portions of this work relating to the behavior of high purity multicrystals was carried out under the sponsorship of the U.S. Air Force Office of Scientific Research. Other portions of the work were carried out under a Grant from the U.S. Army, Army Research Office. The author would again like to express his appreciation to his research assistant, John Suckling.

### References

[1] BELL, J. F.: J. Appl. Phys. 27, 1109 (1956).
[2] BELL, J. F.: J. Appl. Phys. 31, 277 (1960).
[3] BELL, J. F.: J. Appl. Phys. 32, 1982 (1961).
[4] BELL, J. F., and W. M. WERNER: J. Appl. Phys. 33, 2416 (1962).
[5] BELL, J. F.: J. Appl. Phys. 34, 134 (1963).
[6] BELL, J. F.: Proc. IUTAM Symposium on 2nd Order Effects, Haifa, 1962.
[7] The Scientific Papers of Sir Geoffrey Ingram Taylor, G. K. BATCHELOR, editor, Vol. 1, Cambridge University Press 1958, Nos. 21, 22, 27, 47.

[8] BISHOP, J. F. W., and R. HILL: Phil. Mag. 42, 414, 1298 (1951).
[9] BELL, J. F.: Manuscript submitted for publication.
[10] BELL, J. F.: The Initial Development of an Elastic Strain Pulse Propagating in a Semi-Infinite Bar, Tech. Rep. No. 6, U.S. Army Ballistics Research Laboratories, Aberdeen, Contract No. DA-36-034-509-ORD-7 RD; The Johns Hopkins University, 1960.
[11] BELL, J. F.: J. Appl. Phys. 31, 2188 (1960).
[12] BELL, J. F., and J. SUCKLING: Proc. 4th U.S. Nat. Congress Appl. Mech., Berkeley, 1962.
[13] BELL, J. F.: J. Mech. Phys. Solids 9, 261 (1961).
[14] FILBEY, G. L.: Tech. Report No. 8, U.S. Army Ballistics Research Laboratories, Aberdeen, Contract No. DA-36-034-21 X 4992. 509-ORD-3104 RD; The Johns Hopkins University, 1961.
[15] TRUESDELL, C.: Archive Rat. Mech. Anal. 8, 263 (1961).
[16] BELL, J. F.: J. Mech. Phys. Solids 9, 1 (1961).
[17] JOHNSON, J. E., D. S. WOOD and D. S. CLARK: J. Appl. Mech. 9, 253 (1953).
[18] CLARK, D. S.: Trans. Am. Soc. Metals 45, 34 (1953).
[19] KOLSKY, H., and L. S. DOUCH: J. Mech. Phys. Solids 10, 195 (1962).
[20] HAUSER, F. E., J. A. SIMMONS and J. E. DORN: Proc. AIME Conf. on Response of Materials to High Velocity Deformation, 1960, 93.

# Visualisation of Wave Propagation in Impulse-Loaded Bars

By **Erwin David**, **Rudi Schall**, and **Hubert Schardin**

Franco-German Research Institute, St. Louis (Haut-Rhin), France,
and Ernst-Mach-Institut, Freiburg, Germany

## Abstract

Various experimental techniques which can be used for the visuali-sation of stress waves are considered and their applicability to the study of wave propagation in bars is discussed. Some experimental results which were obtained by means of shadow photography are given. These were obtained with bars immersed in water and show the existence of precursor waves which travel with the velocity of propagation in an unbounded medium. A qualitative theory of the wave system is outlined and it is shown that the concept of group velocity in terms of the POCH-HAMMER-CHREE treatment leads to some difficulties in this problem.

A second investigation using a different method of visualisation, namely flash radiography is also described. This technique has here been used for studying the passage of intense plastic waves in rods.

## 1. General Remarks on Methods

A plane plate provides the simplest example for the visualization of stress waves. In this connection, reference should be made to SCHARDIN's paper [1]. For transparent plates of such materials as glass or lucite the following methods may be used with transmitted light:

  (a) schlieren optics,
  (b) shadow optics with direct or indirect shadow image, and
  (c) photo-elasticity.

For specimens with one optically plane reflecting surface,
  (d, e) schlieren and shadow optics with reflected light, and
  (f) photo-elastic films bonded to the surface
can be used, and techniques suitable for intense plastic waves are

  (g) changes in the reflectivity of the surface marking the wave front, and

  (h) flash radiography.

All these methods are suitable not only for plates but also for rods of rectangular or square cross section, especially if the surfaces are of optical quality, though rods of this kind have little practical importance. Cylindrical bars of metals, glass or plastics are of greater interest for experiment and application, but with them certain difficulties arise. Thus, for bars of glass or transparent plastics that are immersed in a liquid of the same refractive index, methods (a) and (b) are not in general applicable on account of insufficient optical homogeneity and method (c) should be more satisfactory. For methods (d) and (e) the cylindrical surface presents a serious difficulty even it is reflecting and of good precision. Method (f) needs waves of some intensity, and methods (g) and (h) are only suitable for very intense plastic waves.

These difficulties may be overcome by immersing the bar in a liquid ([1], Fig. 28) and observing waves in the liquid by methods (a) or (b). This immersion does not distort the waves seriously, it only causes a slight attenuation of the waves in the bar. As is well known, the ratio of transmitted to reflected wave energy is roughly proportional to the ratio of acoustic impedances, that is to the products of acoustic wave velocity and density. This ratio is approximately 8 for water to aluminium or glass and more than 20 for water to heavy metals. Moreover, the bar waves of interest to us are waves which have smaller amplitudes in the lateral than in the longitudinal direction. Consequently, a very small fraction of the wave energy is radiated laterally into the liquid.

As the ambient medium, a liquid (which in most cases will be water) is obviously preferable to a dense gas or vapor. Liquids have convenient optical refractivities and suitable sonic velocities. In the optical arrangement, the liquid may simply be contained in a box with two windows of selected plate glass.

Most bar waves of interest are axially symmetric. Only bending waves are of antisymmetric type. Waves of still higher symmetry are hardly excited.

In schlieren optics the image contrast is to a first approximation proportional to the differences in the angles that the light rays have been bent in the sonic field. The major part of the bending takes place where the light rays are parallel or tangential to a wave front. The curvature of the rays there is proportional to the first derivative of the refractive index, which is in turn proportional to the first derivative of the density (or pressure). The optical selection of those parts of the sonic field, where light rays and sonic wave fronts are parallel, especially in our case of axial symmetry and rather weak waves, makes the image practically a cross section in a plane normal to the light rays through the axis of the bar. This also applies to shadow optics, but there the concentration or dilution of the light beam by a gradual change of the refracted angle is responsible for the contrast in the image, and the first derivative of the refractive index is then replaced by the second derivative of the refractive index, density, or pressure.

On account of these considerations, the schlieren, and even more the shadow, method favor the visualization of short wave lengths in the sonic field. We denote the sonic wavelength by $\Lambda$, the mean light wavelength (approx. $0.5 \cdot 10^{-3}$ mm) by $\lambda$, the distance between the cross sectional plane of the wave field and the shadow plane by $a$, and a characteristic dimension of the wave field, for instance the diameter of the bar, by $D$.

To resolve the smallest wavelength $\Lambda_{min}$ optically, the distance $a$ must satisfy

$$a \ll \Lambda_{min}^2/\lambda. \tag{1}$$

With $\Lambda_{min} = 1$ mm, we have $a \ll 2$ m. The distance $a = 50$ cm actually used in the experiment should therefore be permissible.

A wave of intensity (energy density) $I$ in the bar causes waves in the surrounding water which give rise to a contrast in the shadow photograph

$$\text{Contrast} \simeq \text{const.} \; (I a^2 D/\Lambda^3)^{1/2}. \tag{2}$$

With a wavelength ratio of 10 to 1 between bar waves and precursor waves, the wave energies will have the ratio 1000 : 1 if the image contrast of both wave systems is the same, and these figures are representative of the experimental conditions.

A decrease in the bar diameter $D$ diminishes the visualization contrast if the sonic intensity $I$ is held constant and if $\Lambda_{min}$ and $a$ are supposed to be independent of $D$. In fact, however, $\Lambda_{min}$ tends to decrease slightly with $D$. This necessitates a decrease of the distance $a$ in accordance with (1) and leads to a severe decrease in the image contrast by (2).

## 2. Visualization of Pressure Waves in Bars

The classical methods of observation of bar waves (condensers, induction coils, piezo crystals, strain gages) showed the normal bar waves in agreement with the theory. If the equipment was sufficiently sensitive, some faint ripples appeared on the record. These travelled along the bar with a speed comparable to that of longitudinal waves in unbounded material.

Fig. 28 in Reference [1] does not show any precursor waves, presumably because the bar was too thin and TOEPLER's schlieren method was used. As mentioned above, shadow photography is more sensitive for short waves.

Our first visualization of precursor waves was similar to Fig. 1. Long before the arrival of the bar wave, the end of the bar sends out waves that behave quite curiously. They propagate straight on, in apparent contradiction to Huyghens' principle. On the other hand, the waves excited in the water when the bar wave arrives at the end of the bar, are quite in agreement with this principle (see Fig. 28 in Reference [1]). The second curious effect is the apparent absence of any lateral effects. (The bar shown in Fig. 1 had only a diameter of 4 mm.)

Fig. 1. Aluminum bar, diameter 4 mm, length 60 cm. Bar wave running along bar and precursor waves at left end of bar.

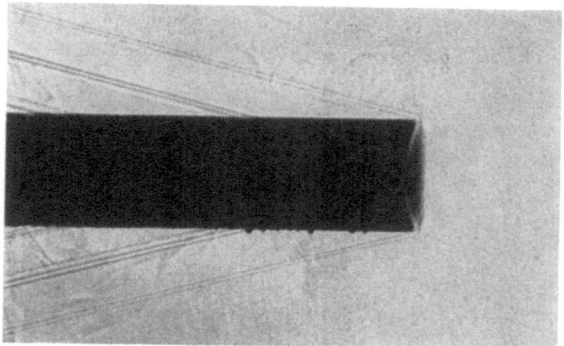

Fig. 2. Iron bar, diameter 32 mm, length 60 cm. First precursor group just arrives at end.

Figs. 2 and 3 taken with a bar with a diameter of 32 mm readily explain these effects. The intense waves sent out at the end are in reality groups of waves of rather short wavelength. The bending of the light rays in these waves is so intense that it destroys the correct optical image, but on the lateral edges of the waves the true wavelength is approximately visible. For this small wavelength there is no contradiction to Huyghens' principle. Also the conical waves drawn laterally are not entirely missing, but the first of them is rather faint though the longitudinal wave corresponding to it is of considerable intensity.

Fig. 4 shows another example. Here the grouping of the waves is less pronounced. Especially later on, a continuous series of waves follows with increasing wavelengths.

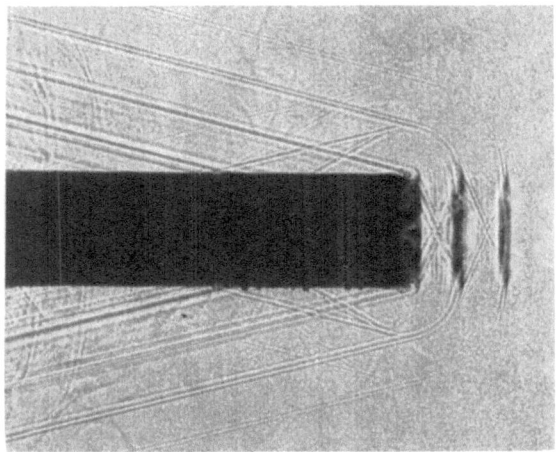

Fig. 3. Iron bar, diameter 32 mm, length 60 cm. Two precursor wave groups have left bar.

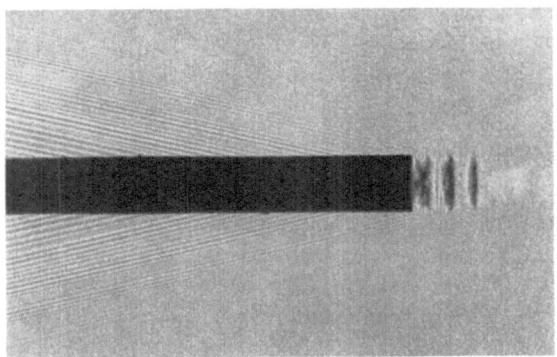

Fig. 4. Aluminum bar, diameter 16 mm, length 60 cm. Example of more continuous wave train.

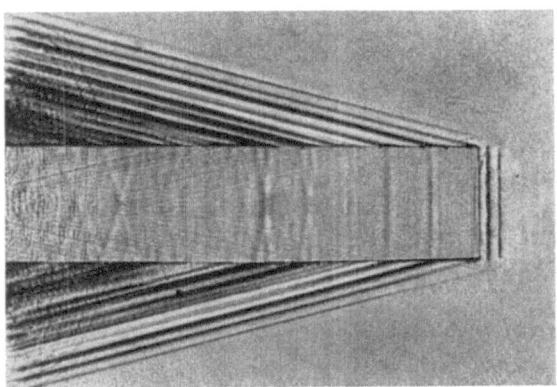

Fig. 5. Glass strip, 34 × 9 mm, length 40 cm. Waves in interior and in surrounding water.

Fig. 5 shows somewhat similar waves in a glass strip that is 34 mm high and 9 mm thick in the light direction. The ratio between the inner wave-spacings in the glass and the outer ones in the water is about 4 : 1. This ratio is of course the ratio of the velocities of sound and is otherwise equal to the sine of the angle between lateral waves in the water and the rim of the strip.

## 3. Interpretation of Precursor Observations

In the records obtained by C. W. Curtis ([2], pp. 32, 33) the precursor appears in the form of small ripples. It is a hard task for the theorist to determine the nature of the two-dimensional wave geometry from a poor one-dimensional record.

In contrast our Fig. 3 clearly reveals the significant features of this geometry, which is by no means simple.

The bar diameter is here 32 mm, the ratio of the longitudinal wave velocity and wave dimensions in the bar to the wave velocity and dimensions in the water is about 4. From the image of the water waves we draw the following conclusions regarding the waves in the bar. Groups of approximately plane longitudinal waves follow each other. They consist of 2, 3, 4 and later on more waves in the bar of about 4 mm wavelength. The distance between the groups is about 49 mm. These groups are connected by waves which appear as comparatively faint crosses on the image. In three dimensions they are double cones. The trace velocity of these waves at the surface of the bar is equal to the longitudinal wave velocity. Therefore, in view of the aforementioned height of the cone, the velocity of these waves is slightly more than 0.5 of the longitudinal velocity which is the velocity of shear waves.

To explain these phenomena, we remember that a reflected longitudinal wave and a drawn transverse (shear) wave are produced when a plane longitudinal wave hits a plane surface in oblique incidence. This simple picture fails for an incident wave with a front that is normal to the surface, but this case does not arise in practice. We observe firstly that the plane longitudinal waves do not seem to be real plane waves of equal amplitude from the axis to the surface of the bar. The amplitude decreases towards the surface. Secondly, in Fig. 4, we easily detect an increase of apparent wavelength backwards along the bar. This increase is accompanied by an increase of velocity, more clearly visible on the images with 8 mm bars. The true speed of longitudinal waves cannot be exceeded. The observed speed must therefore be a trace speed between somewhat oblique longitudinal waves and the surface. Because of axial symmetry the form of the wave fronts of longitudinal waves must be a cone or multiple cone similar to that of the shear waves. But the cones are extremely flat at the beginning and rather flat further on. Mathematically, we take a plane wave oblique to the axis, rotate it around the bar axis, and superpose these waves.

## 4. Theoretical Treatment

In the theory, the problem was always attacked by use of the concept of group velocity (compare C. W. CURTIS [2]). The question arises whether this concept applies to the wave system observed. The answer is distinctly in the negative.

In the mathematical formulation of the concept of group velocity, the starting point is an infinite train of waves. These waves propagate without change of shape with a definite phase velocity. By FOURIER-integral-superposition of infinite wave trains having slightly different wave lengths, the wave group is formed with a smoothly varying amplitude. The single wave in the group is of the same shape as in the infinite wave trains. The group propagation speed is, as is well known, the derivative of the frequency with respect to the wave number.

Our observed precursor with its complicated combination of longitudinal and transverse waves has no single waves corresponding to one single POCHHAMMER-CHREE-wave.

Further details are obtained by physical reasoning. A given wave has a certain energy density and also a certain energy stream density which transports the energy. If we divide the stream density by the density itself, we get the speed with which the energy is transported. This is the group velocity.

In our precursor wave system we have on the one hand the longitudinal waves. As explained above they may consist of plane waves the normals of which form small angles $\alpha$ with the axis of the bar. The direction of the energy transport deviates from the axial direction by this angle. Accordingly, the component of the energy transport speed in the axial direction is the *longitudinal wave velocity* multiplied by cos $\alpha$. The trace speed, that is the speed of the intersection of oblique wave front with the surface of the bar is the *longitudinal wave velocity* divided by cos $\alpha$. This agrees well with Fig. 4. The first waves have small angles $\alpha$, and the energy transport speed is only slightly lower and the trace speed only slightly higher than the longitudinal speed.

The later waves have greater angles $\alpha$; the transport speeds are lower and the trace speeds higher than the longitudinal velocity.

On the other hand, conditions are entirely different for the shear waves. The angles $\beta$ between the axis and the normals of the shear wave adjust themselves so that the trace speed, that is the *shear wave velocity* divided by cos $\beta$, is equal to that of the longitudinal waves. In consequence the component of the energy transport speed in the axial direction, that is the *shear wave velocity* multiplied by cos $\beta$, turns out to be only about one quarter of that of the longitudinal waves.

Longitudinal and shear waves are necessarily connected in the precursor. But in contradiction to the concept of group velocity, they have quite different energy transport velocities. Consequently, the precursor wave does not behave like a correct group that proceeds with little or no change in form.

The very first waves of the precursor reach the observation point with almost exactly the speed of longitudinal waves, but energy is lost continually into the shear waves, which feed the succeeding wave system. The intensity of the very first group therefore declines, presumably in an exponential way.

Longitudinal waves are not only multiplied by intermediate shear waves. As was mentioned, the longitudinal waves consist of components of small obliqueness $\alpha$. Their direct reflection as longitudinal waves at the surface leads to the building up of groups of a few longitudinal waves in Figs. 2 and 3 and of a practically continuous train of longitudinal waves in Fig. 4. New waves created by multiple direct reflection of longitudinal waves and others created by intermediate shear waves give rise to random interference phenomena which cause more or less irregular phase and amplitude changes.

The head of the precursor nearly always moves with approximately longitudinal wave speed, even if in large distances one head group after the other gradually disappears. Quite different from the head, the tail of the precursor waves obviously moves rather slowly. The spreading speed of the precursor group is therefore comparable to its mean propagation speed. It is questionable whether a mean propagation speed has a definite meaning at all, but whatever definition we use, the mean speed is surely smaller than the bar velocity.

We conclude: with our simple analysis in terms of plane waves we can understand all the features of the precursor. One question remains, however: does the inapplicability of the concept of group velocity exclude the usual Pochhammer-Chree-expansion? It does not because both concepts are independent. It is convenient if within the frame of Pochhamer-Chree-theory the simplifying concept of group velocity may be used. If not, many Pochhamer-Chree functions of many different branches must be superposed point for point in the two-dimensional coordinates of the bar volume. This will be a difficult task, even when a large computer is used, and it presumably would not be worth the effort.

### 5. Flash Radiography of Plastic Waves

A method of shock visualisation which applies only to very intense waves is the method mentioned as (h) in Section 1. Fig. 6a shows a typical set-up to produce a plane detonation shock in an inert specimen and Fig. 6b a flash radiograph (exposure time $3 \cdot 10^{-7}$ sec) of a shock wave in perspex obtained with this arrangement [3].

The attractive feature of this technique is that it affords a direct measurement of the density behind the shock. The density $\varrho_1$ is determined *densitometrically* with a precision of 1% comparing film blackening

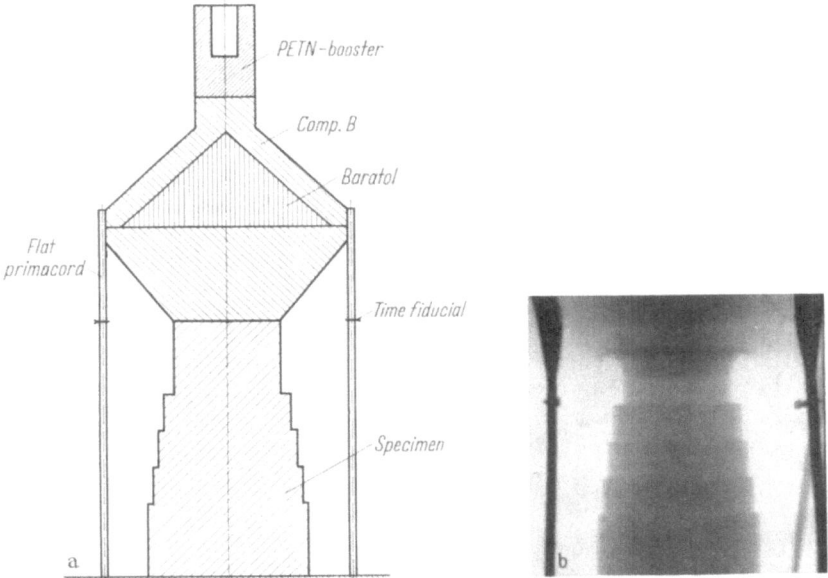

Fig. 6a and b. Flash radiographic measurement of dynamic compression in perspex.
a) Experimental setup; b) Flash radiography 2.2 μs after entrance of shock waves into specimen.

behind the shock front with that of rods of greater absorber thickness. Two primacords radiographed with the phenomenon provide the time measurement necessary to determine the wave speed $V$. From $\varrho_1$ and $V$, the shock pressure $p = \varrho_0 V^2 (1 - \varrho_0/\varrho_1)$ and the HUGONIOT pressure $p(v)$ can be deduced, $v$ being the specific volume.

The main shortcoming of this technique is that release waves starting at the rod walls cause the shock front to be curved and that the $X$-ray beam measures average density across the diameter. Near the high explosive for larger rods and corresponding greater explosive charges

(200 gr) the influence of the free surface should however be small. HUGONIOT curves measured with the radioflash method agree substantially with those obtained following the free surface method for some materials (e.g. for perspex); for other materials (e.g. Mg, Al), there is a certain discrepancy in so far as radiographic results seem to indicate a lower compressibility than has been determined by the cinematic free surface measurements.

## Acknowledgements

The shadow photography experiments on bars, basic for the major part of this paper, have been performed by H. NASDALA, and the flash radiography and the equipment used for the experiments of Section 5 have been made by G. THOMER both of the French-German-Research-Institute, St. Louis (Haut-Rhin), France.

### References

[1] SCHARDIN, H.: International Symposium on Stress Wave Propagation in Materials, New York: Interscience Publishers 1960, pp. 289—302.
[2] CURTIS, C. W.: ibid., pp. 15—43, especially p. 33.
[3] SCHALL, R., and G. THOMER: Technical Report, AF EOAR 61—52 (July 1962). This research has been sponsored in part by the Air Force Special Weapons Center through the European Office, Aerospace Research, United States Air Force.

# Transient Stress Wave Boundary Interactions[1]

## By John S. Rinehart

Mining Research Laboratory, Colorado School of Mines, Golden, Colorado, U.S.A.

## 1. Introduction

Impulsive loads such as are produced by explosions and impacts, on introducing transient stresses into the materials against which they act, start a series of events. At first, fractures and highly localized internal stresses may develop; later, rod, plate, LOVE, RAYLEIGH, and other similar waves may appear. The character of these transitory and evolutionary processes is controlled in a large measure by the interactions which take place both among the transient stress waves themselves and between the stress waves and the boundaries or interfaces which they encounter. These interactions can exert a strong attenuating effect on a wave by disorganizing it and splitting its energy up into diverse modes. The attainment of a full appreciation of the attenuation processes requires a complete understanding of the nature of the interactions themselves, an understanding which may be hampered by the fact that the interaction processes are quite complicated, difficult, and extremely tedious to treat in a rigorous manner. However, some of the more important interactions are amenable to relatively simple treatment. The purpose of this paper is to describe a few interactions that have not heretofore been discussed and to present some new experimental data bearing on them. The specific interactions are: motion associated with the encounter of a wave with an obliquely inclined free surface; fracturing caused by two interfering obliquely inclined waves; the influence of a non cohesive interface on the passage of a wave; the progress of a radial fracture through dissimilar materials; the development of a corner fracture; and some diffraction effects.

## 2. Motion at Inclined Free Surface

When an elastic longitudinal wave strikes a free surface obliquely, as indicated in Fig. 1, the energy of the incident wave, $A$, is redistributed, appearing partly in the form of a shear wave, $B$, and partly in the form

---

[1] Work supported in part by the National Science Foundation and in part by the Office of Naval Research.

of a longitudinal wave, $C$, the relative amounts of energy in the newly created waves being dependent on the angle with which the front of the incident wave strikes the surface and on Poisson's ratio. Specifically, the stress levels $\sigma_B$ and $\tau_C$, respectively, of the reflected longitudinal wave and the newly created shear wave will, at any time during the interaction of the wave with the surface, be given by

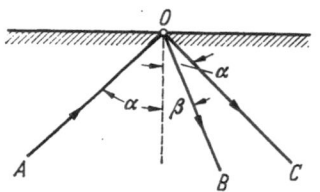

Fig. 1. Generation of shear and longitudinal waves at free surface.

$$\sigma_B = -R\sigma_A$$

and

$$\tau_C = [(R + 1)\cot 2\beta]\sigma_A$$

where $\sigma_A$ is the stress level of the incident wave at the same instant and $R$ is the reflection coefficient given by

$$R = (\tan\beta \tan^2 2\beta - \tan\alpha)/(\tan\beta \tan^2 2\beta + \tan\alpha).$$

The angles $\alpha$ and $\beta$ are, respectively, the angle of incidence of the longitudinal wave and the angle which the normal to the wave front of the shear wave makes with the normal to the surface. The two angles are related by the equation

$$\sin\alpha/\sin\beta = [2(1 - \nu)/(1 - 2\nu)]^{1/2}$$

here $\nu$ is Poisson's ratio.

During the interaction of the transient stress wave with the surface, the latter is set into motion, the direction and magnitude of its movement being governed by Poisson's ratio and the angle of incidence, $\alpha$, of the original wave. The angle $\bar{\alpha}$ which the trajectory of the point $O$ (Fig. 1) makes with the normal to the free surface can be computed from the equation

$$\sin(\bar{\alpha}/2) = [(1 - 2\nu)/2(1 - \nu)]^{1/2}\sin\alpha.$$

It is important to note that surface reaction in the case of an elastic solid is very different from the reaction in the case of a fluid. At the free surface of a fluid, a shear wave is not generated, the surface moving perpendicularly to itself regardless of the angle of incidence. A point on the free surface of an elastic solid experiences lateral as well as forward motion. It does not, however, except for the case of normal incidence, move perpendicularly to the incident wave front, the deviation becoming greater with increasing angle of incidence. By observing the direction of motion of the surface, it is easily possible to determine whether a material is behaving as an elastic solid or as a fluid under the action of an obliquely transient stress wave.

Some experiments have been carried out in the Mining Research Laboratory of the Colorado School of Mines simply to see whether the motion is actually that predicted. A small Plexiglas pellet was rigidly affixed to a Plexiglas block in the manner illustrated in Fig. 2. The motion of the pellet was then observed when a plastic blasting cap, position as indicated, was detonated. This technique is completely described in [1]. The wave, on reaching the pellet, entered it, and was reflected, causing the pellet to undergo substantially the same movement as a point on the surface would. By making the bond between the pellet and the surface of the block considerably weaker than the Plexiglas itself, the bond was broken by the tension developed by the reflected longitudinal wave, permitting the pellet to detach itself from the block and fly off.

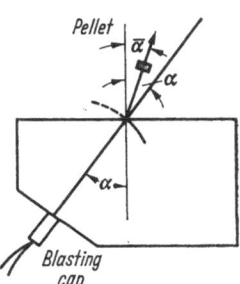

Fig. 2. Experimental arrangement used to project pellet from free surface.

Table 1. *Comparison between Observed and Calculated Values of $\bar{\alpha}$*

| Angle of incidence $\alpha$ (degrees) | Calculated $\bar{\alpha}$ (degrees) | Experimentally[1] observed $\bar{\alpha}$ (degrees) |
|:---:|:---:|:---|
| 15 | 12 | 11, 12, 12, 13 |
| 30 | 24 | 21, 22, 23, 24, 25 |
| 45 | 34 | 28, 29, 30, 31, 32 |
| 60 | 42 | 32, 38, 40, 45 |
| 75 | 46 | 35, 41, 52 |

The results of a number of observations made in this way using Plexiglas blocks ($\nu = 0.40$) are listed in Table 1. It is possible with these data to compare calculated and observed values of the angle of emergence, $\bar{\alpha}$, for different angles of incidence, $\alpha$. The agreement between calculated and observed values of $\bar{\alpha}$ is close, indicating that the wave is surely elastic and that the elastic equations given above are indeed the correct ones to use to describe the motions imparted to a free surface by an obliquely inclined transient stress wave.

Viewed somewhat differently, these experiments provide a novel, practical, and useful experimental method for determining POISSON's ratio dynamically. With $\alpha$ known an $\bar{\alpha}$ observed, the above equation can be used to compute POISSON's ratio [2].

---

[1] Several tests were run for each value of $\alpha$. Each number corresponds to one test.

## 3. Intersection of Two Inclined Waves

Fractures, such as corner fractures and spalls, produced by the high localized tensile stresses accompanying the interference of two stress waves, are frequently composed of an array of short individual fractures having a well defined orientation, especially when the stress is near the fracture threshold. It seems reasonable to assume that each small fracture is oriented perpendicularly to the highest principal tensile stress generated by the superposition of the two interfering waves causing the fracture. It is easy to show that the direction, $\varphi$, of the principal stress established by the two waves will be given by

$$\tan 2\varphi = [(\sigma_1 - \sigma_2)/(\sigma_1 + \sigma_2)] \tan 2\alpha$$

where $2\alpha$ is the angle between the two wave fronts and $\varphi$ is the angle between the principal stress and the bisector of the angle between the two wave fronts. The stresses, $\sigma_1$ and $\sigma_2$, are the respective intensities of the two waves.

In order to observe qualitatively the general validity of this analysis, an electric detonator (point $O$, Fig. 3) was exploded in contact with a small block of Plexiglas, the dimensions of the block being judiciously chosen so that the stress producing the fractures would be near the threshold for spalling. Under these circumstances, the spall was not a continuous fracture but consisted of a number of small disconnected fractures, these being the fractures whose orientations were examined. Spall fractures occur by superposition of the front of the reflected wave, $W_R$, which is a tensile wave, and the tail of the incident wave, $W_T$, which is a compressive wave. The two waves differ in intensity and are, except along the axis, inclined to one another. At the time of their intersection, the principal stress developed by their superposition will not lie along the bisector of the angle between the two wave fronts because of their differing intensities, but will be at an angle $\varphi$ with the bisector.

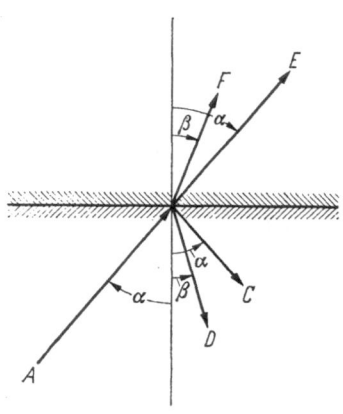

Fig. 3.
Subfractures forming spall surfaces.

For a spherical impinging wave of the type generated by the blasting cap, the angle between the incident and reflected waves increases with increasing off-axis distance. If it is assumed that the plane of each small fracture will be oriented perpendicular to the principal stress developed at the point where the fracture occurs, then the orientation of each fracture should change with off axis distance in the manner indicated in Fig. 3. Experimentally such a change in orientation was observed, the orientations of the fractures following the pattern of the figure, each fracture becoming more inclined to the surface with increasing distance from the axis.

## 4. Inclined Non-Cohesive Interface

Theoretical considerations indicate that a non-cohesive interface, being incapable of supporting shear, would, when struck by an obliquely incident, longitudinal, compressive stress wave, modify the wave, with additional shear and longitudinal components developing in the neighborhood of the interface. As a consequence, momentum will be extracted from the initially incident disturbance, reducing the intensity of the longitudinal wave that propagates beyond the interface.

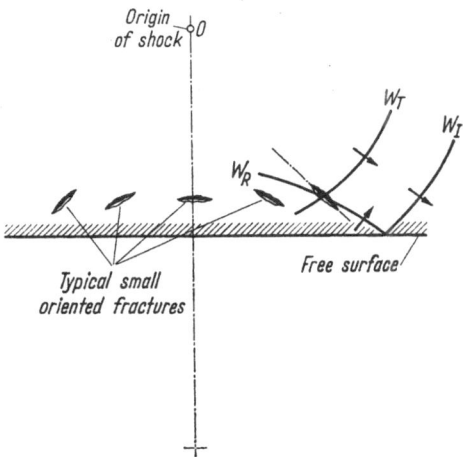

Fig. 4. Generation of transmitted and reflected longitudinal and shear waves at noncohesive interface.

The conditions which must be satisfied at such a boundary are continuity of displacement normal to the interface, continuity of stress normal to the interface, and vanishing of shearing stresses on both of the surfaces forming the interface. It can be shown that these conditions will only be satisfied if the four new waves indicated in Fig. 4 develop. Two of these, a longitudinal wave, $C$, and a transverse wave, $D$, comprise the reflected part of the incident wave; the other two waves, $E$ and $F$, also a longitudinal and a transverse wave, respectively, comprise the transmitted portion. The relative intensities of these waves are functions of the angle of incidence, $\alpha$, of Fig. 4, and POISSON's ratio, $\nu$, and are given specifically by the expressions

$$\frac{C}{A} = \frac{\sin 2\alpha \sin 2\beta}{K^2 \cos^2 2\beta + \sin 2\alpha \sin 2\beta},$$

$$\frac{D}{A} = \frac{K \cos 2\beta \sin 2\alpha}{K^2 \cos^2 2\beta + 2\sin 2\alpha \sin 2\beta},$$

$$\frac{E}{A} = \frac{K^2 \cos^2 2\beta}{K^2 \cos^2 2\beta + 2\sin 2\alpha \sin 2\beta},$$

and

$$\frac{F}{A} = \frac{K^2 \cos^2 2\beta}{K^2 \cos^2 2\beta + \sin 2\alpha \sin 2\beta}$$

where

$$K = [2(1-\nu)/(1-2\nu)]^{1/2} = \sin \alpha / \sin \beta,$$

and where $A$ is the level of stress of the incident wave at any instant, and $C$, $D$, $E$, and $F$ are the corresponding stresses in the newly generated waves.

These relative intensities are plotted for Plexiglas ($\nu = 0.40$) in Fig. 5 as a function of the angle of the angle of incidence $\alpha$. The curves indicate that while the bulk of the energy remains in the transmitted

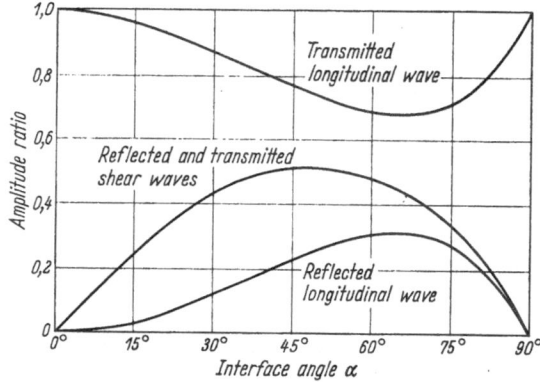

Fig. 5. Theoretical values of amplitude of transmitted and reflected longitudinal and shear waves.

longitudinal wave, two transverse waves and an additional longitudinal wave of appreciable intensity develop, the important exception being for $\alpha = 0$ where the wave is transmitted through the interface with undiminished intensity. The intensities of the two shear waves which are identical for all angles of incidence rise to a maximum equal to about 0.5 A at $\alpha = 45°$ decaying then symmetrically to zero at 0° and 90°. The intensity of the transmitted longitudinal wave decreases with increasing $\alpha$, reaching a minimum of about 0.7 A at 70° after which it starts to increase fairly rapidly. The intensity of the reflected longitudinal wave is a mirror image of the intensity of the transmitted wave, reaching a maximum of about 0.3 A at 70°.

The theoretical predictions regarding variation of transmitted longitudinal stress have been verified experimentally through a series of tests in which electric blasting caps were used to generate transient stress pulses in Plexiglas. The experimental arrangement is illustrated in Fig. 6. A typical specimen used in the tests consisted of two pieces of

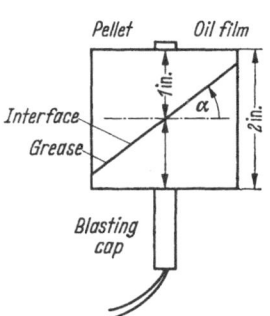

Fig. 6. Experimental arrangement for measuring stress transmitted through noncohesive interface.

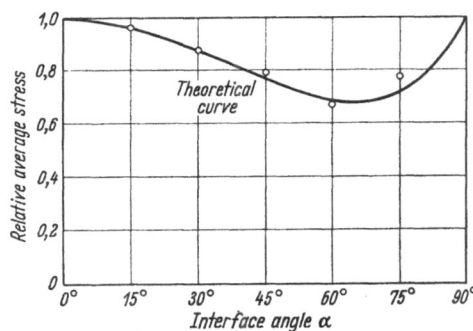

Fig. 7. Relative average stress as a function of interface angle $\alpha$. Solid curve is theoretical; each point represents average of three readings.

Plexiglas having polished surfaces juxtaposed with only a thin film of grease separating them. In each successive test, the Plexiglas block against which the cap was detonated was cut and shaped in such a way that the front of the stress pulse struck the interface at an appropriate preselected angle. The transmitted impulse (or stress) was determined by measuring the velocity with which a small Plexiglas pellet, previously affixed to the second piece, was propelled when the impulse of the transmitted stress wave became trapped in it. Some representative experimental results are plotted in Fig. 7 in which relative stress is plotted against angles of incidence equal, respectively, to 15, 30, 45, 60, and 75°. Each point is the average of three tests. The agreement between theory and experiment is seen to be exceedingly good. The thin film of grease acts as a non-cohesive interface which transmits energy in accordance with the equations derived from elastic theory.

## 5. Fracture Path in Dissimilar Materials

The tensile hoop stress that develops along the front of a divergent transient wave frequently generates fractures. Such fracturing is usually referred to as radial fracturing since the radiating fractures emanating from a localized impact arise in this way. In a homogeneous isotropic material, the fracture is a straight line, lying along a radius and perpendicular to the maximum hoop stress and terminating when the stress reaches a level below which it is too weak to break the material. When the materials of a body is heterogeneous, the fracture no longer follows a straight line, running instead a zig-zag course which changes direction as it moves from one constituent material to the next.

For the simple case of two juxtaposed plates of dissimilar materials, it has been shown experimentally that the change in direction of the fracture is associated with a change in curvature of the wave front as it passes from one material to the other. In the experiments, small slabs of Plexiglas and rock were bonded together with a thin layer of Duco cement and clamped at right angles to their bonded interface to insure efficient transmission of the stress wave across the interface. The impact of a 0.22-caliber lead bullet was used to generate the radial fractures. In the case of two similar materials, Plexiglas-Plexiglas, no change in fracture orientation was observed across the interface. A change in fracture orientation did occur, however, as predicted, for an array of Plexiglas and limestone and for an array of Plexiglas and plaster of Paris where the longitudinal wave velocities differed, being 9070, 11000, and 11200 ft/sec, respectively, for Plexiglas, limestone, and plaster of Paris.

The directional change to be anticipated for the case of a spherical wave moving from a low velocity material to a high velocity material (Fig. 8), is easily predicted by application of SNELL's law, being given by

$$\sin \varphi_2 = (v_2/v_1) \sin \varphi_1$$

where $\varphi_1$ is the angle between the incident fracture and the normal to the interface; $\varphi_2$ is the angle between the refracted fracture and the same normal; and $v_1$ and $v_2$ are the respective longitudinal velocities of the materials. The experimental and the theoretically deduced values of the refraction angle $\varphi_2$ for the two arrays of different materials are compared in Table 2 for several values of $\varphi_1$ and are seen to agree closely, indicating that SNELL's law can be realistically applied to this situation.

The two component, multilayered structure, shown in Fig. 9, is a somewhat more complex array of dissimilar materials. In such an array, the fracture will change direction each time it passes from one layer to the next, being bent away from the normal in going from the low velocity material to the high velocity material; and back again, toward the normal, when it goes from the high velocity material to the low

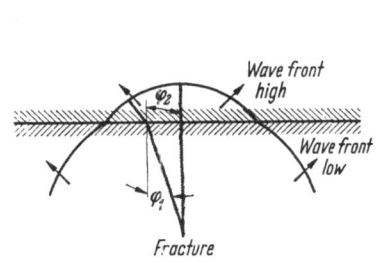

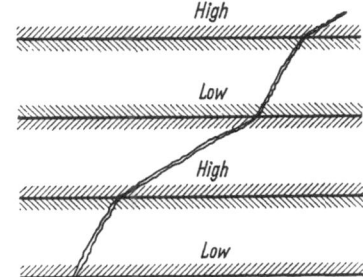

Fig. 8. Diagram illustrating change in curvature of spherical wave front in passing from low velocity material to high velocity material.

Fig. 9. Path of fracture which moves from low velocity material through high velocity material to low velocity material.

velocity material. The fracture travels a zig-zag course, changing direction at each interface, its path being easily determined theoretically by applying SNELL's law.

Table 2. *Comparison of Observed and Predicted Fracture Angles (degrees)*

| Material Combination | Measured $\varphi_1$ | Predicted $\varphi_2$ | Measured $\varphi_2$ |
|---|---|---|---|
| Plexiglas-limestone | 50 | 66 | 59 |
|  | 40 | 50 | 56 |
|  | 24 | 29 | 32 |
|  | 22 | 27 | 28 |
|  | 3 | 4 | 4 |
| Plexiglas-plaster of Paris | 40 | 52 | 51 |
|  | 22 | 27 | 32 |
|  | 19 | 24 | 28 |
|  | 17 | 21 | 23 |

## 6. Corner Fracturing

When a compressive transient stress wave becomes trapped in a corner, the two release or tension waves that develop and eat in simultaneously from the two inclined free surfaces soon meet, generating a fracture described as a corner fracture. The inclination of this fracture to the free surface is easily and accurately predictable. There is an unusual feature of fractures of this type occurring in plastics, metals, and rocks which has not been previously reported: the fact that the true corner fracture does not extend clear to the tip of the corner. It generally stops short of it. It may then bend off toward one surface or the other, running up to that surface. Occasionally it will split and run to both surfaces.

The four most common situations are illustrated in Fig. 10. Which of the situations develops seems to depend upon the inclination of the wave front with respect to the surfaces, upon the intensity of the wave, and to some extent upon its duration. It has been demonstrated experimentally, using blasting caps to generate transient waves in small Plexiglas blocks, that any one fracture system is quite reproducible. Further, when the fracture breaks through to a surface, it does so along the shortest path. When the shortest path is indeterminate, as in the case of a fracture inclined at 45° to both surfaces (Fig. 10a), the breaks

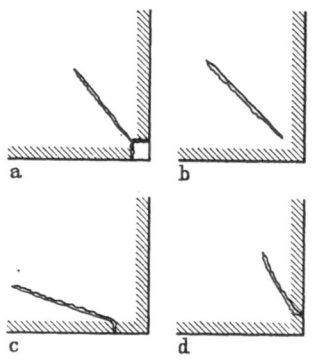

Fig. 10 a — d. Paths of representative corner fractures.

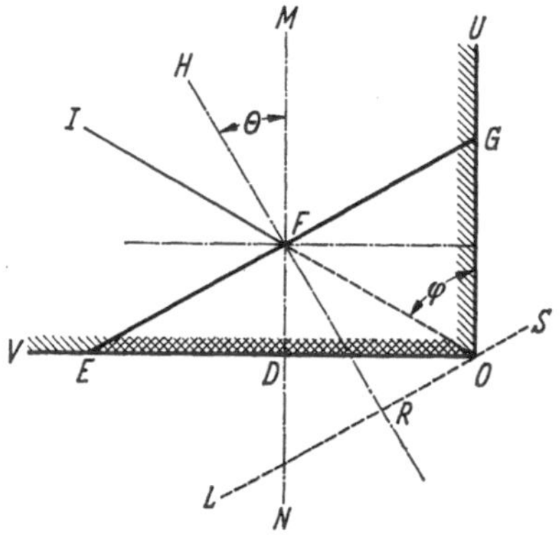

Fig. 11. Geometry of corner fracture.

go to both surfaces, or if the wave is weak, to neither (Fig. 10b). The deviatory part of the fracture is less long for intense waves than for weak ones, the principal corner fracture extending close to the tip of the corner.

In order to explain in detail the formation of the corner fracture, it is necessary to introduce into the analysis the notion of the finite duration of the wave, it being this aspect which accounts for the inability of the corner fracture to run right up to the tip of the corner. The case of a plane wave entering a 90° corner is illustrated in Fig. 11. It is assumed for simplicity that the wave is sharp fronted, flat topped, and of short duration. The drawing illustrates the situation obtaining at the exact instant that the wave fronts are combining to form the corner fracture. The line $SL$ is the position that the wave front would have reached had it not been reflected and the line $GE$ is the position of the rear of the wave. The distance, $FR$, corresponds to the duration of the disturbance. The wave front makes an angle $\theta$ with the lower surface, $\overset{,}{V}O$, of the corner. The fracture will be expected to develop along the line $OI$, where the fronts of the two waves meet, making an angle $\varphi$ with the side surface of the corner, where $\varphi = 90 - \theta$. From the point $F$ inward toward the point $I$ along the fracture line $OI$, the fronts of the waves are meeting and interfering in a region which is unstressed because all of the incident wave has passed beyond it, the rear of the wave at the time of the encounter lying along the line $EG$. From the point $O$ to the point $F$ along the same line, the situation is quite different. Along this portion of the fracture line, the material is under compression immediately before the two waves meet. Thus the tensile stress, tending to generate a fracture, is much higher from the point $F$ toward the point $I$ than it is from $O$ to $F$. It is entirely possible that the tensile stress along $OF$ will be too low to generate a fracture whereas beyond $F$, where the stress is higher, a fracture will develop. In fact, the tensile stress is very likely to be too low to generate a fracture since the two interfering waves are much less intense than the incident wave as a consequence of the energy partitioning occurring during the oblique reflections of the waves. Once a fracture has taken place along the line $FI$, the fracture may then extend itself to either of the two surfaces, to both, or terminate at $F$, as mentioned previously (Fig. 10).

Even though the case described here is highly idealized, observations on Plexiglas, rock, and metal specimens have shown that in general the fracture runs to the bottom surface of the corner when the distance, $RD$, from the tip of the corner fracture to the bottom surface is less than the distance $FC$ to the side surface; it runs to the side surface when $FD$ is greater than $FC$; it goes to both surfaces when $FD$ is approximately equal $FC$ and the wave is strong; and it goes to neither surface when the wave is weak. All of these observations are in good qualitative agreement with predictions.

Most real situations are considerably more complex than the simple one analyzed here. The usual transient wave is divergent rather than plane and decays in intensity quickly after its initial rapid rise. However, the same general considerations must apply and the final fracture pattern should be substantially the same.

## 7. Diffraction Effects

Transient waves moving through bodies containing openings can be strongly influenced by the presence of these openings, the wave undergoing severe transformation in character as a consequence of reflections from the surfaces of these openings. Oblique reflections of a longitudinal wave from free surfaces will soon transform the wave into transverse and longitudinal waves, these in turn giving rise to a complex pattern of highly localized stress inhomogeneities which may or may not sunder the body. Other and exceedingly significant and interesting effects not heretofore described are the transformations that accompany the progress of an abruptly terminated wave front into an unstressed material, effects analogous in certain ways to optical diffraction effects.

Such a terminated wave front, complete with its attendant transformed waves is shown in Fig. 12. Here a plane, sharp-fronted wave having its wave front perpendicular to the line $DLOK$ has just passed by an open corner. When it reaches the corner, a portion (not shown) of the wave is reflected, the remainder moving on forward unobstructed. This cutting of the wave leaves a free end, $O$, and provides a point through which some of the energy of the wave can escape into the unstressed material to its side. As a consequence, there develops around $O$ the pattern of wave and wavelets evident in Fig. 12. The two principal new waves are shear waves: one, $OB$, entering the unstressed region to the left; and the other, $OA$, moving into the region behind the wave front. Both are inclined to the normal to the wave front, $OD$, at the angle whose sine is equal to the ratio of the transverse wave velocity to the dilatational wave velocity. In addition, there develop two other regions, included, respectively, between the lines $OF$ and $OB$ and between the lines $OE$ and $OA$. The curve $EOF$ is the arc of a circle having its center at the point $G$, the tip of the corner. Neither $OF$ nor $OE$ is a coherent wave front since the source, $O$, feeding energy to the regions $EOC$ and $BOF$, is moving with the wave front $OC$, making it impossible to construct an envelope to the individual wavelets. The curves, $OE$ and $OF$, simply represent the limits to which the influence of the terminal point $O$ can be felt. The region, $FOB$, lying to the left of $O$, contains compressional strain energy introduced into it by the presence of the transient wave. The region $EOA$, to the right of $O$, is a region from which strain energy has been extracted. In these two regions it is appropriate to describe the transient state in terms of energy or momentum density rather than wave fronts, a crude measure of the density being the spatial concentration of wavelets. In addition, during passage of the original wave, the intensity of its front, $OC$, degrades in the neighborhood of

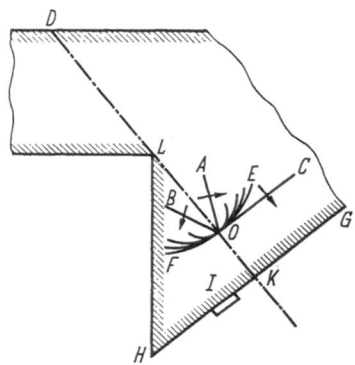

Fig. 12. Wave patterns associated with diffraction of plane wave which passes by re-entering corner.

the point, $O$, resulting in the development of a front along which the intensity is no longer uniform.

A complete and rigorous analytical description of the process is exceedingly difficult. However, the general features of the situation can be assessed qualitatively using the simple ideas put forward above. Furthermore, it has been possible to obtain experimental data bearing on the matter. In this instance, a blasting cap placed in contact with a small specially cut block of Plexiglas was detonated, the cap being located at the point $D$ of Fig. 12. The motion was observed of a small Plexiglas pellet placed at a preselected point, $I$, on the free surface, $GH$. By making the distance from the blasting cap to the surface large ($DK$ in the figure, equaling 2 inches), the wave front ($OC$) was sufficiently flat to be treated as if it were plane.

In these experiments the thickness of the pellet and its position along the line $GH$ were both varied, permitting momentum from the several regions of the wave complex to be trapped and evaluated. The distance, $DL$, from cap to corner was kept constant at 1 inch.

The results are plotted in Fig. 13 in which momentum trapped is plotted against position, the distance $IK$, on the surface $HG$ for each of four different pellet thicknesses, 1/32, 1/16, 3/32, and 1/8 inch, the

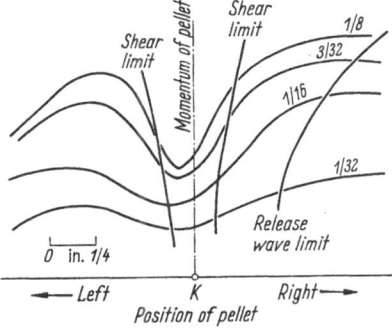

Fig. 13. Momentum of pellet as function of position on surface $HO$ of Fig. 12 (Number on each curve is thickness of pellet in inches).

diameter of each pellet being 1/4 inch. Three additional lines have been drawn on the graph. These lines delineate the areas of influence of the two shear waves, $OA$ and $OB$, and the release wave, $OE$, as far as any action they might have on the pellets is concerned. For example, a pellet placed to the left of the release wave limit will leave the surface before the release wave arrives. Similarly a pellet placed to the left of the left hand shear limit will leave the surface before the shear wave on that side arrives and hence before it could influence the momentum trapped by the pellet. Each pellet, having a finite area, averages the stress over that area, making the resolution not especially high. However, the true momentum versus distance curves must have the approximate shapes indicated in the Fig. 13.

Each of the curves exhibits a dip lying roughly between the two shear wave limits with the dip, as might be anticipated, being more pronounced for the thicker pellets than for the thin ones. The presence of the dip is an indication that energy has been transformed from the initial longitudinal wave to the shear wave, complicating momentum entrapment in the pellet.

Perhaps the most surprising feature of the curves is the degree to which stress is sustained to the left of the edge of the initial wave. It is, for example, at a distance of one inch from the border line, still 70 to 80 percent of the original stress in the unaffected portion of the wave front. It is also seen that the decrease with distance is less rapid for the thinner pellets than for the thicker pellets, an aspect which is likely attributable to the decrease in energy density with movement to the left into the region $BOF$ (Fig. 13).

## 8. Conclusion

This paper has described several wave and boundary interactions which play important roles in the fracturing of bodies subjected to impulsive loads and which form important links in the chain of events involved in the metamorphosis of a transient stress disturbance into a more complex wave form in a homogeneous material, as well as the progress of a wave through a heterogeneous material. The treatment is direct and simple and based on the elementary and fundamental notion that any change in the shape of a body can be resolved into two components, a distortion and a dilatation, the two being transmitted at distinctly different rates. A number of experimental results have been presented which confirm the qualitative correctness and usefulness of the treatment. While much is lacking in the experiments with respect to rigor and exactness, nonetheless the solutions provide extremely useful tools for the development of an understanding of the phenomenology of the wave and boundary interaction processes and in this way contribute to a better appreciation of attenuation mechanics.

In conclusion, I wish to acknowledge the contributions of ULRICH LORBER, CHARLES HAAS, JAIME EISEN, JOSE DEL SOLAR, WILLIAM McCLAIN, ROBERT AVEZOU, and JEAN-PIERRE FORTIN, all students at the Colorado School of Mines, who performed the experiments described here.

### References

[1] RINEHART, J. S., and W. C. McCLAIN: J. Appl. Phys. **31**, 1809 (1960).
[2] RINEHART, J. S.: Nature **194**, 369 (1962).

# Reflection of a Plastic Wave at an Obstacle

By N. V. Zvolinsky and O. V. Rykov

Institute of Geophysics, Moscow, U.S.S.R.

## 1. Introduction

The propagation of a plastic wave and its interaction with an obstacle have been investigated by several authors. References [2] and [3] are concerned with the propagation, and references [4] and [5] with the interaction problem. In these papers it was assumed that the density of the material remained constant during unloading and linear or piece-wise linear approximations were used to the laws of loading.

The present paper shows that the interaction of a plane wave with an obstacle is readily treated when the stress during loading is supposed to be proportional to a power of the strain. A power law of this kind is in good agreement with experiments over a wide range of stress.

The results of the present investigation are useful in the treatment of wave phenomena in soft soil.

## 2. Preliminary Remarks

For the study of plane waves, the properties of the material are adequately specified by two pieces of information, namely

(a) the law of cubical compression

$$\sigma = \sigma(\theta) \tag{2.1}$$

and

(b) the yield condition

$$|\sigma_x - \sigma_y| = -m\sigma + m'. \tag{2.2}$$

Here, $\sigma_x$ and $\sigma_y$ are the extreme principal stresses, $\sigma$ is the mean normal stress, $\theta$ the cubical compression, and $m$ and $m'$ are positive constants. Note that only phenomena involving plastic deformation are discussed in this paper.

The law of cubical compression is assumed to be different during loading (line $AB$ in Fig. 1) and unloading (line $BD$): the mean normal stress $\sigma(\theta)$ is given by $\sigma(\theta) = f_1(\theta)$ for loading $(d\sigma/dt > 0)$, while it is indeterminate for unloading $(d\sigma/dt < 0)$.

If we restrict ourselves to the discussion of a single plane plastic wave, we only need the uniaxial equivalent of (2.1) and (2.2), which we assume to be

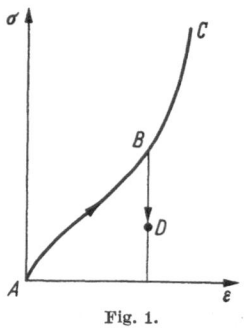

Fig. 1.

$$\sigma_x = \sigma^0 |\varepsilon_x|^n, \qquad \sigma^0 < 0, \qquad n > 1, \qquad (2.3)$$

where $\sigma^0$ and $n$ are constants.

In agreement with experimental facts, we assume that on reloading (after unloading of the type represented by the line $BD$ in Fig. 1) the increase of $\sigma_x$ is not accompanied by a change in density until $\sigma_x$ has reached the value (represented by the ordinate of $B$) at the beginning of the last unloading process. The behavior of the material under further loading is represented by the continuation $BC$ of the original loading diagram.

For simplicity, the index $x$ will be omitted in the following, and $\sigma$ and $\varepsilon$ will be used instead of $\sigma_x$ and $\varepsilon_x$. As we only intend to discuss plastic phenomena, we may use the law (2.3) which disregards an initial elastic part of the loading diagram.

## 3. Propagation of a Plane Plastic Wave Generated by Exterior Action

Let the uniform stress at a certain material plane be given as a function of time. Waves will propagate from this plane in both normal directions. We consider one of these waves and let the $x$ axis be parallel to the line of propagation. The phenomenon is conveniently described by using the Lagrangian coordinate $h$ and the time $t$ as independent variables and writing the Eulerian coordinate $x(h, t)$ in the form $x(h, t) = h + u(h, t)$, where $u(h, t)$ is the displacement of the particle $h$.

We take $u(h, 0) = 0$, that is, we suppose that the medium is initially unperturbed. At the material plane $h_0$, the stress $\sigma_0$ is prescribed as a function of time; its absolute value is assumed to increase instantaneously at $t = 0$ and then to decrease. These assumptions concerning the character of the excitation are essential for the following treatment. It is natural to expect that the instantaneous loading produces a shock front. Let $h^*$ be the Lagrangian coordinate of this front (Fig. 2). Particles with Lagrangian coordinates $h$ satisfying $h_* < h < h_0$ are in a region of unloading, where the following equations apply:

(a) equation of motion

$$\frac{\partial \sigma}{\partial h} - \varrho_0 \frac{\partial v}{\partial t} = 0; \qquad (3.1)$$

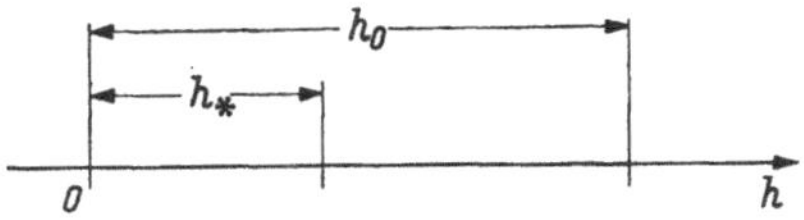

(b) equation of continuity

$$\frac{\partial x}{\partial h} = \frac{\varrho_0}{\varrho(h)} = 1 + \varepsilon(h). \qquad (3.2)$$

Fig. 2.

Here, $v(h, t)$ is the velocity of the particle $h$, and $\varrho_0$ and $\varrho(h)$ are the values of the density at this particle before and after the passage of the shock front:

It follows from (3.1) and (3.2) that

$$x(h, t) = \int_{h_0}^{h} \frac{\varrho_0 d\eta}{\varrho(\eta)} + x_0(t), \qquad (3.3)$$

$$v(h, t) = \frac{\partial x}{\partial t} = x_0'(t), \qquad (3.4)$$

$$\sigma(h, t) = \varrho_0 x_0''(t)(h - h_0) + c(t). \qquad (3.5)$$

While the stress $\sigma_0$ at the section $h = h_0$ is in principle given as a function of time, it may also be regarded as a function of $h_*$ which is a monotonically increasing function of time. Since this way of formulating the problem facilitates its solution, we shall assume that the function

$$\sigma_0 = \sigma_0(h_*)$$

is given. Once the problem has been solved, the function $h_*(t)$ is known and the stress $\sigma_0$ can be determined as a function of time.

The conditions at the section $h = h_0$ and at the shock front together with the law of loading (2.3) furnish the four equations

$$\sigma(h_0, t) = \sigma_0(h_*), \qquad (3.6)$$

$$\sigma(h_*, t) = \sigma^0 |\varepsilon_*|^n, \qquad (3.7)$$

$$\sigma_0(h_*, t) = \varrho_0 \varepsilon_* h_*'^2, \qquad (3.8)$$

$$v(h_*, t) = -\varepsilon_* h_*', \qquad (3.9)$$

in which the subscript * indicates values at the shock front.

Eliminating $x_0$, $c$, and $\varrho$ from Eq. (3.3), (3.4), and (3.5) and setting

$$Z = (-\varrho_0 h_*'^2/\sigma^0)^{n/(n-1)},$$

we obtain the following differential equation for $h_*$:

$$\frac{dZ}{dh_*} + \frac{2n}{n+1} \frac{Z}{h_* - h_0} + \frac{2n}{n+1} \frac{\sigma_0(h_*)}{\sigma^0(h_* - h_0)} = 0.$$

Being linear, this equation is readily integrated to yield

$$Z = \frac{c}{(h_0 - h_*)^{2n/(n+1)}} + \frac{2n}{n+1} \frac{1}{(h_0 - h_*)^{2n/(n+1)}} \int_{h_*}^{h_0} \frac{\sigma_0(\xi)}{\sigma^0} (h_0 - \xi)^{(n-1)/(n+1)} d\xi.$$

For $h_* \to h_0$, the first term on the right-hand side of this equation increases indefinitely whereas the second term remains finite. The particle velocity is finite even at $t = 0$, so that the constant of integration must be given the value 0. With

$$\Phi(h) = \left[ \int_h^{h_0} \frac{\sigma_0(\xi)}{\sigma^0} (h_0 - \xi)^{(n-1)/n+1} d\xi \right]^{(n-1)/2n},$$

we therefore have

$$- h'_* = \left[ \frac{-\sigma^0}{\varrho_0} \right]^{1/2} \left[ \frac{2n}{n+1} \right]^{(n-1)/2n} \Phi(h_*)(h_0 - h_*)^{(1-n)/(1+n)}. \quad (3.10)$$

The time at which the shock front reaches a given position $h_*$ is

$$t = \left[ \frac{-\varrho_0}{\varrho_0} \right]^{1/2} \left[ \frac{n+1}{2n} \right]^{(n-1)/2n} \int_{h_*}^{h_0} (h_0 - \xi)^{(n-1)/(n+1)} \Phi^{-1}(\xi) d\xi, \quad (3.11)$$

and the strain immediately behind the shock front is given by

$$- \varepsilon_* = \left[ \frac{2n}{n+1} \right]^{1/n} \Phi(h_*)]^{2/(n-1)} (h_0 - h_*)^{-2/(n+1)}. \quad (3.12)$$

The Eulerian coordinate of the particle $h$ at the generic instant $t$ is

$$x(h, t) = h + \int_{h^*}^{h} \varepsilon(\eta) d\eta. \quad (3.13)$$

The velocity $v(h, t)$ of this particle and the stress $\sigma(h, t)$ are readily determined by means of (3.4), (3.5), (3.12), and (3.13). For the stress, one finds

$$\sigma(h, t) = \frac{1}{h_* - h_0} [\sigma_0(h_*)(h_* - h) - \varrho_0 \varepsilon_* h'^2_*(h - h_0)]. \quad (3.14)$$

The formulas (3.10) through (3.14) completely describe the wave generated by the stress applied at the section $h_0$; in particular, (3.11) and (3.14) enable us to express the stress $\sigma_0(h_*)$ as a function of time.

We stress the following consequences of the results obtained above.

1. Even if the external stress were only applied to the section $h_0$ during a finite interval of time, the stress wave would propagate indefinitely with a monotonically decreasing velocity that tends asymptotically to zero. (The treatment presented above is only justified for sufficiently high stress intensity at the shock front.)

2. Between the section $h_0$ and the shock front, the stress $\sigma(h, t)$ is a linear function of $h$.

3. In this interval, the particle velocity is independent of $h$.

## 4. Reflection of Wave at Immobile Obstacle

The wave studied in the preceding section will be called the *incident wave* and quantities referring to it will be given the subscript 1. We next have the discuss the form in which this wave is reflected at an obstacle. Quantities referring to the *reflected wave* will be given the subscript 2.

To solve the problem of nonlinear reflection, we introduce certain assumptions regarding the reflected wave. Specifically, we assume that the stress is increased by the reflection and that the reflected wave also has a shock front (Fig. 3). The region between the obstacle and the front of the reflected wave will be called Region II. We first discuss the conditions in this region and postpone the discussion of the stress intensity ahead of the front of the reflected wave. According to our assumptions, the particle at the reflected front experiences a stress jump, the intensity of which depends on the stress-strain diagram. Accordingly, the following equations apply in Region II:

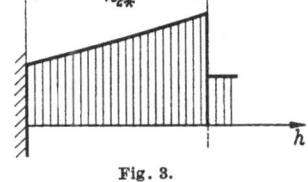

Fig. 3.

$$\frac{\partial \sigma_2}{\partial h} - \varrho_0 \frac{\partial v_2}{\partial t} = 0, \qquad (4.1)$$

$$\frac{\partial x_2}{\partial h} = \frac{\varrho_0}{\varrho_2(h)} = 1 + \varepsilon_2(h). \qquad (4.2)$$

Taking the coordinate origin at the immobile obstacle, we have the following conditions at the obstacle:

$$x_2(0, t) = 0, \qquad v_2(0, t) = 0. \qquad (4.3)$$

It follows from (4.2) that

$$x_2(h, t) = \int_0^h \frac{\varrho_0 d\eta}{\varrho_2(\eta)} + x_{20}(t), \qquad (4.4)$$

$$v_2(h, t) = x'_{20}(t), \qquad (4.5)$$

and (4.3) shows that

$$x_2(t) = 0, \qquad v_2(h, t) = 0, \tag{4.6}$$

$$\sigma_2(h, t) = \sigma_2(t). \tag{4.7}$$

Let us now investigate the conditions ahead of the front of the re-flected wave. When the incident wave reaches the obstacle, the reflected wave begins to advance into the region through which the incident wave has passed. The reflected wave therefore propagates in an already perturbed region. The question arises whether the reflected wave sends some perturbation ahead of itself or simply propagates in the field left behind by the incident wave. The answer to this question depends on the stress-strain law for unloading. According to our assumption, there are no density changes during unloading and subsequent reloading as long as the highest stress $\sigma_B$ reached at the front of the incident wave is not surpassed (point $B$ in Fig. 2). In this regime, perturbations travel with infinite speed, and the reflected wave must therefore send a signal ahead. In fact, the front of the reflected wave is preceded by a wave that travels at infinite (or, practically, at very great) speed. In principle, this wave may also be regarded as part of the reflected wave. In treating this wave we further assume that the stress $\sigma'$ ahead of the shock front equals, for each particle, the highest stress $\sigma_B$ to which this particle has been subjected[1]. It follows from this assumption that ahead of the front of the reflected wave a spatially constant though time-dependent stress superimposes itself on the stress in the incident wave. The intensity of this additional stress is formed from the condition that the stress $\sigma_B$ is attained ahead of the reflected front. The density and particle velocity ahead of the front of the reflected wave are those of the incident wave.

The preceding remarks enable us to determine the parameters of the reflected wave. Indicating quantities in the region ahead of the reflected front by a prime, we express conservation of mass and momentum by the relations

$$(D - v')\varrho' = (D - v_2)\varrho_2,$$

$$\sigma' - \sigma_2 = (D - v_2)\varrho_2(v_2 - v_1),$$

in which $D$ denotes the velocity of propagation of the reflected front. The stress-strain law furnishes the additional relation

$$\sigma_2 = \sigma^0(1 - \varrho_2/\varrho_{2*})^n, \tag{4.8}$$

---

[1] This assumption represents an independent hypothesis in so far as all pre-viously made assumptions could be accepted even if another choice for $\sigma'$ were made. In this case, however, the shock front would certainly be unstable.

and we have $v_2 = 0$,

$$D = \frac{dx_2(h_{2*})}{dt} = \frac{\varrho_0}{\varrho_2(h_{2*})}\, h'_{2*}\,.$$

Substituting these values into the relations established for the reflected front and setting

$$\varrho' = \varrho_1(h)\,, \qquad v' = v_1(t)\,, \qquad \sigma' = \sigma_B(h)$$

in accordance with our assumptions, we find

$$v_1 = \left[\frac{\varrho_0}{\varrho_2} - \frac{\varrho_0}{\varrho_1}\right]_{h=h_{2*}} h'_{2*}\,,$$

$$\sigma_B - \sigma_2 = -\varrho_0 \left[\frac{\varrho_0}{\varrho_1} - \frac{\varrho_0}{\varrho_2}\right]_{h=h_{2*}} h'^2_{2*}\,.$$

Eliminating the functions $\varrho_2(h)$ and $\sigma_2(t)$ from these equations and (3.8), and noting that

$$v_1 = -\varepsilon_1(h_{1*})h'_{1*} = -\frac{\sigma_1(h_{1*})}{\varrho_0 h_{1*}}\,,$$

we obtain

$$\left[1 - \frac{\varepsilon_1(h_{1*})}{\varepsilon_1(h_{2*})}\, \frac{h'_{1*}}{h'_{2*}}\right]^n = 1 - \left[\frac{\varepsilon_1(h_{1*})}{\varepsilon_1(h_{2*})}\right]^n \frac{h'_{2*}}{h'_{1*}}\,. \qquad (4.9)$$

Integration of this nonlinear differential equation under the initial condition $h_{2*}(0) = 0$ furnishes the function $h_{2*}(t)$. The density and the stress are then obtained from

$$\varrho_2(h) = \varrho_1(h) \left[1 + \frac{\varrho_1(h)}{\varrho_0}\, \frac{v_1(t)}{h'_{2*}(t)}\right]^{-1}\,,$$

$$t = t(h) \equiv t(h_{2*})\,; \qquad (4.10)$$

$$\sigma_2(t) = \sigma_B(h) + \varrho_0 v_1(t) h'_{2*}(t)\,,$$

$$h = h(t) \equiv h_{2*}(t)\,. \qquad (4.11)$$

To facilitate the integration of (4.9), we may use the following approximation. Physical considerations show that

$$\frac{\varepsilon_1(h_{1*})}{\varepsilon_1(h_{2*})} < 1\,, \qquad \left|\frac{h'_{1*}}{h'_{2*}}\right| < 1\,.$$

In view of these inequalities, we may develop the left-hand side of (4.9) into a power series. If we are justified in retaining only the two first terms of these series, we have

$$\frac{h'_{2*}}{h'_{1*}} = -n^{1/2} \left[\frac{\varepsilon_1(h_{2*})}{\varepsilon_1(h_{1*})}\right]^{(n-1)/2}\,.$$

Integration of this equation furnishes the following implicit relation between $h_{2*}$ and $h_{1*}$:

$$\int_0^{h_{2*}} \frac{dh}{|\varepsilon_1(h)|^{(n-1)/2}} + n^{1/2} \int_0^{h_{1*}} \frac{dh}{|\varepsilon_1(h)|^{(n-1)/2}} = 0. \qquad (4.12)$$

The quantities in (4.9) are readily interpreted in terms of a fictitious incident wave that propagates beyond the obstacle. In Fig. 4 the stress in the incident wave (and in its continuation in the absence of the

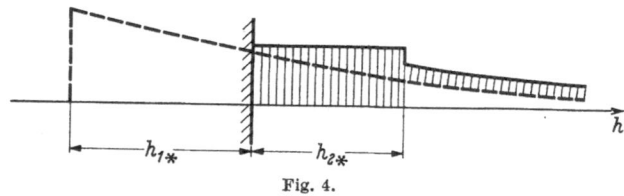

Fig. 4.

obstacle) is shown by the dashed line. The front of this wave determines the values $h_{1*}$ and $\varepsilon_1(h_{1*})$, while the front of the reflected wave is at a particle with the coordinate $h_{2*}$ and the strain $\varepsilon_1(h_{2*})$.

The results obtained above yield the following conclusions.

(a) At each instant, the velocity of propagation of the reflected wave exceeds that of the incident wave.

(b) In accordance with our assumptions, we have

$$\varrho_2 > \varrho_1, \qquad |\sigma_2| > |\sigma_B|.$$

(c) At the instant when the incident wave reaches the obstacle $(h_{1*} = h_{2*} = 0)$, the ratio $q = h'_{2*}/h'_{1*}$ is given by

$$\left(1 + \frac{1}{q}\right)^n = 1 + q.$$

Values of $q$ calculated for various values of $n$ are given in Table 1.

Table 1

| $n$ | 1.0 | 1.5 | 2.0 | 2.5 | 3.0 |
|---|---|---|---|---|---|
| $q$ | 1.0 | 1.32 | 1.62 | 1.84 | 2.15 |

Since $\sigma_{1*} = \sigma_B$ when the incident wave reaches the obstacle, we find from (4.11) that

$$(\sigma_2/\sigma_B)_{t=t_0} = 1 + q. \qquad (4.13)$$

The quotient on the left of (4.13) may be called the coefficient of reflection. In the present example, it depends only on a single constant of the material, namely the exponent $n$; it is independent of the intensity of the incident wave. Since $q > 1$, the coefficient of reflection is greater than 2.

## 5. Special Types of Loading

(a) Let us assume that the applied stress $\sigma_0(h_{1*})$ instantaneously rises to the value $\sigma_0$ and thereafter remains constant. The quantities characterizing the incident and reflected waves then vary in a similar manner with time. For the incident wave, we have

$$h'_{1*} = -(-\sigma^0/\varrho_0)^{1/2}(\sigma_0/\sigma^0)^{(n-1)/2n},$$

$$h_0 - h_{1*} = (\sigma_0/\sigma^0)^{(n-1)/2n}(-\varrho_0/\sigma^0)^{1/2}t,$$

$$\varepsilon_{1*} = -(\sigma_0/\sigma^0)^{1/n},$$

$$\varrho_1/\varrho_0 = [1 - (\sigma_0/\sigma^0)^{1/n}]^{-1},$$

$$\sigma_1(h, t) = \sigma_0.$$

For the reflected wave, we have

$$h'_{2*} = -q\,h'_{1*}, \qquad h_{2*} = -q\,h_{1*},$$

$$\varrho_2(h) = \varrho_0\left[1 - \frac{q+1}{q}\left(\frac{\sigma_0}{\sigma^0}\right)^{1/n}\right]^{-1},$$

$$\sigma_2(t) = (1 + q)\,\sigma_0,$$

where $q$ is a positive root of the equation

$$\left(1 + \frac{1}{q}\right)^n = 1 + q.$$

(b) Consider the applied stress

$$\sigma_0(h_{1*}) = \begin{cases} 0 & \text{for} \quad h_{1*} > h, \\ \sigma_0 & \text{for} \quad h_0 - l < h_{1*} < h_0, \\ 0 & \text{for} \quad h_{1*} < h_0 - l. \end{cases}$$

The incident wave is represented by the functions

$$h'_{1*} = (-\sigma^0/\varrho_0)^{1/2}(\sigma_0/\sigma^0)^{(n-1)/2n}[l/(h_0 - h_{1*})]^{(n-1)/(n+1)},$$

$$t = \left[-\frac{\varrho_0}{\sigma_0}\right]^{1/2}\left[\frac{\sigma^0}{\sigma^0}\right]^{(n-1)/2n} l\left\{1 + \frac{n+1}{2n}\left[\left(\frac{h_0 - h_{1*}}{l}\right)^{2n/(n+1)} - 1\right]\right\},$$

$$\varepsilon_1(h) = -(\sigma_0/\sigma^0)^{1/n}[l/(h_0 - h)]^{2/(n+1)},$$

$$\frac{\varrho_1(h)}{\varrho_0} = \left\{1 - \left[\frac{\sigma_0}{\sigma^0}\right]^{1/n}\left[\frac{l}{h_0 - h}\right]^{2/(n+1)}\right\}^{-1},$$

$$\sigma_1(h, t) = \sigma_0\frac{h - h_0}{h_{1*} - h_0}\left[\frac{l}{h_0 - h_{1*}}\right]^{2n/(n+1)}.$$

These five formulas hold for $h_{1*} < h_0 - l$. For $h_0 - l < h_{1*} < h_0$, the stress remains constant, and the incident wave is identical with that of the preceding example.

For the reflected wave, we have

$$h_{2*}' = -n^{1/2} \left[ \frac{h_0 - h_{1*}}{h_0 - h_{2*}} \right]^{(n-1)/(n+1)} h_{1*}'$$

$$= -n^{1/2} \left[ \frac{-\sigma^0}{\varrho_0} \right]^{1/2} \left[ \frac{\sigma_0}{\sigma^0} \right]^{(n-1)/2n} \left[ \frac{l}{h_0 - h_{2*}} \right]^{(n-1)/(n+1)},$$

$$\frac{\varrho_2(h)}{\varrho_0} = \frac{\varrho_1}{\varrho_0 + \varrho_1(v_1/h_{2*}')},$$

$$\sigma_2 - \sigma_B = \frac{\sigma_0 n^{1/2} l^{2n/(n+1)}}{(h_0 - h_{1*})(h_0 - h_{2*})^{(n-1)/(n+1)}}.$$

## 6. Interaction of Wave and Obstacle

Let us consider the reflection of a wave at an obstacle that is imbedded in the medium. Since we only consider plane waves, we must assume that the obstacle is bounded by two parallel planes. Without loss in generality, we assume the obstacle to be an indefinitely thin plate possessing a certain density per unit of surface area.

When a wave impinges normally on this plate, the latter begins to move and transmits a wave to the medium on the other side (transmitted wave). Let us restrict the discussion to a comparatively short interval of time for which one may make the following assumptions.

1. The stress in the reflected wave is greater than that in the incident wave (shock front in the reflected wave).

2. The plate moves only in one direction, the direction of propagation of the incident wave.

3. By its motion, the plate generates a compression wave, in which the stress intensity at the plate increases monotonically.

The solution of the problem will show that these assumptions are justified.

On the other side of the obstacle a transmitted wave is formed, which depends on the parameters of the incident wave and the surface density of the obstacle. Let us first consider the case where the transmitted wave is weak and amenable to treatment by the linear theory. This corresponds to a massive obstacle. The region of the transmitted wave will be indicated by the subscript 4.

The displacement $u_4$ satisfies

$$\frac{\partial^2 u_4}{\partial h^2} - \frac{1}{a^2} \frac{\partial^2 u_4}{\partial t^2} = 0, \tag{6.1}$$

where

$$a^2 = \frac{\varkappa}{\varrho_0}, \qquad \varkappa = \lambda + 2\mu,$$

$\lambda$ and $\mu$ being the LAMÉ constants. The transmitted wave, which propagates in the negative $h$ direction, is of the form

$$u_4(h, t) = f(h + at).$$

We must determine the reflected and transmitted waves and the motion of the obstacle from the following conditions. At the shock front of the reflected wave,

$$(D - v_2)\varrho_2 = (D - v_1)\varrho_1,$$

$$\sigma_B - \sigma_2 = (D - v_2)\varrho_2(v_2 - v_1).$$

At the obstacle,

$$x_4(0, t) = x_{20}(0, t) = U(t),$$

$$v_2(h, t) = x'_{20}(t) = U'(t).$$

The motion of the obstacle is governed by

$$\sigma_2(0, t) - \sigma_4(0, t) = M U''(t), \tag{6.2}$$

where $M$ is the surface density and $U$ the displacement of the obstacle. As before, we have for the reflected wave

$$x_2(h, t) = \int_0^h \frac{\varrho_0 \, d\eta}{\varrho_2(\eta)} + x_{20}(t),$$

$$\sigma_2(h, t) = \varrho_0 x''_{20}(t) h + \sigma_2(0, t).$$

The velocity of propagation of the reflected wave can be determined from

$$D = \frac{dx_{2*}}{dt} = \frac{\varrho_0}{\varrho_2(h_{2*})} h + v_2.$$

Elimination of all unknown functions except $h_{2*}$ and $v_2$ yields an equation that is similar to (4.9):

$$\left[1 - \beta \frac{\varepsilon_1(h_{1*})}{\varepsilon_1(h_{2*})} \frac{h'_{1*}}{h'_{2*}}\right]^n = 1 - \left[\frac{\varepsilon_1(h_{1*})}{\varepsilon_1(h_{2*})}\right]^n \beta \frac{h'_{2*}}{h'_{1*}} \tag{6.3}$$

where

$$\beta = 1 + \frac{v_2}{\varepsilon_1(h_{1*})h'_{1*}} = 1 - \frac{v_2}{v_1}.$$

Eq. (6.3) contains two unknown functions, so that we must add the following not previously used equation concerning the motion of the obstacle:

$$\frac{dv_2}{dt} + P(t)v_2 = Q(t),$$

where

$$P(t) = \frac{(\varkappa/a) + \varrho_0 h'_{2*}}{M + \varrho_0 h_{2*}}, \qquad Q(t) = \frac{\sigma_B + \varrho_0 h'_{2*} v_1}{M + \varrho_0 h_{2*}}. \qquad (6.4)$$

The solution of this equation for which $v_2 = 0$ when the incident wave reaches the obstacle ($t = t_0$) is given by

$$v_2(t) = \frac{B}{M + \varrho_0 h_{2*}} \int_{t_0}^{t} \left[1 - \left|\frac{\varepsilon_1(h_{1*})}{\varepsilon_1(h_{2*})}\right|^n \frac{h'_{2*}}{h'_{1*}}\right] exp\left[\frac{-\varkappa}{aM} \int_{\tau}^{t} \frac{d\tau'}{1 + (\varrho_0 h_{2*}/M)}\right] d\tau'$$

$$(6.5)$$

Between them, Eqs. (6.3) and (6.5) determine the functions $h_{2*}$ and $v_2$, in terms of which the other functions are expressed.

The stress exerted by the reflected wave on the obstacle is found to be

$$\sigma_2(0, t) = \varrho_0 \frac{(\varkappa h_{2*}/a M) - h'_{2*}}{M - \varrho_0 h_2{}^*} v_2(t)$$

$$+ \sigma_B \left[1 - \left|\frac{\varepsilon_1(h_{1*})}{\varepsilon_1(h_{2*})}\right|^n \frac{h'_{2*}}{h'_{1*}}\right]\left[1 + \frac{\varrho_0 h_{2*}}{M}\right]^{-1}, \qquad (6.6)$$

where $v_2(t)$ is given by (6.5).

When the reflected wave first appears ($t = t_0$), we have $h_{1*} = h_{2*} = 0$, $v_2 = 0$, $\beta = 1$, and

$$\frac{h'_{2*}}{h'_{1*}} = q.$$

The coefficient of reflection $q$ is obtained from (6.6). One has

$$\frac{\sigma_2(0, t)}{\sigma_B} = 1 + q,$$

exactly as for an immobile obstacle. In fact, at the instant $t = t_0$, the obstacle is immobile. The stress on the other face of the obstacle is

$$\sigma_4(0, t) = \frac{\varkappa}{a} v_2.$$

The solution of Eqs. (6.3) and (6.5) would require the use of numerical methods. Alternatively, an approximate form of Eq. (6.5) may be used in the same way as in Section 3. From physical considerations,

$$0 \le \beta \le 1, \qquad 0 < \frac{\varepsilon_1(h_{1*})}{\varepsilon_1(h_{2*})} \le 1, \qquad \frac{h'_{1*}}{h'_{2*}} < 1.$$

Let us assume that

$$\left| \beta \frac{\varepsilon_1(h_{1*})}{\varepsilon_1(h_{2*})} \frac{h'_{1*}}{h'_{2*}} \right| < 1.$$

Developing the left-hand side of (6.3) into a power series and retaining only the first two terms, we set

$$\frac{h'_{2*}}{h'_{1*}} = -h^{1/2} \left| \frac{\varepsilon_1(h_{1*})}{\varepsilon_1(h_{2*})} \right|^{(n-1)/2}.$$

In this approximation, the value of $h_{2*}/h_{1*}$ can be determined in the same way as for the immobile obstacle.

Let us finally treat, in greater detail, an incident wave of the type discussed in Section 4, Example (a). The formulas simplify considerably in this case, and the problem may even be solved in closed form when a further approximation is introduced.

When a step-wave impinges on the obstacle, we have

$$\sigma_1(h, t) = \sigma_0, \qquad \sigma_B = \sigma_0,$$

$$h'_{1*} = -(-\sigma^0/\varrho_0)^{1/2}(\sigma_0/\sigma^0)^{(n-1)/2n}, \qquad \varepsilon_{1*}(h) = -(\sigma_0/\sigma^0)^{1/n}.$$

Eq. (4.16) then has the following simple form:

$$\left(1 - \beta \frac{h'_{1*}}{h'_{2*}}\right)^n = 1 - \beta \frac{h'_{2*}}{h'_{1*}},$$

where $v_1$ is constant and $\beta = 1 - (v_2/v_1)$. The ratio $h'_{1*}/h'_{2*}$ is then approximately equal to $q$ (its value at $t = t_0$). This approximation enables us to derive explicit and simple formulas for the stress $\sigma_2(0, t)$ and the velocity $v_2$. Introducing the dimensionless variables

$$\tau = \frac{(t - t_0)\varkappa}{a\,M}, \qquad\qquad v_2^0 = v_2/a, \qquad\qquad v_1^0 = v_1/a,$$

$$-c = h'_{1*}\varrho a/\varkappa = h'_{1*}/a,$$

we obtain, from (6.5),

$$v_2^0 = \frac{\sigma_0}{\varkappa} \frac{1+q}{1+qc} [1 - (1 + qc\tau)^{-(qc+1)/qc}]. \tag{6.7}$$

For the incident wave, we have

$$v_1^0 = \sigma_0/(\varkappa c).$$

Eq. (6.7) is represented graphically in Fig. 5. The stress exerted on the obstacle by the incident wave is found from (6.6):

$$\frac{\sigma_2(0,\, t)}{\sigma_0} = \frac{1+q}{1+qc\,\tau}\left[1 - \frac{qc(1-\tau)}{1+qc}\right][1 - (1 + qc\,\tau)^{-(qc+1)/qc}]. \qquad (6.8)$$

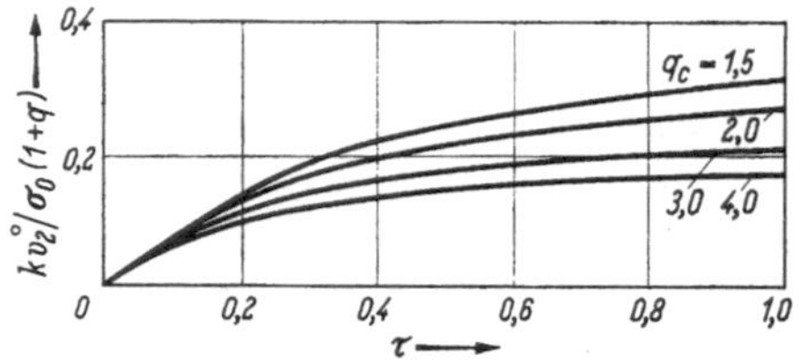

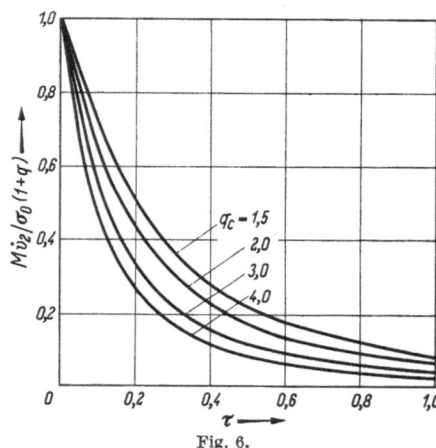

Fig. 5.

When the incident wave impinges on the obstacle, the stress rises instantaneously to the value $\sigma_0(1+q)$, as in the case of an immobile obstacle, and then decreases as shown in Fig. 6.

At the front of the reflected wave, we have the stress

$$\frac{\sigma_2(h_{2*},\, t)}{\sigma_0} = \frac{1+q}{1+qc\tau}\left[1 + qc(1 + qc\tau)^{-(qc+1)/qc}\right]. \qquad (6.9)$$

Fig. 6.

The acceleration of the obstacle is given by

$$\frac{dv_2^0}{dt} = -\frac{\sigma_0(1+q)}{M}(1 + qc\,\tau)^{-(2qc+1)/qc} \qquad (6.10)$$

(see Fig. 7). The maximum absolute value of the acceleration, namely

$$w = \sigma_0(1+q)/M,$$

occurs when the incident wave reaches the obstacle.

It should be kept in mind that Eqs. (6.7) through (6.10) are based on the following assumptions: the deformation of the continuum is uniaxial, the stress at the front of the reflected wave is greater than that

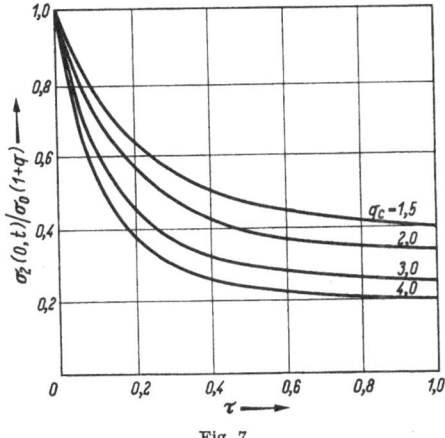

Fig. 7.

in the incident wave, and no unloading has as yet occurred in the transmitted wave. The suitability of these assumptions must be verified for every specific problem.

The case $c < 1$ can not be treated along these lines because we have assumed that the incident wave propagates in an unperturbed medium. If $c < 1$, the plastic wave is preceded by an elastic wave. This case must be studied independently.

### References

[1] ALEKSEENKO, V. C., S. S. GRIGORIAN, A. F. NOVGORODOV, and G. V. RYKOV: Dokl. Akad. Nauk SSSR. 133, 1311 (1960).
[2] KALISKI, S., and G. OSIECKI: Proc. Vibration Problems (Warsaw) 1, 49 (1959).
[3] PERZYNA, P.: Contribution to Non-homogeneity in Elasticity and Plasticity, London: Pergamon Press 1958, p. 431.
[4] LYAKHOV, G. M., and N. I. POLAKOVA: Izv. Akad. Nauk SSSR., Otd. Tehn. Nauk, Mehanika i Mashinostroenie, No. 2, 73 (1959).
[5] LYAKHOV, G. M., and N. I. POLAKOVA: ibid., No. 3, 99 (1960).

# A Simple Viscoelastic Analogy for Stress Waves[1]

By **Chi-chang Chao** and **Jan D. Achenbach**[2]

Department of Aeronautics and Astronautics, Stanford University, Stanford, California, U.S.A.

## 1. Introduction

The subject of wave propagation in a linearly viscoelastic medium has received considerable attention in recent years. The problems that have been considered are, however, mainly of a one-dimensional nature (see, for example, BERRY and HUNTER [1] and MORRISON [2]). Many other contributions on one-dimensional wave propagation have been reviewed in survey articles by KOLSKY [3], LEE [4], and HUNTER [5]. In the case of three-dimensional wave propagation, the problem of the characterization of the material is more complicated. For small strains, the mechanical behavior of a homogeneous isotropic material is completely defined by the behavior of the material in bulk and in shear.

Every statement about the fundamental behavior in bulk or in shear should be supported by experimental evidence. Most of the experimental techniques for the determination of the mechanical properties of viscoelastic solids, however, are carried out either in shear or under uniaxial conditions of stressing. Experiments in shear, yield creep or relaxation curves, or values for the complex modulus on the basis of which a proper representation for the shear behavior can be selected. If the creep curve obtained from such an experiment is approximated by the sum of exponential functions with constant coefficients, the corresponding shear behavior of the material is characterized by a model consisting of a network of springs and dashpots. In fact, all idealized models can be viewed as merely the consequences of curve fittings of the experimental creep curves. These idealized solids were used extensively and satisfactorily in the study of quasi-static viscoelastic problems. However, in the case of dynamic response of viscoelastic materials, these models pose great difficulties in the mathematical analysis as was pointed out in [5]. With an eye on the subsequent mathematical analysis and at

---

[1] The work presented here was supported by the U.S. Army Research Office (Durham) under grant DA-ARO(D)31-124-G238.

[2] Now at Northwestern University, Evanston, Illinois, U.S.A.

the same time fitting the general creep behavior of most viscoelastic materials, a constitutive relation more suitable for dynamic problems was proposed in [6] and is adopted in this paper.

Since very few test results are available for the dilatation behavior of viscoelastic solids, none of the theoretical assumptions on the fundamental dilational behavior are supported by appropriate evidence.

The most widespread assumptions are:

(a) Incompressibility,

(b) Elastic behavior in bulk,

(c) Identical viscoelastic behavior in bulk and in shear.

The first assumption (a) is commonly adopted for fluids. It leads to a simple relation between the LAPLACE transforms of the viscoelastic equivalents of YOUNG's modulus and the shear modulus. Some test results are available to support (a) for low frequency type disturbances. In the present study, where the interest is focussed on the solid type of viscoelastic materials, (a) is not applicable.

The assumption of elastic behavior in bulk was first suggested by BOLTZMANN, and has been generally accepted by many other investigators. REINER [7], however, points out that every material will show, under isotropic stress, a retarded elasticity. According to KOLSKY and SHI [8], assumption (b) is probably realistic for a crystalline solid for which hydrostatic pressure merely changes interatomic distances.

If bulk viscosity is recognized, reasons of mathematical expedience lead to assumption (c). This assumption is equivalent to the assumption of a constant POISSON's ratio. It is obvious that the assumption does not make sense for liquid type materials. For physical reasons the final change of volume of an element has to be small, consequently, the viscoelastic effect in bulk can only be minute, as opposed to that in shear, which can be of much larger magnitude. Thus, for very large viscoelastic effects in shear it seems to be advisable to apply assumption (b). If, on the other hand, the final shear deformations are bounded (retarded elasticity) and not too large, it is possible to use assumption (c). From the mathematical point of view, this assumption has great advantages over the others as will be seen later in this paper. Even if the viscoelastic shear effects do not remain very small, this assumption still gives results which can be considered as an approximation to the real behavior. A general approach to wave propagation in a viscoelastic medium with a constant POISSON's ratio was given by SHEMYAKIN [9].

In the present work, the application of the LAPLACE transform technique to viscoelastic wave propagation problems is discussed. The well-known correspondence principle is considered first. It is then shown that under the restricting condition of constant POISSON's ratio, many viscoelastic problems can be solved provided that the solution of the corresponding elastic problem is known.

By adopting the proposed constitutive relation of [6], the visco-elastic solutions are expressed in terms of a convolution integral with the elastic solution and some known functions as its integrands. Two illustrative examples are given at the end of this paper. In the first example, the displacement components in the radial and the vertical directions on the surface of a viscoelastic half-space loaded suddenly by a vertical force of constant magnitude are evaluated and plotted. The second example gives the stress in the radial direction for the problem of the expanding spherical cavity in an infinite visco-elastic medium.

## 2. Three-Dimensional Stress-Strain Relations for a Viscoelastic Material

In the theory of elasticity it has been shown that for small strains, and under the assumptions that the solid is homogeneous and isotropic, two independent constants completely define the stress-strain relations.

If the shear modulus $\mu$ and the bulk modulus $K$ are chosen for this purpose, the relation between stresses and strains can be written

$$s_{ij} = 2\mu e_{ij}, \tag{2.1}$$

$$\sigma_{ii} = 3K\varepsilon_{ii}, \tag{2.2}$$

where $s_{ij}$ is the stress deviator, $e_{ij}$ the strain deviator. The tensors $s_{ij}$ and $e_{ij}$ are defined as

$$s_{ij} = \sigma_{ij} - \frac{1}{3}\,\delta_{ij}\sigma_{kk}, \tag{2.3}$$

$$e_{ij} = \varepsilon_{ij} - \frac{1}{3}\,\delta_{ij}\varepsilon_{kk}. \tag{2.4}$$

According to the usual summation convention $\sigma_{ii}/3$ is the mean normal stress and $\varepsilon_{ii}/3$ the mean extension. Substitution of (2.1), (2.2) and (2.4) into (2.3) gives the usual stress-strain relations of the form:

$$\sigma_{ij} = \left(K - \frac{2}{3}\,\mu\right)\varepsilon_{kk}\delta_{ij} + 2\mu\varepsilon_{ij}. \tag{2.5}$$

The relations between stresses and strains in a homogeneous and isotropic linearly viscoelastic solid are time dependent. Instead of the algebraic relations of the elastic material, there are now the following integral expressions

$$2\mu e_{ij} = s_{ij} + \int\limits_{-\infty}^{t} \psi_1(t - \tau) \frac{ds_{ij}}{d\tau}\, d\tau, \tag{2.6}$$

$$3K \varepsilon_{ii} = \sigma_{ii} + \int\limits_{-\infty}^{t} \psi_2(t - \tau) \frac{d\sigma_{ii}}{d\tau}\, d\tau, \tag{2.7}$$

where $\psi_1$ and $\psi_2$ are the creep functions governing the shear and dilatational behavior of the medium.

Expressions of the type (2.6) and (2.7) are not easily handled in theoretical calculations. By introducing the LAPLACE transform and assuming, as usual, that no disturbances are present for $t < 0$, the equations are replaced by much simpler algebraic relations connecting the transformed values of stresses and strains. Application of the LAPLACE transform to (2.5) and (2.6) and some additional manipulation yields:

$$\bar{\sigma}_{ij} = \left[ \overline{K}(p) - \frac{2}{3}\, \overline{\mu}(p) \right] \bar{\varepsilon}_{kk} \delta_{ij} + 2\overline{\mu}(p)\, \varepsilon_{ij}, \tag{2.8}$$

where

$$\overline{\mu}(p) = \frac{\mu}{1 + p\bar{\psi}_1(p)} = \mu\alpha(p), \tag{2.9}$$

$$\overline{K}(p) = \frac{K}{1 + p\bar{\psi}_2(p)} = K\gamma(p). \tag{2.10}$$

## 3. Equations of Motion and Correspondence Principle

The LAPLACE transforms of the stress equations of motion of a continuum in the absence of body force are:

$$\bar{\sigma}_{ij,j} = \varrho p^2 \bar{\mu}_i \tag{3.1}$$

where $\varrho$ is the density and the continuum is assumed initially at rest.

Substitution of (2.8) into (3.1) yield the governing differential equations for the transformed displacements of a viscoelastic medium.

$$(\overline{K} + \frac{1}{3}\, \overline{\mu})\bar{u}_{j,\,ji} + \overline{\mu}\bar{u}_{i,\,jj} = \varrho p^2 \bar{u}_i. \tag{3.2}$$

Decompose the displacements into equivoluminal and irrotational parts according to:

$$\bar{u}_i = \bar{v}_i + \bar{w}_i, \tag{3.3}$$

15  Kolsky/Prager, Stress Waves

where

$$\bar{v}_{i,i} = 0, \tag{3.4}$$

$$\overline{w}_{i,j} = \overline{w}_{j,i}. \tag{3.5}$$

The transformed equations of motion (3.2) will then be satisfied if:

$$\bar{v}_{i,jj} = \frac{p^2}{\bar{c}_1^2} \, \bar{v}_i, \tag{3.6}$$

$$\overline{w}_{i,jj} = \frac{p^2}{\bar{c}_2^2} \, \overline{w}_i, \tag{3.7}$$

where

$$\bar{c}_1^2 = \frac{\overline{K}(p) + \dfrac{4}{3}\,\bar{\mu}(p)}{\varrho}, \tag{3.8}$$

$$\bar{c}_2^2 = \frac{\bar{\mu}(p)}{\varrho}. \tag{3.9}$$

The same manipulation can be applied for the elastic medium if Eq. (2.5) is used instead of (2.8). The analogous equations to (3.6) and (3.7) for the elastic case are obtained if $\overline{K}(p)$ and $\bar{\mu}(p)$ in (3.8) and (3.9) are replaced by the elastic modulus $K$ and $\mu$, respectively. The following conclusion can then be drawn: the LAPLACE transforms of the solutions for a viscoelastic wave propagation problem can be obtained from the LAPLACE transforms of the solutions for the elastic problem with the same boundary and initial conditions by replacing $\overline{K}$ by $K(p)$ and $\mu$ by $\bar{\mu}(p)$. This conclusion is known as the correspondence principle [10]. The problem of obtaining the displacement solutions for the viscoelastic body is now reduced to a problem of inverting LAPLACE transforms. For certain choices of $\alpha(p)$ and $\gamma(p)$—(2.9) and (2.10)— the inversion may become very difficult [5]. If a certain type of viscoelasticity is required it is advisable to choose $\alpha(p)$ and $\gamma(p)$ so that not only the mechanical behavior is represented in a satisfactory way, but also such that it is possible to obtain the inverse LAPLACE transforms.

## 4. The Displacement and Stress Fields Inside a Body of Viscoelastic Material of Constant Poisson's Ratio

It is convenient to introduce the following discussion by a particular example. We will consider the displacements in an infinite viscoelastic medium due to a time-dependent concentrated force which is applied at a given point (the origin, Fig. 1).

General expressions for the displacement components in a purely elastic infinite medium due to a time dependent force $\chi(t)$ were first obtained by STOKES. Formulas equivalent to those evaluated by STOKES were given by LOVE [11]. If a single force of magnitude $\chi(t)$ acts at the origin in the direction of the z-axis, we have the following expression for the displacement in the radial direction of a system of cylindrical coordinates:

$$u_r = \frac{c_2^2}{4\pi\mu} \frac{3zr}{R^5} \int_{R/c_1}^{R/c_2} \tau\,\chi(t-\tau)\,d\tau + \frac{1}{4\pi\varrho}\frac{zr}{R^3}\left[\frac{1}{c_1^2}\chi\left(t-\frac{R}{c_1}\right) - \frac{1}{c_2^2}\chi\left(t-\frac{R}{c_2}\right)\right].$$
$$(4.1)$$

The LAPLACE transform of (4.1) is

$$\bar{u}_r(r,z,p) = \frac{Fa(p)}{\mu}\,f_1(r,z,h,k),$$
$$(4.2)$$

where $Fa(p)$ is the LAPLACE transform of $\chi(t)$ and

$$f_1(r,z,h,k) = \frac{1}{4\pi}\frac{zr}{R^3}\left[\frac{h^2}{k^2}\left(1+\frac{3}{Rh}+\frac{3}{R^2h^2}\right)e^{-hR} - \left(1+\frac{3}{Rk}+\frac{3}{R^2k^2}\right)e^{-kR}\right],$$
$$(4.3)$$

$$h^2 = p^2/c_1^2, \qquad (4.4)$$

$$k^2 = p^2/c_2^2. \qquad (4.5)$$

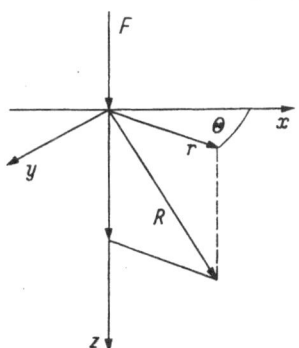

Fig. 1. Force in an infinite solid.

Let $\chi(t) = Fa(t)$ now be applied in a viscoelastic infinite medium. The expression (4.2) can then be considered as the LAPLACE transform of the corresponding elastic solution. According to the correspondence principle, the LAPLACE transforms of the viscoelastic solution can then be written as

$$(\bar{u}_r)_{vi} = \frac{Fa(p)}{\mu\alpha(p)}\,f_1(r,z,\bar{h},\bar{k}) = \frac{Fa(p)}{\mu\alpha(p)}\,f_1(r,z,p/\bar{c}_1,p/\bar{c}_2). \qquad (4.6)$$

Expressions for $\bar{c}_1(p)$ and $\bar{c}_2(p)$ are given by (3.8) and (3.9) respectively.

In order to determine the inverse LAPLACE transforms of (4.6) we will first consider a function $f(p)$ of the type

$$f(p) = h(p)g[\xi(p)]. \qquad (4.7)$$

The functions $h(p)$ and $g[\xi(p)]$ are assumed to have the following properties:

(a) $h(p)$ and $\xi(p)$ are analytic functions of $p$ with possible singularities in the left half plane;

(b) $g[\xi(p)]$ is an analytic composite function of $p$ for $Re(p) > 0$;

(c) The inverse LAPLACE transform of $g(p)$ exists and is defined as $G(t)$, thus

$$G(t) = L^{-1}[g(p)], \tag{4.8}$$

$$g(p) = \int\limits_0^\infty G(t)e^{-pt}dt; \tag{4.9}$$

(d) $\lim\limits_{p\to\infty}[\xi(p) - p] = A$,     where     $A$ is real and $A \geq 0$;  (4.10)

(e) $$\lim\limits_{p\to\infty} h(p) = 1. \tag{4.11}$$

The inverse LAPLACE transform of $f(p)$ can formally be written as

$$F(t) = \frac{1}{2\pi i}\int\limits_{\delta-i\infty}^{\delta+i\infty} f(p)e^{pt}dp. \tag{4.12}$$

It can be shown that the assumptions (a), (b), (c), (d), and (e) are sufficient for convergence of (4.12).

From (4.9) it is evident that

$$g[\xi(p)] = \int\limits_0^\infty G(s)e^{-\xi(p)s}ds, \tag{4.13}$$

hence

$$f(p) = \int\limits_0^\infty h(p)G(s)e^{-\xi(p)s}ds. \tag{4.14}$$

Substitution of (4.14) into the integral representation (4.12) for $F(t)$ yields

$$F(t) = \frac{1}{2\pi i}\int\limits_{\delta-i\infty}^{\delta+i\infty} \left[\int\limits_0^\infty G(s)h(p)e^{-\xi(p)s}ds\right] e^{pt}dp. \tag{4.15}$$

It is convenient to rearrange the integrand by writing

$$e^{As}h(p)e^{-\xi(p)s} = \{h(p)e^{-[\xi(p)-A]s} - e^{-ps}\} + e^{-ps} \tag{4.16}$$

and to define

$$h(p, s) = h(p)e^{-[\xi(p)-A]s} - e^{-ps}. \tag{4.17}$$

The expression (4.16) can then be written as

$$h(p)e^{-\xi(p)s} = e^{-As}e^{-ps} + e^{-As}h(p, s).$$ (4.18)

Substituting (4.18) into (4.15), we write

$$F(t) = I_1 + I_2,$$ (4.19)

where

$$I_1 = \frac{1}{2\pi i} \int_{\delta-i\infty}^{\delta+i\infty} \left[ \int_0^\infty G(s)e^{-As}e^{-ps}ds \right] e^{pt}dp,$$ (4.20)

$$I_2 = \frac{1}{2\pi i} \int_{\delta-i\infty}^{\delta+i\infty} \left[ \int_0^\infty G(s)e^{-As}h(p, s)ds \right] e^{pt}dp.$$ (4.21)

Inspection of (4.20) shows that

$$I_1 = e^{-At}G(t).$$ (4.22)

To evaluate $I_2$ the order of integration for $I_2$ is reversed. It can be shown that this operation is admissible because of the assumptions (a), (b), (c), (d), and (e) [12]. Thus

$$I_2 = \int_0^\infty G(s)e^{-As}H(t, s)ds,$$ (4.23)

where

$$H(t, s) = L^{-1}[h(p, s)].$$ (4.24)

The function $h(p, s)$ was defined in (4.17). Substitution of (4.22) and (4.23) into (4.19) yields the result

$$F(t) = L^{-1}\{h(p)g[\xi(p)]\} = e^{-At}G(t) + \int_0^\infty G(s)e^{-As}H(t, s)ds$$ (4.25)

where $A$, $G(s)$ and $H(t, s)$ are given by (4.10), (4.9) and (4.24).

Let us now return to the expression (4.6). The function $f_1(r, z, p/\bar{c}_1, p/\bar{c}_2)$ is a composite function of $p$. From the expressions (3.8) and (3.9) it is seen that $p/\bar{c}_1$ and $p/\bar{c}_2$ can be written as

$$p/\bar{c}_1 = \frac{p\varrho^{1/2}}{\left[K\gamma(p) + \frac{4}{3}\mu\alpha(p)\right]^{1/2}} = \frac{p}{c_1\sqrt{\beta(p)}},$$ (4.26)

$$p/\bar{c}_2 = \frac{p\varrho^{1/2}}{[\mu\alpha(p)]^{1/2}} = \frac{p}{c_2\sqrt{\alpha(p)}}.$$ (4.27)

In order to apply the procedure outlined through the Eqs. (4.8) to (4.25) to $(\bar{u}_r)_{vI}$, it is thus in the first place necessary to assume that, apart from a multiplying constant, $p/\bar{c}_1$ and $p/\bar{c}_2$ are identical functions of $p$. The equations (4.26) and (4.27) show that this is only possible if

$$\alpha(p) = \beta(p) = \gamma(p). \qquad (4.28)$$

The transformed POISSON's ratio is given as

$$\bar{\nu} = \frac{3\bar{K}(p) - 2\bar{\mu}(p)}{2[3\bar{K}(p) + \bar{\mu}(p)]} = \frac{3K\gamma(p) - 2\mu\alpha(p)}{2[3K\gamma(p) + \mu\alpha(p)]}. \qquad (4.29)$$

From (4.29) it is then seen that the assumption (4.28) implies a constant POISSON's ratio. Under the assumption (4.28) the function $(\bar{u}_r)_{vl}$ is of the type $f(p)$ as given by (4.7), namely, the product of a function of $p$ and a composite function of $p$.

The expression (4.6) for $(\bar{u}_r)_{vl}$ can now be written as

$$(\bar{u}_r)_{iv} = \frac{Fa(p)}{\mu\alpha(p)} f_1\left[r, z, p/c_1\sqrt{\alpha(p)}, p/c_2\sqrt{\alpha(p)}\right]. \qquad (4.30)$$

Comparing (4.30) with (4.7) one can choose

$$\xi(p) = p/\sqrt{\alpha(p)}, \qquad (4.31)$$

$$g[\xi(p)] = \frac{Fa[p/\sqrt{\alpha(p)}]}{\mu} f_1\left[r, z, p/c_1\sqrt{\alpha(p)}, p/c_2\sqrt{\alpha(p)}\right], \qquad (4.32)$$

$$h(p) = \frac{1}{\alpha(p)} \frac{a(p)}{a[p/\sqrt{\alpha(p)}]}. \qquad (4.33)$$

Inspection of (4.2), (4.32), and (4.13) shows that $G(s)$ is the solution of the corresponding elastic problem

$$G(s) = [u_r(x, y, z, s)]_{el}. \qquad (4.34)$$

The composite function $g[\xi(p)]$ thus satisfies the assumptions (b) and (c). The inverse LAPLACE transform of $(\bar{u}_r)_{vl}$ can be determined if $\xi(p)$ and $h(p)$ satisfy the assumptions (4.10), (4.11) and (a).

According to (4.10) we have to satisfy

$$\lim_{p \to \infty} [\xi(p) - p] = A \qquad \text{where} \qquad A \geq 0.$$

With the use of (4.31), this requirement becomes

$$\lim_{p \to \infty} \left[\frac{p}{\sqrt{\alpha(p)}} - p\right] = A \geq 0. \qquad (4.35)$$

Eq. (4.35) will be satisfied by any viscoelastic material with initial elasticity. Therefore one can write for large values of $p$

$$\frac{p}{\sqrt{\alpha(p)}} \approx p + A \quad \text{and} \quad \frac{1}{\sqrt{\alpha(p)}} \approx 1 + \frac{A}{p}.$$

The substitution of these into (4.33), confirms the validity of equation (4.11). It can then be stated that if (4.35) is satisfied and the singularities of $1\sqrt{a(p)}$ are in the left-half plane (physically, these singularities correspond to the fact that the material has an initial elasticity and possesses a reasonable creep behavior), the solution for the viscoelastic problem is

$$(u_r)_{\text{vi}} = e^{-At}[u_r(x, y, z, t)]_{\text{el}} + \int_0^\infty e^{-As}[u_r(x, y, z, s)]_{\text{el}} H(t, s)\, ds, \quad (4.36)$$

where

$$H(t, s) = L^{-1}\left\{ \frac{1}{\alpha(p)} \frac{a(p)}{a[p/\sqrt{\alpha(p)}]} \exp\left(\{-[p/\sqrt{\alpha(p)}] + A\}s\right) - e^{-ps}\right\}. \quad (4.37)$$

The technique discussed above for the particular example can be generalized and applied whenever the LAPLACE transforms of the corresponding elastic solutions can be expressed in the form (4.2). This is indeed the case for a number of elastic problems for which solutions are known. Examples are the problems of the normally applied step force to an elastic half space [13], the tangentially applied force to a half space [14], and the problem of the expanding spherical cavity [15]. In these cases an application of the correspondence principle and the assumption of constant POISSON's ratio will yield functions in $p$ of the type $f(p) = h(p)g[\xi(p)]$, which can be inverted as shown in this section.

The conclusion is that for certain wave propagation problems in a viscoelastic solid with a constant POISSON's ratio, the displacement solutions can be expressed in terms of the displacement solutions of the corresponding elastic problem as

$$u_{\text{vi}}(x, t, z, t) = e^{-At}u_{\text{el}}(x, y, z, t) + \int_0^t e^{-As}u_{\text{el}}(x, y, z, s) H(t, s)\, ds. \quad (4.38)$$

The upper limit of the integrand can be taken as $t$, since the integrand will vanish for $s > t$.

A result similar to (4.38) can now easily be obtained for the stresses. In general, the LAPLACE transforms of the stresses will have the form

$$\bar{\sigma}_{\text{vi}} = \bar{\lambda}(p) L_1(\bar{u}) + \bar{\mu}(p) L_2(\bar{u}) \quad (4.39)$$

where $L_1$ and $L_2$ are linear operators on the displacements which only involve derivatives with respect to the space coordinates. With (4.28) one can write

$$\bar{\sigma}_{vi} = \alpha(p)[\lambda L_1(\bar{u}) + \mu L_2(\bar{u})]. \qquad (4.40)$$

For each individual element of $L_1(u)$ and $L_2(u)$ the technique outlined in the Eqs. (4.8) to (4.25) can be applied. It is easy to see that the general result is

$$\sigma_{vi}(x, y, z, t) = e^{-At}\sigma_{el}(x, y, z, t) + \int_0^t e^{-As}\sigma_{el}(x, y, z, s) K(t, s)ds \qquad (4.41)$$

where

$$K(t, s) = L^{-1}\left\{\frac{a(p)}{a[p/\sqrt{\alpha(p)}]} \exp\left(-\{[p/\sqrt{\alpha(p)}] - A\}s\right) - e^{-ps}\right\}. \qquad (4.42)$$

The functions $H(t, s)$ and $K(t, s)$ depend on the particular properties of the viscoelastic material and the time dependence of the applied load.

## 5. Evaluation of $u_{vi}(x, y, z, t)$ and $\sigma_{vi}(x, y, z, t)$ for a Three-Parameter Viscoelastic Solid

Any further studies of the Eqs. (4.38) and (4.41) require the specific descriptions of the stress-strain relations of the viscoelastic medium in question. A three-parameter model discussed in [6] is now considered. It was pointed out in [6] that the creep function and the complex modulus of this model are very similar to those of the standard linear solid, while the stress-strain relations are such that solutions of dynamic problems can be obtained more readily.

The transformed shear modulus of this three-parameter linearly viscoelastic solid is

$$\mu(p) = \mu\frac{(pT + m)^2}{(pT + \sqrt{m})^2} = \mu\alpha(p). \qquad (5.1)$$

where $\mu$ is the modulus of initial elasticity, $m(0 < m < 1)$ is the relaxation ratio, and $T$ is a relaxation time. The function $\xi(p)$ is

$$\xi(p) = \frac{p}{\sqrt{\alpha(p)}} = \frac{p(pT + \sqrt{m})}{pT + m}. \qquad (5.2)$$

The constant $A$, which was defined in Eq. (5.14) is

$$A = \lim_{p \to \infty} \frac{p(pT + \sqrt{m})}{pT + m} - p = \frac{\sqrt{m} - m}{T}. \qquad (5.3)$$

With Eq. (4.38) the solution for a displacement can be expressed in terms of the solution of the corresponding elastic problem as

$$u_{\mathrm{vi}} = \exp\left(-\left[\sqrt{m} - m\right]s/T\right) u_{\mathrm{el}}(x, y, z, t)$$

$$+ \int_0^t \exp\left(-\left[\sqrt{m} - m\right]s/T\right) u_{\mathrm{el}}(x, y, z, s) H(t, s)\, ds. \tag{5.4}$$

After substitution of (5.2) and (5.3), $H(t, s)$ is given by (4.37).

We will now consider the case where the force is applied in a manner described by a step function $Fa(t) = FH(t)$, and thus $a(p) = 1/p$. The expression for $h(p, s)$ can then be written as:

$$h(p, s) = \frac{(pT + \sqrt{m})^3}{(pT + m)^3} \exp\left(-\left[\frac{p(pT + \sqrt{m})}{pT + m} - \frac{\sqrt{m} - m}{T}\right] s\right) - e^{-ps} \tag{5.5}$$

or

$$h(p, s) = h_1(p) + h_2(p) + h_3(p) + h_4(p), \tag{5.6}$$

where

$$h_1(p) = e^{-ps}\left[\exp\left(\frac{m[\sqrt{m} - m]s}{T[pT + m]}\right) - 1\right], \tag{5.7}$$

$$h_2(p) = e^{-ps}\frac{3(\sqrt{m} - m)}{pT + m}\exp\left(\frac{m[\sqrt{m} - m]s}{T[pT + m]}\right), \tag{5.8}$$

$$h_3(p) = e^{-ps}\frac{3(\sqrt{m} - m)^2}{(pT + m)^2}\exp\left(\frac{m[\sqrt{m} - m]s}{T[pT + m]}\right), \tag{5.9}$$

$$h_4(p) = e^{-ps}\frac{(\sqrt{m} - m)^3}{pT + m)^3}\exp\left(\frac{m[\sqrt{m} - m]s}{T[pT + m]}\right). \tag{5.10}$$

The following inversions are known from tables of LAPLACE transforms [16]

$$L^{-1}\left(e^{\frac{a}{p}} - 1\right) = a^{1/2}t^{1/2}I_1(2a^{1/2}t^{1/2}), \tag{5.11}$$

$$L^{-1}\left[p^{-\nu-1}e^{\frac{b}{p}}\right] = b^{-\frac{\nu}{2}}t^{\frac{\nu}{2}}I_\nu(2b^{1/2}t^{1/2}), \tag{5.12}$$

$$L^{-1}[f(p + c)] = e^{-ct}F(t), \tag{5.13}$$

$$L^{-1}[e^{-dp}f(p)] = F(t - d)H(t - d), \tag{5.14}$$

where $I_1$ and $I_\nu$ represent the modified BESSEL functions of order one and $\nu$ and where $H(t - d)$ is the HEAVISIDE unit function.

Without difficulties the inverse LAPLACE transforms of (5.7), (5.8), (5.9), and (5.10) can be obtained as:

$$L^{-1}[h_1(p)] = \exp\left(-m[t - s]/T\right)\frac{ks^{1/2}}{(t - s)^{1/2}}I_1(\delta_s)H(t - s), \tag{5.15}$$

where

$$k = \frac{(m\sqrt{m} - m^2)^{1/2}}{T} \tag{5.16}$$

$$\delta_s = 2 k s^{1/2} (t - s)^{1/2}, \tag{5.17}$$

$$L^{-1}[h_2(p)] = \frac{3(\sqrt{m} - m)}{T} \exp\left(-m[t-s]/T\right) I_0(\delta_s) H(t - s), \tag{5.18}$$

$$L^{-1}[h_3(p)] = \frac{3(\sqrt{m} - m)^2}{T^2 k} \frac{(t - s)^{1/2}}{s^{1/2}} \exp\left(-m[t-s]/T\right) I_1(\delta_s) H(t - s), \tag{5.19}$$

$$L^{-1}[h_4(p)] = \frac{(\sqrt{m} - m)^3}{T^3 k^2} \frac{(t - s)}{s} \exp\left(-m[t-s]/T\right) I_2(\delta_s) H(t - s). \tag{5.20}$$

The function $H(t, s)$ is equal to the sum of the expressions (5.15), (5.18), (5.19), and (5.20). With the recurrence formula

$$I_1(x) = \frac{x}{2} I_0(x) - \frac{x}{2} I_2(x), \tag{5.21}$$

the function $H(t, s)$ can be expressed as a combination of $I_0(\delta_s)$ and $I_2(\delta_s)$ only. The result (5.4) can next be rewritten in a non-dimensional form by introducing the dimensionless variables

$$v = \frac{cs}{d}, \tag{5.22}$$

$$\theta = \frac{d}{cT}, \tag{5.23}$$

$$\tau = \frac{ct}{d}, \tag{5.24}$$

where $d$ and $c$ are appropriately chosen characteristic length and velocity parameters.

After some rearranging the final result is

$$u_{\mathrm{vi}} = e^{-(\sqrt{m}-m)\theta\tau} u_{\mathrm{el}}(\tau)$$

$$+ \theta(\sqrt{m} - m) \int_0^\tau \exp\left(-[\sqrt{m} - m]\theta v\right) \exp\left(-m\theta[\tau - v]\right)$$

$$\times \left[m\theta v + 3 + 3(\sqrt{m} - m)\theta(\tau - v)\right] I_0(\delta_v) u_{\mathrm{el}}(v) dv$$

$$- \theta(\sqrt{m} - m) \int_0^\tau \exp\left(-[\sqrt{m} - m]\theta v\right) \exp\left(-m\theta[\tau - v]\right)$$

$$\times \left[m\theta v + 3(\sqrt{m} - m)\theta(\tau - v) - \frac{\sqrt{m} - m}{m}\frac{\tau - v}{v}\right] I_2(\delta_v) u_{\mathrm{el}}(v) dv \tag{5.25}$$

where with (5.16), (5.17), (5.22), and (5.23)

$$\delta_v = 2(m\sqrt{m} - m^2)^{1/2}\theta v^{1/2}(\tau - v)^{1/2}. \tag{5.26}$$

The expression for $\sigma_{\mathrm{vl}}$ can be derived in a similar way;

$$\sigma_{\mathrm{vl}} = \exp.\left(-[\sqrt{m} - m]t/T\right)\sigma_{\mathrm{el}}(x, y, z, t)$$

$$+ \int_0^t \exp\left(-[\sqrt{m} - m]t/T\right)\sigma_{\mathrm{el}}(x, y, z, s)K(t, s)ds \tag{5.27}$$

where, with (4.42) and for $a(p) = 1/p$:

$$K(t, s) = L^{-1}\left\{\frac{pT + \sqrt{m}}{pT + m}\exp\left(-\left[\frac{p(pT + \sqrt{m})}{pT + m} - \frac{\sqrt{m} - m}{T}\right]s\right) - e^{-ps}.\right. \tag{5.28}$$

The function $K(t, s)$ is now easily obtained with the aid of (5.11), (5.12), (5.13), and (5.14). With the recurrence formula (5.21), $K(t, s)$ can be expressed as a combination of $I_0(\delta_s)$ and $I_2(\delta_s)$ only. After introduction of the dimensionless variables $v$ (5.22), $\theta$ (5.23), and $\tau$ (5.24), the final result can be written as

$$\sigma_{\mathrm{vl}} = e^{-(\sqrt{m} - m)\theta\tau}\sigma_{\mathrm{el}}(\tau)$$

$$+ \theta(\sqrt{m} - m)\int_0^\tau \exp\left(-[\sqrt{m} - m]\theta v\right)\exp\left(-m\theta[\tau - v]\right)$$

$$\times (m\theta v + 1)I_0(\delta_v)\sigma_{\mathrm{el}}(v)dv$$

$$- \theta^2 m(\sqrt{m} - m)\int_0^\tau \exp\left(-[\sqrt{m} - m]\theta v\right)\exp\left(-m\theta[\tau - v]\right)$$

$$\times vI_2(\delta_v)\sigma_{\mathrm{el}}(v)dv, \tag{5.29}$$

where $\delta_v$ is given by (5.26).

The substitution of $m = 1$ or $T = \infty$ reduce (5.25) and (5.29) to elastic solutions, in which the coefficients $\lambda$ and $\mu$ of initial elasticity appear as the elastic coefficients. These solutions will be called the first elastic solutions. The second elastic solutions are obtained by taking $m\lambda$ and $m\mu$ as elastic coefficients. The second elastic solution provides the displacements and stresses to which (5.25) and (5.29) approach when $\tau$ increases.

The multiplying factor of the elastic solution, $\exp\left[-\left(\sqrt{m} - m\right)\theta\tau\right]$, provides a measure of the decay of the wave front as the pulse moves through the medium. There is a relation between this decay and the slope of the creep curve at $t = 0$. The integrals in (5.25) and (5.29) indicate the viscoelastic aftereffect; they influence the magnitude of the displacements and the stresses immediately after the passage of the wave front.

## 6. Wave Propagation in a Viscoelastic Half-Space

In the study of wave propagation in a medium bounded by an infinite plane surface — a "half-space"— one will find that in addition to the previously discussed pressure and shear waves there exist wave-like effects with depthwise rapidly decaying amplitudes. The possibility of such a wave in an elastic medium was first recognized by RAYLEIGH, after whom it is named. The disturbance generated in a semi-infinite medium by a suddenly applied vertical force was more recently studied by PEKERIS [13]. He obtained closed-form expressions for the displace-

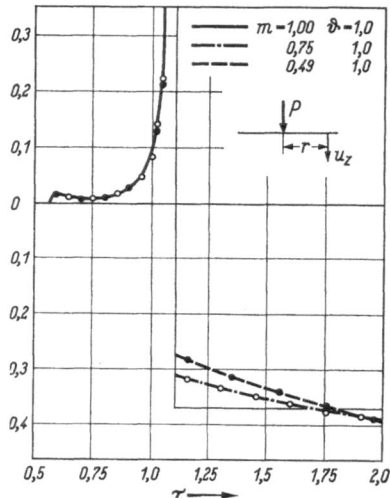

Fig. 2. Vertical displacement $u_r$ at the surface $z = 0$ as a function of $z = \dfrac{c_2 t}{r}$ and $\vartheta = \dfrac{r}{c_2 t}$.

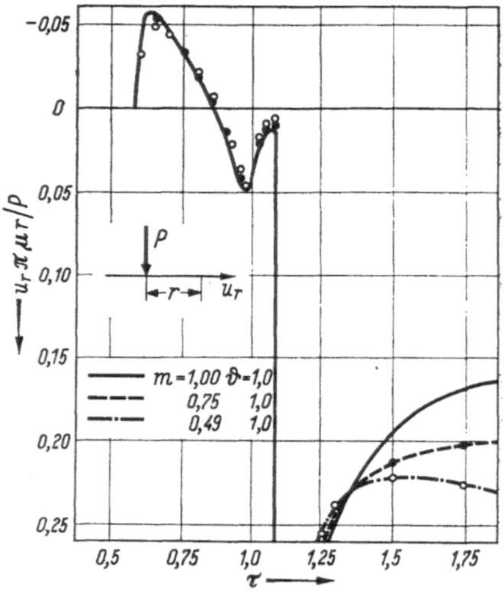

Fig. 3. Radial displacement $u_r$ at the surface $z = 0$ as a function of $z$ and $\vartheta$.

ments in the radial and the vertical direction on the surface $z = 0$. Inspection of the PEKERIS solutions shows that there exist indeed three distinct wave fronts traveling with velocities

$$c_1 = \left(\frac{\lambda + 2\mu}{\varrho}\right)^{1/2}, \qquad c_2 = \left(\frac{\mu}{\varrho}\right)^{1/2} \qquad \text{and} \qquad c_R = \frac{c_2}{\gamma}$$

respectively. The arrival of the RAYLEIGH waves is marked by an infinite discontinuity in both $u_r$ and $u_z$.

The analogue of PEKERIS' problem in the viscoelastic medium of the previous section will now be considered. Solutions for this medium can be obtained by substituting the PEKERIS' elastic solutions in the expression (5.25) and (5.29).

In PEKERIS' results $\lambda = \mu$ was used, a simplification commonly employed in geophysics. The dimensionless time parameter was $\tau = c_2 t/r$. It can easily be checked that the wave fronts of the pressure, shear and RAYLEIGH waves then are at values of $\tau = 1/\sqrt{3}$, $\tau = 1$ and $\tau = \gamma = \frac{1}{2} \left(3 + \sqrt{3}\right)^{1/2}$ respectively. Examination of the elastic solution of this problem shows that the distance $r = \sqrt{x^2 + y^2}$ and the shear wave velocity $c_2$ are the most advantageous choices for the characteristic length and velocity parameters $d$ and $c$ in (5.22)—(5.24). The expression (5.25) is evaluated for values of $m = 0.5$, $m = 0.75$ and $m = 1.0$ (first elastic case). The parameter $\theta$ was taken as $\theta = 1$. The results are plotted in Fig. 2 and 3.

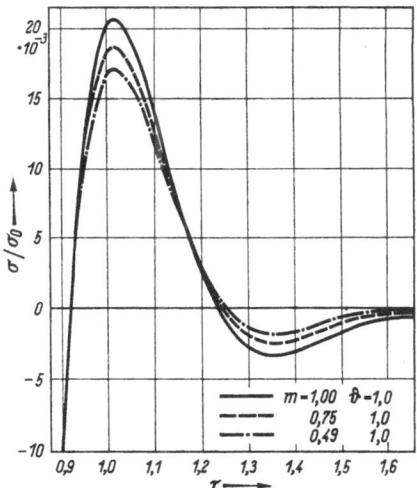

Fig. 4.
Numerical results.

# 7. The Dynamical Expansion of a Spherical Cavity

The problem arises if a pressure $p(t)$ is applied to the surface of a spherical cavity of radius $a$ in an infinite medium. The particular problem, which will be discussed here for the viscoelastic solid defined in Section V, is the one where the pressure varies as the HEAVISIDE unit function $p(t) = p_0 H(t)$.

It may be mentioned that the problem of the dynamical expansion of spherical cavities in other types of material, especially ductile metals, has been surveyed by HOPKINS [17]. The elastic solution for the present problem was taken from that review article.

For $\lambda = \mu$ the displacement potential is

$$\Phi(\tau', R) = - \frac{a^3 p_0}{4\mu} \frac{1}{R} \left[1 - \left(\frac{3}{2}\right)^{1/2} \exp\left(-\frac{2\tau'}{3}\right) \sin\left(\frac{2\tau'\sqrt{2}}{3} + \alpha\right)\right], \quad (7.1)$$

where

$$\tau' = \frac{c_1}{a}\left(t - \frac{R-a}{c_1}\right), \tag{7.2}$$

$$\tan \alpha = \sqrt{2}, \tag{7.3}$$

$$c_1 = \left(\frac{\lambda + 2\mu}{\varrho}\right)^{1/2}. \tag{7.4}$$

As

$$\sigma_R = (\lambda + 2\mu)\frac{\partial^2 \Phi}{\partial R^2} + \frac{2\lambda}{R}\frac{\partial \Phi}{\partial R} \tag{7.5}$$

one has, with the aid of (7.1),

$$\sigma_R = \frac{a^3 p_0}{2R^3}\left\{-2 + \exp\left(-\frac{2\tau'}{3}\right)\left[\left(\sqrt{2} - \frac{2R\sqrt{2}}{a} + \frac{R^2\sqrt{2}}{a^2}\right)\sin\left(\frac{2\tau'\sqrt{2}}{3}\right)\right.\right.$$
$$\left.\left. + \left(2 - \frac{2R^2}{a^2}\right)\cos\left(\frac{2\tau'\sqrt{2}}{3}\right)\right]\right\}. \tag{7.6}$$

The radial distance $R$ and the dilatational wave velocity $c_1$ are here chosen as the characteristic length and velocity parameters $d$ and $c$ in (5.22)—(5.24). For the particular value of $R/a = 10$, the solution of the viscoelastic problem is again obtained by substituting (7.6) into (5.29). Numerical results are plotted in Fig. 4.

### References

[1] BERRY, D. S., and S. C. HUNTER: J. Mech. Phys. Solids 4, 72 (1956).
[2] MORRISON, J. A.: Quart. Appl. Math. 14, 153 (1956).
[3] KOLSKY, H.: Proc. Conf. Stress Propagation, Penn. State Univ., 1959 (ed. by N. DAVIDS), New York: Interscience Publ. Co. 1960, p. 59.
[4] LEE, E. H.: Proc. Conf. Stress Propagation, Penn. State Univ., 1959 (ed. by N. DAVIDS), New York: Interscience Publ. Co. 1960, p. 199.
[5] HUNTER, S. C.: Progress in Solid Mechanics (ed. by I. N. SNEDDON and R. HILL), Vol. I, Amsterdam: North-Holland Publ. Co. 1960, Chap. I.
[6] ACHENBACH, J. D., and C. C. CHAO: J. Mech. Phys. Solids 10, 245—252 (1962).
[7] REINER, M.: Lectures on Theoretical Rheology, Amsterdam: North-Holland Publ. Co. 1960.
[8] SHI, Y. Y., and H. KOLSKY: Brown University Report No. 562, (14)/8, 1960.
[9] SHEMYAKIN, YE. I.: Prikl. Math. Mekh. 22, 289 (1958).
[10] BLAND, D. R.: The Theory of Linear Viscoelasticity, Oxford: Pergamon Press 1960.
[11] LOVE, A. E. H.: The Mathematical Theory of Elasticity, 4th ed., New York: Dover Publications 1946.
[12] ACHENBACH, J. D.: Stanford University, SUDAER Report No. 132, 1962.
[13] PEKERIS, C. L.: Proc. Nat. Ac. Sci. (Wash) 41, 469—480 (1955).
[14] CHAO, C. C.: J. Applied Mechanics 27, (1960) p. 559
[15] ERINGEN, A. C.: Quart. J. Mech. 10, 257 (1957).
[16] ERDELYI, A., et al.: Tables of Integral Transforms, Bateman Manuscript Project, Vol. 1, New York: McGraw-Hill Book Co. 1954.
[17] HOPKINS, H. G.: Progress in Solid Mechanics (ed. by I. N. SNEDDON and R. HILL), Vol. I, Amsterdam: North-Holland Publ. Co. 1960, Chap. III.

# An Example of the Influence of Yield on High Pressure Wave Propagation

By Erastus H. Lee[1] and David T. Liu

Stanford University and Lockheed Missiles and Space Company respectively

## Abstract

The propagation of plane waves in an elastic-plastic medium with displacement normal to the wave front is considered. A closed form solution for moderate unloading behind a shock-wave is developed, which determines the shock-wave attenuation. Yield strength and the influence of temperature change are taken into account. This solution enables the influence of various material characteristics to be assessed. The elastic-plastic solution is compared with the compressible fluid model which neglects yield influences, and with the rigid-plastic approximation in shear. In the particular problem treated, the former determines an appreciable difference in attenuation, while the latter provides only a poor approximation, since the elastic resilience in shear has an appreciable influence on the unloading characteristics.

## Introduction

The remarkable measurements carried out by the groups at the Los Alamos Scientific Laboratory and Stanford Research Institute (see for example [1—5], and the review [6]) which characterise the propagation in solids of plane waves generated by contact explosives, provide data on deformation at high strain rates by means of a system which is amenable to relatively precise analysis. Measurements are taken in the plane part of the wave-front away from the influence of edge effects, so that the displacement is constrained to be normal to the wave front, and the equations of motion can be expressed as a single scalar equation. Thus the influence of lateral motion which complicates the interpretation of measurements of wave propagation down a rod [7] is eliminated. Since the pressure generated by explosive loading is generally large compared with the yield stress of the metals investigated, most of the analyses have been based on the neglect of the influence of yield strength

---

[1] Consultant to Lockheed Missiles and Space Company.

on the process, which is equivalent to representing the solid by a compressible fluid model. While this provides a satisfactory approximation to some overall characteristics of the system, there may be certain aspects of the wave interactions which are particularly sensitive to the influence of yield strength. This effect will, of course, become more dominant when situations are considered for which the maximum pressure is smaller. In initiating a study of the influence of yield strength in these phenomena, we sought to analyse a situation which could be expected to be particularly affected by the yield strength of the material, and which lends itself to relatively simple but applicable approximate analysis, so that this influence can be assessed by comparing solutions based on different approximations. The attenuation of a shock wave due to unloading stresses behind it is such an aspect. The interaction between the shock front and the unloading waves is governed by material characteristics prescribing the response to reduction of the compressive stress normal to the wave front following a large impulsive increase in that stress component. As detailed in the next section, plasticity theory determines an elastic unloading region during which the shear stresses in the specimen are reversed in sign, followed by plastic flow in recovery with strain increments opposite in sign to those generated in the shock wave. That strain increments can be decomposed into independent dilatation and shear components in the state of high compressive, approximately hydrostatic, stress is suggested by BRIDGMAN's work on the insensitivity of plastic flow to hydrostatic compression [8], and confirmed by the interpretation of shock wave measurements in 2024 aluminum by FOWLES [5].

The stress-strain relations governing the initiation of unloading are discussed in the next section, and in the subsequent one are applied to the unloading analysis through the use of characteristic conditions. A particular example is evaluated for a shock wave in 2024 aluminum alloy, and the calculated attenuation behind the shock front is compared with that deduced from the compressible liquid approximation. Comparison is also made with a rigid-plastic unloading approximation, which provides a contrasting approximation to the liquid one.

## Stress-Strain Relations for Unloading

As mentioned in the introduction, we shall consider the deformation to be the combination of isotropic compression due to the average hydrostatic pressure, and elastic-plastic distortion due to the stress deviator or shear stress influence. Only a brief development of the theory is given, since it has already been presented in the literature [5, 9—11]. Let $x$ be the Cartesian coordinate normal to the plane wave surface, then by symmetry the principal stresses, taken positive in compression, are: $\sigma_x$, $\sigma_y = \sigma_z$. These can be defined as true stresses to simplify plasticity relations, but the $\sigma_x$ component will also be the corresponding nominal stress component since lateral expansion is prevented. Since this is the only stress component which appears in the equation of motion, the advantage of using the nominal stress in plastic wave theory [12] is retained. Since no lateral motion occurs, the principal strain components are: $\varepsilon_x$, 0, 0, again $\varepsilon_x$ being positive for compressive strain. It is convenient to use LAGRANGE type coordinates with the displacement component $u$ considered as a function of the time $t$ and the position coordinate $x$ in the undeformed body. Then

$$\varepsilon_x = -\partial u/\partial x \tag{1}$$

is the nominal strain, and is equal to $(V_0 - V)/V_0$ of the physical literature (see, for example [1]) where $V$ is the specific volume and $V_0$ its value for the undeformed state.

For the determination of the initial attenuation of a shock wave by unloading waves following behind it, we shall be concerned with comparatively small changes in stress and strain from the highly compressed state generated by the shock wave. We shall assume a non-linear elastic type compressibility relation, including the influence of temperature changes, as deduced from shock wave experiments [1]. For stress and strain increments $\Delta\sigma_x$ and $\Delta\varepsilon_x$ following the passage of the shock-wave, the compressibility relation governing averaged normal stresses and dilatation takes the form

$$(\Delta\sigma_x + 2\Delta\sigma_y) = 3k\Delta\varepsilon_x; \tag{2}$$

$\Delta\sigma_x$ and $\Delta\sigma_y$ are to be considered finite, but small compared with the shock stress; $k$, the gradient of the pressure-compression curve, will be a function of the average stress, but can be considered constant during the moderate unloading $\Delta\sigma_x$, to a satisfactory order of approximation, as is clear, for example, from a study of the relevant pressure-compression curves in [1] and [3], to be discussed with the numerical results at the end of this paper. In effect, the unloading adiabat is replaced by its tangent at the maximum stress value, and $k$ is chosen accordingly as the gradient there of this adiabat. Eq. (2) contains the assumption that anisotropy of the stress and strain tensors does not modify the compressibility relation based on averaged values of the normal components. This is likely to be a good approximation for pressures large compared with the yield stress, for then contribution to the internal energy from shear components will be comparatively small.

Since the hydrostatic component of the stress tensor is dominating, and since there is evidence [5] of only minor influence of this and strain-rate on the deviator stress-strain relations governed by the elastic-plastic laws, we shall assume elastic-ideally plastic behavior, with the yield stress $Y_0$ in simple tension. The introduction of yield clearly overrides the particular work hardening law which applies. Then the plasticity limit, either according to the VON MISES or TRESCA laws, becomes

$$|\sigma_x - \sigma_y| = Y_0, \tag{3}$$

and since for a given strain, compressibility prescribes the average hydrostatic pressure

$$p(\varepsilon_x) = (\sigma_x + 2\sigma_y)/3, \tag{4}$$

(3) and (4) give

$$\sigma_x = p \pm \frac{2Y_0}{3} \tag{5}$$

if plastic flow is occurring in loading or unloading respectively. It should be noted that the approximation of perfect plasticity is assumed to apply when the temperature rise due to irreversible compression in the shock wave is taken into account. In the example considered in this paper of 2024 aluminium subjected to a maximum stress of 200 kilobars, the temperature rise is about 220 °C [1], and for short time stressing this has only a minor influence on the yield stress [13].

For elastic unloading when $|\sigma_x - \sigma_y| < Y_0$, increments of strain satisfy elastic relations with the compressibility $k$ appropriate for the hydrostatic stress magnitude and the thermodynamic characteristics of the unloading. For the latter, as mentioned above, the influence of the shear stresses and strains is neglected, since in the situations considered they will modify such variables as internal energy only slightly. We shall assume that the elastic shear modulus $G$ based on increments in natural strain is independent of the hydrostatic pressure. BRIDGMAN [14] showed that the increase of $G$ with hydrostatic pressure is much less for most metals than that of the compressibility $k$, and, following MORLAND [10], we shall take $G$ to be constant. For the highly compressed state during initial unloading the principal values of the increment of natural strain are: $\Delta\varepsilon_x/(1 - \varepsilon_x)$, $0, 0$, since $\Delta\varepsilon_x$ was defined as a nominal strain component. Shear behavior can be conveniently expressed in terms of stress and strain differences between components in the $x$ and $y$ directions:

$$\Delta\sigma_x - \Delta\sigma_y = 2G\Delta\varepsilon_x/(1 - \varepsilon_x). \tag{6}$$

Elimination of $\Delta\sigma_y$ between (2) and (6) gives:

$$\Delta\sigma_x/\Delta\varepsilon_x = k + 4G/3(1 - \varepsilon_x). \tag{7}$$

Since $\Delta\varepsilon_x$ is small compared with $\varepsilon_x$, $\varepsilon_x$ can be considered constant in this expression. Eqs. (3) and (6) show that $\Delta\varepsilon_x = -Y_0(1 - \varepsilon_x)/G$ for the change from plastic flow in loading to plastic flow in unloading, hence the corresponding stress reduction is determined by (7):

$$\Delta\sigma_x = -Y_0[4/3 + k(1 - \varepsilon_x)/G]. \tag{8}$$

Fig. 1 shows the configuration of stress-strain relations for unloading from the state $\sigma_x^{max}$, $\varepsilon_x^{max}$ according to the expressions developed above. The broken curve represents the hydrostatic adiabat, and the full lines

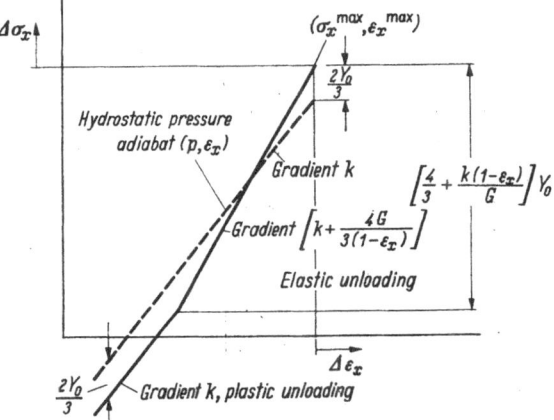

Fig. 1. Stress-strain relations for initial unloading.

the corresponding elastic-plastic unloading path. Unloading stress increments $\Delta\sigma_x$ are considered small compared with $\sigma_x$, so that the region of the hydrostatic adiabat utilized can be adequately replaced by its tangent although appreciable curvature can occur over its whole range down to zero stress. This opportunity to use linear stress-strain relations in the investigation of the initial attenuation of a shock wave forms the basis for the comparatively simple analysis given below.

The assumptions utilized in the above derivation, of the superposition of plastic properties in shear on the hydrostatic compression configuration, and the neglect of strain rate effects, were confirmed for 2024 aluminium through shock wave experiments by G. R. Fowles [5] up to a pressure of 50 kilobars. The assumption of constant shear modulus behind the shock wave was not, however, checked in these experiments, since elastic wave velocities in front of the shock-wave only were measured.

## Wave Propagation Solution

We shall consider the problem of a body with a plane surface subjected to suddenly applied uniform pressure, monotonically decreasing in magnitude after the initial discontinuous rise. For most materials this causes a shock wave to propagate into the body from the surface, followed by a tail of continuously decreasing stress. The latter causes the shock-wave to be attenuated because unloading waves from this tail overtake the shock-wave and interact with it. The analysis developed below provides a closed form solution to this problem for applied pressure decreasing linearly with time, thus presenting a convenient vehicle for assessing properties of the solution and comparing solutions based on different approximations to the stress-strain characteristics.

For moderate unloading $\Delta\sigma_x$, as analysed in the previous section, the stress-strain relation comprises two straight lines, one for elastic and one for plastic unloading as depicted in Fig. 1. The compressible liquid model, corresponding to the broken line in Fig. 1, can also be taken to be linear in the range considered. The theory of plastic wave propagation in one dimension using the LAGRANGE coordinate $x$ takes on a particularly simple form for linear stress-strain characteristics using nominal values of these variables [12]. With

$$c^2 = \frac{1}{\varrho} \frac{d\sigma}{d\varepsilon}\Big|_x, \tag{9}$$

the characteristics, and characteristic relations are

$$\frac{dx}{dt} = \pm c, \qquad \varrho c v \pm \sigma = C_1, \tag{10}$$

where $\varrho$ is the initial density, $\sigma$ is written for $\sigma_x$, $v$ is the particle velocity, and $C_1$ a constant for each characteristic. BOHNENBLUST [15] showed (10) to be correct even when the linear stress-strain relation changes with the position coordinate $x$, as long as its gradient, and hence $c$, remain constant. For an attenuating shock, $\sigma^{max}$ and $\varepsilon^{max}$ in Fig. 1 vary continuously with $x$ as different sections of the plate are considered, and hence unloading occurs for different sections over a set of distinct lines. Since the changes in $\sigma^{max}$ through the plate for the attenuation region considered are of the order of $\Delta\sigma_x$, the gradient of the compressibility curve can be assumed constant for unloading at all sections, and hence (10) holds in both the elastic and plastic unloading regions, assuming the appropriate value of $c$.

In order to determine the conditions immediately behind the shock, $\sigma^{max}(x)$ and $\varepsilon^{max}(x)$, which form the initial values for the unloading solution, we must use the HUGONIOT conditions across the shock (see, for example [1]). Using subscript "$b$" to denote before passage of the shock, and "$a$" after, the HUGONIOT relations take the form

$$\sigma_a - \sigma_b = \varrho c_2 (v_a - v_b),\tag{11}$$

where $c_2$ is the shock velocity in the LAGRANGE sense, given by

$$c_2 = \sqrt{\frac{1}{\varrho}\frac{\sigma_a - \sigma_b}{\varepsilon_a - \varepsilon_b}}.\tag{12}$$

For the problem under consideration, an elastic wave precedes the shock, but its influence can be eliminated from the analysis by considering differences in stress and particle velocity from the corresponding values, $\sigma_0$ and $v_0$, when the shock was first initiated at the face of the plate. Thus for constant unspecified conditions ahead of the shock, subtraction of (11) from the corresponding equation with ($\sigma_a$, $v_a$) replaced by ($\sigma_0$, $v_0$) gives

$$\sigma_0 - \sigma = \varrho c_2 (v_0 - v),\tag{13}$$

where the subscript $a$ is dropped, and thus ($\sigma$, $v$) are the stress and velocity behind the shock. (13) is based on the assumption that $c_2$ is constant, and this is shown below to provide a satisfactory approximation.

The stress-strain relation governing each phase of the motion depends on the thermodynamic conditions pertaining. Thus stress increase through the passage of a shock involves anisentropic deformation which is governed by the HUGONIOT curve, whereas unloading is governed by

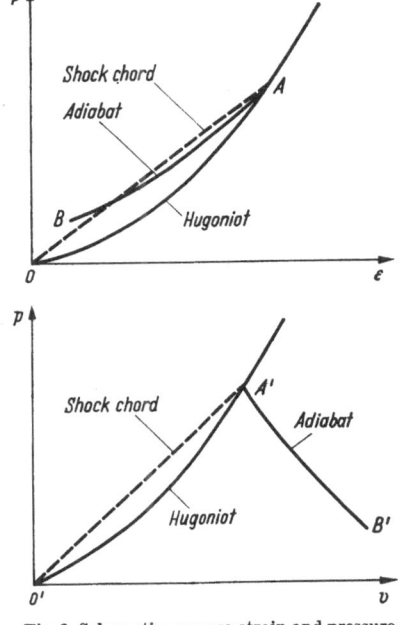

Fig. 2. Schematic pressure-strain and pressure-particle velocity curves.

an adiabat along which entropy is conserved. Since the shock pressures we analyse are many times the yield stress, we shall discuss the choice of stress-strain relation in terms of the corresponding hydrostatic situation. Moreover, the yield influence is partially eliminated, since we express stress by its change from $\sigma_0$. Fig. 2 shows the pressure vs. strain and pressure vs. particle velocity curves for a hypothetical material, which for expository purposes over-emphasises the curvature of the HUGONIOT curves and the differences between possible material histories. The HUGONIOT curve gives the conditions behind shocks of different pressure magnitudes. For a particular point $A$, the gradient of the shock chord $OA$ determines the shock velocity according to (12), with $p$ replacing $\sigma$. The gradient of the shock chord $O'A'$ in the $p,v$ diagram is $\varrho c_2$ by (11). The $p - v$ adiabat shown, $A'B'$, is of the type discussed in [1, 2] and [3] since we shall use values for 2024 aluminum from [3] for the particular solution evaluated. $A'B'$ represents a velocity increase due to unloading by a wave moving in the opposite direction to the shock wave. For unloading from behind the shock which causes attenuation, the particle velocity is reduced rather than increased along the adiabat. However, for the type of curve depicted, the gradient at $A'$ gives $\varrho c_1$, $c_1$ being the gradient of the characteristics for unloading in the compressible fluid case, which is related to the compressibility $k$ through

$$c_1^2 = k/\varrho. \tag{14}$$

With attenuation the point $A$ moves down the HUGONIOT curve. The shock wave velocity, $c_2$, remains almost constant for moderate changes of $A$, as can be seen from the variation of the gradient of the shock chord $O'A'$ in [3] (the variation suggested by Fig. 2 is greatly over-emphasised). If we assume $c_2$ constant, (13) gives

$$\frac{d\sigma}{dv} = \varrho c_2, \tag{15}$$

which is the gradient of the chord $O'A'$. Clearly, since all values $(\sigma, v)$ lie on the HUGONIOT, $d\sigma/dv$ should be the gradient of the tangent at $A'$ rather than that of the chord. The discrepancy is associated with the actual increase of $c_2$ with $v$, and the correct deduction from (11) should have been

$$\frac{d\sigma}{dv} = \varrho c_2 + \varrho (v - v_b) \frac{dc_2}{dv} = \varrho c_2', \tag{16}$$

where $c_2'$ can be determined directly from the gradient of the tangent to the HUGONIOT at $A'$. Thus a much better approximation to the shock characteristics than (12) (with $\sigma_b = \varepsilon_b = 0$ for the liquid case) and (13), with $c_2$ constant, is the combination

$$\varrho c_2^2 = \sigma/\varepsilon, \tag{17}$$

which is equivalent to

$$\varrho c_2 = \sigma/v \tag{18}$$

by the HUGONIOT conditions, and:

$$\sigma_0 - \sigma = \varrho c_2' (v_0 - v) \tag{19}$$

(18) is based on the gradient of the chord $O'A'$ in Fig. 2, and (19) on the gradient of the tangent to the HUGONIOT at $A'$. With $c_2 \neq c_2'$, (18) and (19) violate HUGONIOT conditions across the shock, but the error can be directly associated with the geometry of the characteristic net through the assumed constancy of $c_2$, and is less than that associated with (15).

The characteristic relations which govern the attenuation of the shock wave can now be expressed in terms of the gradient of the elastic unloading line

$$\varrho c_0^2 = k + 4G/3\,(1 - \varepsilon) \qquad (20)$$

from (9) and (7). With the use of (14) this becomes

$$c_0^2 = c_1^2 + \frac{4}{3}\,c_s^2, \qquad (21)$$

where $c_s$ is the LAGRANGE type shear wave velocity, that is the corresponding velocity of propagation through the undeformed body.

Fig. 3 shows the characteristic field for wave propagation from the surface $x = 0$ on which the pressure $f(t)$ is applied at time $t = 0$,

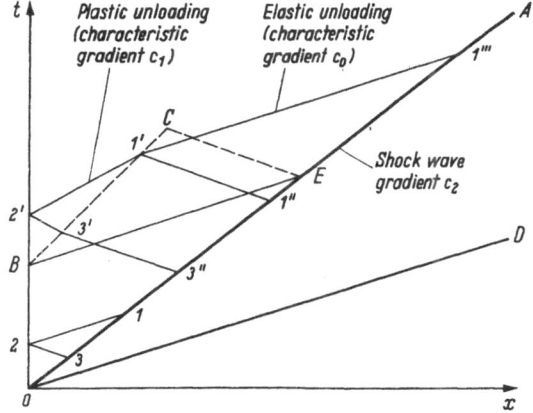

Fig. 3. Configuration of characteristics for an attenuating shock wave.

when the body is undisturbed. $OD$ is the elastic wave front, and $OA$ the shock front. $f(t)$ increases discontinuously to $\sigma_0$ at $t = 0$, and decreases monotonically thereafter. At $B$ the stress has fallen sufficiently to initiate plastic flow in unloading, and the boundary $BC$ separates the elastic and plastic regions.

Relating conditions at the points 1, 2 and 3 on the characteristics through 2 in Fig. 3 determines the initial attenuation. The shock conditions (19) at 1 and 3 give

$$\sigma_0 - \sigma_1 = \varrho\, c_2'(v_0 - v_1), \tag{22}$$

$$\sigma_0 - \sigma_3 = \varrho\, c_2'(v_0 - v_3), \tag{23}$$

and the characteristic relations (10) yield

$$\varrho\, c_0 v_2 + f(t_2) = \varrho\, c_0 v_1 + \sigma_1, \tag{24}$$

$$\varrho\, c_0 v_2 - f(t_2) = \varrho\, c_0 v_3 - \sigma_3. \tag{25}$$

The geometry of the triangle $123$ in Fig. 3 gives

$$t_3 = \frac{c_0 - c_2}{c_0 + c_2}\, t_1, \tag{26}$$

$$t_2 = \frac{c_0 - c_2}{c_0}\, t_1. \tag{27}$$

Elimination of $v_1$, $v_2$, and $v_3$ between (22)—(25) gives the relation:

$$2 f(t_2) = \frac{c_0 + c_2'}{c_2'}\, \sigma_1 - \frac{c_0 - c_2'}{c_2'}\, \sigma_3, \tag{28}$$

which using (26) and (27) expresses the stress $\sigma(t_1)$ immediately behind the shock front as the solution of the finite difference equation

$$\sigma(t) = \frac{c_0 - c_2'}{c_0 + c_2'}\, \sigma\!\left(\frac{c_0 - c_2}{c_0 + c_2}\, t\right) + \frac{2 c_2'}{c_0 + c_2'}\, f\!\left(\frac{c_0 - c_2}{c_0}\, t\right). \tag{29}$$

Since this applies for arbitrary $t$, in the range $OE$, the subscript 1 is dropped. This can be solved by iteration, and in the case of linearly falling surface pressure

$$f(t) = A - Bt \tag{30}$$

has the closed form solution

$$\sigma(t) = A - B\, \frac{c_0^2 - c_2^2}{c_0^2}\, \frac{c_2'}{c_2 + c_2'}\, t. \tag{31}$$

This attenuation of the shock strength will apply over the range $OE$ in Fig. 3, and thereafter the influence of plastic unloading will make itself felt above the characteristic $BE$.

The determination of the boundary $BC$ follows a pattern similar to that adopted above, using the characteristic relations on $1'2'$ and $2'3'$, but replacing the shock conditions by the characteristic relations for $1'1''$ and $3'3''$ since the conditions at $1''$ and $3''$ are known from the previous development. The gradient of the boundary $BC$, denoted by $dx/dt = c_3$, is not known, but the stress values at $1'$ and $3'$ are known in terms of the strength of the shock as it passed these sections, according to (8). Thus the stresses at points on $OA$ and $BC$ in Fig. 3 for the same position coordinate $x$ are related by

$$\sigma_{OA} - \sigma_{BC} = Y_0[4/3 + k(1 - \varepsilon_x)/G] = 2\,Y; \qquad (32)$$

$Y$ thus represents the yield stress for plate compression with lateral expansion prevented. (30) and (32) give

$$t_B = 2\,Y/B. \qquad (33)$$

These conditions on $BC$ enable $c_3$ to be evaluated, and reduce to the form

$$c_1 - c_3 = \frac{c_0 - c_2}{c_0^2\,c_2}\,\frac{(c_0 + c_3)}{(c_1 + c_3)}\,c_1^2\,c_3. \qquad (34)$$

Thus $c_3$ is independent of the particular point $1'$ on $BC$ as long as $1''$ lies below $E$. This establishes that the elastic-plastic boundary $BC$ is linear. If $1''$ lies above $E$, a different stress attenuation function applies along the shock and the boundary will change. This new stress attenuation is determined by the characteristic condition (10) on characteristics of positive slope intersecting $BC$, such as $1'1'''$ in Fig. 3, and the HUGONIOT condition (19) at the shock. The resulting stress behind the shock is given by

$$\sigma(t) = A$$
$$- 2\,Y\left\{\frac{c_2'(c_0 + c_2)}{c_0(c_2 + c_2')} - \frac{c_2'(c_0 - c_2)[2c_2'c_3(c_0 + c_2) + c_2(c_0 - c_2')(c_0 + c_3)]}{c_0c_2(c_0 - c_3)(c_0 + c_2')(c_2 + c_2')}\right\}$$
$$- Bt\,\frac{c_2'(c_0 - c_2)^2\,[2c_2'c_3(c_0 + c_2) + c_2(c_0 - c_2')(c_0 + c_3)]}{c_0^2c_2(c_0 - c_3)(c_0 + c_2')(c_2 + c_2')}. \qquad (35)$$

This applies until the positive characteristic through $C$ intersects the shock front and thereafter it will be necessary to determine the extension of the boundary $BC$. Since the shock stress beyond $E$ is now known, the extension of $BC$ can be carried out by a method similar to that used for $BC$ by using characteristic conditions in the elastic and plastic regions, the unloading condition (32) and the HUGONIOT relation (19) across the shock. Since the stresses are linear along the boundary $x = 0$, and piece-wise linear along the shock front, it is clear that the continuation of the boundary $BC$ will be a polygon, each side determining a new linear attenuation expression. The development of the solution for longer times is shown schematically in Fig. 4. Each boundary segment $CC'$, $C'C''$ ... is determined from the previously calculated attenuation function on $EE'$, $E'E''$ ... Thus there is a band of elastic unloading behind the shock front before the plastic unloading sets in. The boundary pattern will, of course, be modified by the change in form of solution across the characteristic $ECF$, transmitted along $FH$. When this intersets the boundary it will cause a change in gradient, but for the velocities used in the example, this modification only applies after many reflections at $E$, $E'$, $E'' \ldots$

For the liquid model in which yield influences are neglected, the first part of the solution in the region $OBCE$ in Fig. 4, with the hydrostatic velocity $c_1$ (14) replacing $c_0$ (20) in the characteristic relations (10), applies for all time. The effect of yield is thus to cause a speeding up of the unloading wavelets in the elastic band from propagation velocity $c_1$ to $c_0$, and its effect on the attenuation is discussed in the next section.

An approximation to the yield influence which will over-emphasize its effect, in contrast to the liquid assumption, is to assume rigid plastic

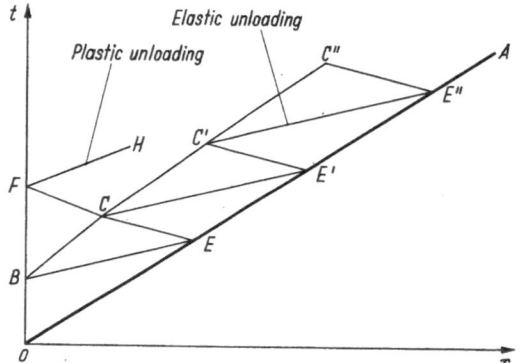

Fig. 4. Extension of the characteristic field.

unloading in shear. This is equivalent to taking $G$ indefinitely large, so that in Fig. 1 the stress drops at constant strain by $4\,Y_0/3$ from $\sigma^{\max}$. The unloading is then parallel to the hydrostatic adiabat in Fig. 1, and becomes coincident with the plastic unloading line corresponding to elastic-plastic theory. With the rigid plastic assumption, the elastic band behind the shock becomes a band of rigid material, and the equations of rigid body motion replace the characteristic relations used above. Fig. 5 indicates the form of the solution which is similar to that detailed in [16] for an analogous problem. The difference equation (29) is replaced by a differential equation, which on $OE_1$ gives the attenuation

$$\sigma(t) = A - Bt\,\frac{c_2'}{c_2 + c_2'}. \tag{36}$$

The boundary $B_1 C_1$ will in general be curved. Its initial gradient, $c_3'$, is given by

$$c_3'^2 + \frac{c_1^2}{c_2}\,c_3' - c_1^2 = 0\,, \tag{37}$$

and the attenuation function beyond $E_1$ also has the gradient given by (36). The stress reduction to yield is $4\,Y_0/3$, so that

$$t(B_1) = 4\,Y_0/3\,B. \tag{38}$$

This is much less than $t(B)$, given by (32) and (33) for the elastic-plastic case shown in Fig. 3.

It is interesting to note that the gradient of the applied pressure, $B$, does not appear in the equations for $c_3$, (34), and $c_3'$, (37), although the characteristic fields are modified by the positions of $B$ and $B_1$ in Figs. 3 and 5 respectively, which depend on the pressure gradient $B$, according to (33), (34), and (38).

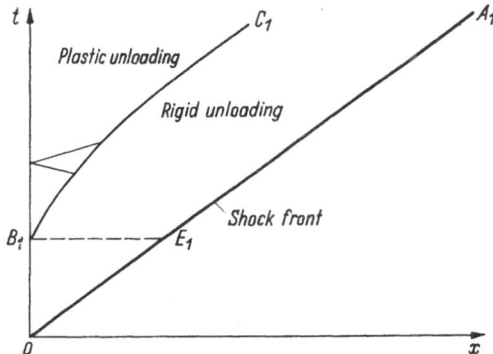

Fig. 5. Wave propagation configuration of a rigid plastic material.

The limit to the extension of the solution through $E$, $E'$, $E'''$ ... in Fig. 4, is determined by the loss of accuracy due to the replacement of the HUGONIOT curve and the adiabats by tangents to these at the point of maximum stress as discussed above, and the geometrical inaccuracy in the characteristic field, caused by taking $c_2$ to be constant. The influence of these assumptions depends on the rate of unloading, and is assessed in the next section for the particular problem evaluated.

## An Example

For a particular example we took the shock-wave results given by WALSH et al. [3] for a maximum shock-wave pressure of 200 kilobars in 2024 aluminum. Approximate values of the relevant constants were taken to provide appropriate relative magnitudes of the various wave velocities which influence the solution. The constants, and the equations in which they are defined are given below:

| | | | |
|---|---|---|---|
| density | | $\varrho$ | $= 2.875 \text{ gm/cm}^5$ |
| maximum strain | | $\varepsilon_x$ | $= 0.16,$ |
| shock wave speed | (18) | $c_2$ | $= 6.77 \text{ km/sec}$ |
| shock wave speed | (19) | $c_2'$ | $= 8.35 \text{ km/sec,}$ |
| fluid and plastic | | | |
| unloading wave | (14) | $c_1$ | $= 8.55 \text{ km/sec,}$ |
| elastic unloading wave | (20) | $c_0$ | $= 9.46 \text{ km/sec,}$ |
| boundary speed | (34) | $c_3$ | $= 6.36 \text{ km/sec.}$ |

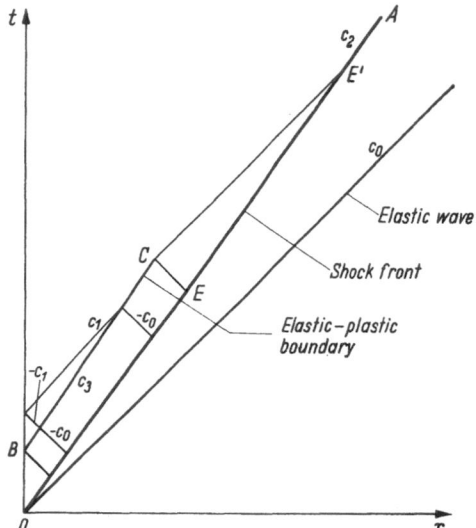

Fig. 6 shows the corresponding wave front configuration in the $x$-$t$ plane. As mentioned in the previous section, the wave speeds, and so the slopes of the lines in Fig. 6 are independent of the attenuation constant $B$. However, a particular value must be selected to fixed the point $B$ at which plastic unloading first occurs.

Fig. 6.
Wave fronts for 2024 aluminum with $\sigma_9 = 200$ KBARS.

The attenuation of the stress at the shock wave according to various assumptions are:

| | |
|---|---|
| fluid case ((31) with $c_1$ replacing $c_0$) | $\sigma(t) = A - 0.205 \text{ Bt,}$ |
| elastic-plastic case on $OE$, Fig. 4, (31) | $\sigma(t) = A - 0.268 \text{ Bt,}$ |
| elastic-plastic case on $EE'$, Fig. 4, (35) | $\sigma(t) = A - 0.362 \ Y - 0.217 \text{ Bt,}$ |
| rigid-plastic case on $OE_1$, Fig. 5, (36) | $\sigma(t) = A - 0.552 \text{ Bt,}$ |

where $Y_0 = 2.4$ kbars and $Y = 3.65 Y_0$. Fig. 7 shows these results plotted on a dimensionless time plot. The points $O$, $E_1$, $E$ and $E'$ correspond to the points designated by the same letters in Figs. 4 and 5.

As mentioned previously, the maximum temperature rise in the shock wave is about 220 °C, which will not produce appreciable reduction in the short time yield strength [13]. The data given in [3] indicate that errors in replacing the HUGONIOT curve and adiabat by their tangents at the 200 kbar points are only a few percent for stress reduction to 150 kbar. The shock wave speed changes by less than 5 percent over the same range, and this will not lead to appreciable distortion of the wave front configuration, Fig. 6. Thus the solution can be expected to be satisfactory down to this stress if the constant $B$ (30) gives this attenuation by the point $E'$ in Fig. 6. For smaller attenuation, extension beyond $E'$ will be satisfactory, but only to an intermediate stress, since geometrical distortion of the characteristic net is then more severe.

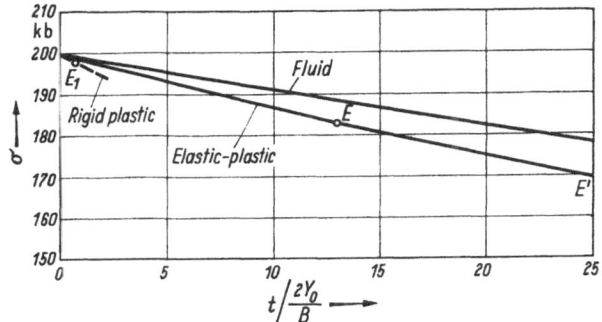

Fig. 7. Shock wave attenuation corresponding to different material properties.

## Discussion

The solution presented above, being of closed form, permits an assessment of the interaction between surface stress reduction and the various wave speeds involved. It provides an example of the convenience of the characteristic method, similar to that given by BOHNENBLUST [15] for elastic unloading, which unfortunately has not appeared in the printed literature. The use of the characteristic relations (10), even when unloading takes place down distinct stress-strain lines at different sections, seems more direct than the wavelet approach given by MORLAND [10]. Using numerical methods, TUPPER [11] evaluated problems with an elastic unloading band transmitting unloading waves from a plastic to a shock front similar to that shown in Fig. 4.

The attenuation curves shown in Fig. 7 indicate the more severe attenuation according to the elastic-plastic solution. Because of the large factor of 3.65 for $Y/Y_0$, the rigid unloading solution over the range $OE_1$ in Fig. 5 is extremely localized in Fig. 7 and determines an appreciably steeper attenuation. It seems that is not likely to be a satisfactory approximation, and this is confirmed by the marked discrepancy in Fig. 1 of the rigid-plastic unloading path from the elastic-plastic one.

It is not possible to compare the results obtained in this paper with attenuation measurements presented in the literature. These have been obtained for much higher pressures, when the high temperature generated by the shock wave is likely to cause appreciable reduction in the yield stress. A comparison of attenuation measurements in the more moderate pressure range could supply information on the influences of pressure on the elastic shear modulus $G$.

## Bibliography

[1] RICE, M. H., R. G. McQUEEN and J. M. WALSH: Compression of Solids by Strong Shock Waves, Solid State Physics, Vol. 6, F. SEITZ and D. TURNBULL, Eds., New York: Academic Press 1958, pp. 1—63.

[2] WALSH, J. M., and R. H. CHRISTIAN: Equation of State of Metals from Shock Wave Measurements, Phys. Rev. 97, 1544 (1955).

[3] WALSH, J. M., M. H. RICE, R. G. McQUEEN and F. L. YARGER: Shock-Wave Compressions of Twenty-Seven Metals, Equations of State of Metals, Phys. Rev. 108, 196 (1957).

[4] DUVALL, G. E.: Some Properties and Applications of Shock Waves, Metallurgical Society Conferences, Vol. 9, P. G. SHEWMON and V. F. ZACKAY, Eds., New York: Interscience Publishers 1961, pp. 165—202.

[5] FOWLES, G. R.: Shock Wave Compression of Hardened and Annealed 2024 Aluminum, J. Appl. Phys. 32, 1475 (1961).

[6] DUVALL, G. E.: Shock Waves in the Study of Solids, Appl. Mech. Rev. 15, 849 (1962).

[7] PLASS, H. J., and E. A. RIPPERGER: Current Research on Plastic Wave Propagation at the University of Texas, Plasticity, Proceedings of the Second Symposium on Naval Structural Mechanics, E. H. LEE and P. S. SYMONDS, Eds., Pergamon Press 1960.

[8] BRIDGMAN, P. W.: Studies in Large Plastic Flow and Fracture, McGraw-Hill Book Co., 1952.

[9] WOOD, D. S.: On Longitudinal Plane Waves of Elastic-Plastic Strain in Solids, J. Appl. Mech. 19, 521 (1952).

[10] MORLAND, L. W.: The Propagation of Plane Irrotational Waves Through an Elastoplastic Medium, Phil. Trans. Roy. Soc. A 251, 341 (1959).

[11] TUPPER, S. J.: On the Propagation of Plane Stress Waves Generated in a Thick Steel Plate by a Surface Explosion, British Technical Report ARDC (B) 12/61, 1 (1961).

[12] LEE, E. H.: A Boundary Value Problem in the Theory of Plastic Wave Propagation, Quart. of Appl. Math. 10, 335 (1953).

[13] MOON, D. P., and W. F. SIMMONS: Selected Short-Time Tensile and Creep Data Obtained Under Conditions of Rapid Heating, Defense Metals Information Center Report 130, Battelle Memorial Inst. 1960.

[14] BRIDGMAN, P. W.: The Physics of High Pressure, London: Bell and Sons 1949, p. 386.

[15] BOHNENBLUST, H. F.: Comments on White and Griffis' Theory of the Permanent Strain in a Uniform Bar Due to Longitudinal Impact, OSRD Report No. 781, 1942.

[16] LEE, E. H., and A. J. WANG: Wave Propagation in an Elastic Rod Exhibiting Internal Coulomb Friction, Considered as a Model for a Ring Spring, J. Appl. Mech. 23, 367 (1956).

# Pulse Propagation in a Viscoelastic Solid with Geometric Dispersion[1]

By Julius Miklowitz

California Institute of Technology, Pasadena, California, U.S.A.

## Abstract

A technique employing double integral transforms is given for treating pulse propagation problems in a viscoelastic solid with geometric dispersion. It is used in connection with a correspondence principle and a linear model. Two examples treated are, taken from the general problem of pulse scattering by a circular cylindrical cavity in the infinite solid. In the corresponding elastic problem RAYLEIGH waves, generated by the pulse-cavity interaction, are spatially non-decaying (in $\theta$ about the cavity), periodic disturbances that predominate for long time. It is shown that their viscoelastic counterparts are spatially attenuated waves.

## Introduction

Recent work employing double integral transforms has proved fruitful in solving transient wave propagation problems in elastic rods and plates. In particular the technique given by the author [1] and LLOYD and the author [2] yields a solution comprised of a series of integrals, each of which is associated with a particular branch of the underlying frequency-wave number spectrum. This enables a study of individual wave types in the transient response to be made. The technique lends itself to viscoelastic problems as the examples treated in the present work will demonstrate.

The examples are taken from the general problem of pulse scattering by a circular cylindrical cavity in the infinite solid, a problem which has had current interest in connection with protective construction. In [3] on the elastic problem the author showed RAYLEIGH waves, generated by the pulse-cavity interaction, are spatially independent (in $\theta$ about the

[1] This work was sponsored by AFSWC, Kirtland Air Force Base, New Mexico, through Contract No. AF 29(601)-5395 with National Engineering Science Company, Pasadena, California.

cavity), nondecaying, and periodic ($2\pi$ in time) disturbances that predominate in the long-time solution. It was therefore natural to investigate the nature of these waves for the viscoelastic medium.

Two problems were treated in [3], one being the case of normal line load excitation of the cavity wall, and the other the case of impingement of a plane compressional pulse on the cavity wall. To investigate the corresponding viscoelastic cases, a model-correspondence principle has been used. Since the RAYLEIGH waves involved are high frequency-short waves, it was natural to employ a model having instantaneous elasticity represented to preserve the hyperbolicity of the governing equations. Towards this end the MAXWELL element suffices, being simple, yet exhibiting the essential feature of what is needed here. As will be shown later, the MAXWELL element introduces spatial decay into the formerly nondecaying elastic RAYLEIGH waves.

## Line Load Problem: Elastic Case

### (a) Statement of Problem

The subject problem is illustrated in Fig. 1. A line load $PF(t)$ is suddenly applied, at time $t = 0$, to the cavity wall $r = a$ at $\theta = 0$; $P$ is a magnitude constant of dimensions force/unit length and $F(t)$ prescribes the time behavior of the input. The governing wave equations are

$$\nabla^2 \varphi = \frac{1}{c_d^2} \frac{\partial^2 \varphi}{\partial t^2}, \quad \nabla^2 \psi = \frac{1}{c_s^2} \frac{\partial^2 \psi}{\partial t^2}, \quad (1)$$

where the potentials $\varphi$ and $\psi$ are related to the displacements through

$$u_r = \frac{\partial \varphi}{\partial r} + \frac{\partial \psi}{r \partial \theta}, \quad u_\theta = \frac{\partial \varphi}{r \partial \theta} - \frac{\partial \psi}{\partial r}; \quad (2)$$

$\nabla^2$ is the Laplacian polar operator, $c_d$

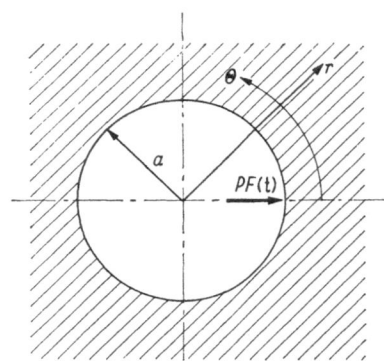

Fig. 1. Problem of cavity surface normal line source.

and $c_s$ are respectively the dilatational and equivoluminal body wave speeds, defined by $c_d^2 = (\lambda + 2\mu)/\varrho$ and $c_s^2 = \mu/\varrho$, where $\lambda$ and $\mu$ are the LAMÉ constants, and $\varrho$ is the material density.

The problem is one of plane strain $(u_z = \partial/\partial z = 0)$, the stresses being given by

$$\sigma_r = \lambda \Delta + 2\mu \varepsilon_r, \qquad \sigma_\theta - \lambda \Delta + 2\mu \varepsilon_\theta,$$
$$\sigma_z = \nu'(\sigma_r + \sigma_\theta), \qquad \sigma_{r\theta} = \mu \varepsilon_{r\theta}, \qquad (3)$$

where the strains and dilatation are given by

$$\varepsilon_r = \frac{\partial u_r}{\partial r}, \qquad \varepsilon_\theta = \frac{u_r}{r} + \frac{\partial u_\theta}{r\,\partial\theta}, \qquad \Delta = \varepsilon_r + \varepsilon_\theta,$$

$$\varepsilon_{r\theta} = \frac{\partial u_r}{r\,\partial\theta} + \frac{\partial u_\theta}{\partial r} - \frac{u_\theta}{r}, \tag{4}$$

and $\nu'$ is POISSON's Ratio.

Boundary conditions (at $r = a$) for the problem are given by

$$\sigma_r(a,\theta,t) = -\frac{P}{a}\,F(t)\,\delta(\theta), \qquad \sigma_{r\theta}(a,\theta,t) = 0, \tag{5}$$

where $\delta(\theta)$ is the symmetrical delta function. The potentials $\varphi$ and $\psi$, hence the displacements and stresses, are required to vanish as $r \to \infty$, i.e.,

$$\lim_{r\to\infty}[\varphi(r,\theta,t),\,\psi,\,u_r,\,u_\theta,\,\text{etc.}] = 0. \tag{6}$$

Initial conditions are taken as

$$\varphi(r,\theta,0) = \psi(r,\theta,0) = \frac{\partial\varphi(r,\theta,0)}{\partial t} = \frac{\partial\psi(r,\theta,0)}{\partial t} = 0 \tag{7}$$

representing quiescence at $t = 0$.

### (b) Formal Solution of Problem

Formal solutions (inversion integrals) in [3] were derived by using the technique given by FRIEDLANDER [4] in his related work on the acoustic wave diffraction problem. From his work it is clear that solutions for the present problems can be obtained in the form of

$$\Phi(r,\theta,t) = \sum_{n=-\infty}^{\infty} \varphi(r,\theta + 2n\pi, t). \tag{8}$$

This gives solution $\Phi$ definition, not only in the physical plane, $r \geq a$, $-\pi \leq \theta \leq \pi$, but also on a RIEMANN surface the sheets of which are given by $(2n-1)\pi < \theta < (2n+1)\pi$, $n = 0, \pm 1, \pm 2, \ldots$ In this formulation a single source, say at $(a, 0)$ is represented by an infinity of sources at $(a, 2n\pi)$. This corresponds physically to the fact that the single source can signal to a receiver, not only through a direct ray, but by rays corresponding to diffracted waves which wrap themselves around the cavity. One can solve for $\Phi$, therefore, by getting a solution for $\varphi(r, \theta + 2n\pi, t)$ in the physical plane, $n = 0$, corresponding to a single source there, and using (8) to get the total solution. This procedure was followed in [3].

[3] employed integral transform pairs to get the formal solutions for the physical plane ($n = 0$), LAPLACE on $t$

$$\bar{\varphi}(r, \theta, p) = \int_0^\infty \varphi(r, \theta, t)e^{-pt}dt, \quad \varphi(r, \theta, t) = \frac{1}{2\pi i}\int_{Br_1} \bar{\varphi}(r, \theta, p)e^{pt}dp \quad (9\,a, b)$$

and FOURIER on $\theta$

$$\tilde{\varphi}(r, \nu, t) = \int_{-\infty}^\infty \varphi(r, \theta, t)e^{i\nu\theta}d\theta, \quad \varphi(r, \theta, t) = \frac{1}{2\pi}\int_{-\infty}^\infty \tilde{\varphi}(r,\nu,t)\, e^{-i\nu\theta}d\nu, \quad (10\,a, b)$$

where $p$ and $\nu$ are the LAPLACE and FOURIER transform parameters, respectively, and $Br_1$ is the well-known contour in the right half $p$-plane.

Application of (9a), (10a), and (7) to (1) gives

$$r^2\frac{d^2\tilde{\bar{\varphi}}}{dr^2} + r\frac{d\tilde{\bar{\varphi}}}{dr} - (r^2 k_d^2 + \nu^2)\tilde{\bar{\varphi}} = 0,$$

$$r^2\frac{d^2\tilde{\bar{\psi}}}{dr^2} + r\frac{d\tilde{\bar{\psi}}}{dr} - (r^2 k_s^2 + \nu^2)\tilde{\bar{\psi}} = 0 \tag{11}$$

the STURM-LIOUVILLE equations for the system, where $\tilde{\bar{\varphi}}$, $\tilde{\bar{\psi}}$ are the transformed displacement potentials, and $k_d = p/c_d$, $k_s = p/c_s$. A solution of (11) satisfying the transform of (6) may have the form

$$\tilde{\bar{\varphi}} = A_\nu(p, \nu)K_\nu(w),$$

$$\tilde{\bar{\psi}} = B_\nu(p, \nu)K_\nu(y), \tag{12}$$

where $K_\nu$ is the modified BESSEL function of the second kind of order $\nu$, and $w = k_d r$ and $y = k_s r$. The related transformed displacements and stresses may be obtained through transformations of (2) and (3) and applying the latter to (12).

Transformation of (5) and application of these to the transformed stresses determines $A_\nu$ and $B_\nu$, hence giving final definition to the transformed potentials, displacements, and stresses. Making use now of (9b) and (10b), the displacements are given formally by

$$\left.\begin{matrix} u_r(r, \theta, t) \\ u_\theta(r, \theta, t) \end{matrix}\right\} = \frac{1}{4\pi^2 i}\int_{Br_1} e^{pt}\left[\int_{-\infty}^\infty \left\{\begin{matrix} \tilde{\bar{u}}_r(r, \nu, p) \\ \tilde{\bar{u}}_\theta(r, \nu, p) \end{matrix}\right\} e^{-i\nu\theta}d\nu\right] dp, \quad (13\,a)$$

where

$$\bar{\bar{u}}_r(r, \nu, p) = \Gamma(p, \nu)\left[k_d A K'_\nu(w) + \frac{\nu}{r} B K_\nu(y)\right], \qquad (13\,\mathrm{b})$$

$$\bar{\bar{u}}(r, \nu, p) = -i\Gamma(p, \nu)\left[\frac{\nu}{r} A K_\nu(w) + k_s B K'_\nu(y)\right], \qquad (13\,\mathrm{c})$$

$$A = c_d^2[K_{\nu+2}(v) + K_{\nu-2}(v)], \qquad B = c_s^2[K_{\nu+2}(u) - K_{\nu-2}(u)],$$

$$\Gamma(p, \nu) = -\frac{P f(p)}{\mu a p^2 K_\nu(\mu) K_\nu(v) C},$$

$$C = \frac{K_{\nu-2}(u)}{K_\nu(u)} \cdot \frac{K_{\nu+2}(v)}{K_\nu(v)} + \frac{K_{\nu+2}(u)}{K_\nu(u)} \cdot \frac{K_{\nu-2}(v)}{K_\nu(v)}$$

$$- \left(1 - \frac{c_d^2}{c_s^2}\right)\left[\frac{K_{\nu+2}(v)}{K_\nu(v)} + \frac{K_{\nu-2}(v)}{K_\nu(v)}\right],$$

and where $u = k_d a$, $v = k_s a$, $f(p)$ is the LAPLACE transform of $F(t)$, and the prime denotes differentiation with respect to the arguments of the $K$ functions (this form of solution restricts $K_\nu(u)$ and $K_\nu(v)$ to non-vanishing values).

### (c) Exact Inversion

(13a) can be inverted exactly through residue theory and contour integration. The details are given in [3] so the following can be brief. Consider the inner integral in (13a). Since $p$ is complex, $C$ may have zeros. Further, for $\theta > 0$, convergence of (13a) requires that these roots of $C$ be restricted to $\mathrm{Im}\,\nu \leq 0$[1]. Hence the inner integral in (13a) is inverted by using residue theory applied to the contour shown in Fig. 2 with the result

$$\left.\begin{matrix} u_r(r, \theta, t) \\ u_\theta(r, \theta, t) \end{matrix}\right\} = \frac{1}{2\pi i}\int_{Br_1} e^{pt}\left[i\sum_{\nu_j}\left\{\begin{matrix} N_r(r, \nu, p)\big/\dfrac{\partial C}{\partial \nu} \\ N_\theta(r, \nu, p)\big/\dfrac{\partial C}{\partial \nu} \end{matrix}\right\}_{\nu=\nu_j(p)} e^{-i\nu_j\theta}\right]dp, \qquad (14)$$

where $\nu_j(p) = (\nu_r - i\nu_i)_j$, in terms of its real and imaginary components[2], and $N_r = C\bar{\bar{u}}_r$, $N_\theta = C\bar{\bar{u}}_\theta$.

---

[1] Since the problem is one of symmetry about $\theta = 0$, it suffices to consider $\theta > 0$.

[2] Note that complex $\nu$ corresponds to a spatially attenuated wave which can be physically realized in a linear elastic system.

Now the LAPLACE transform in (14) may be inverted by termwise contour integration. Completion of $Br_1$ to the left as shown in Fig. 3, for each of the terms in (14), is inherent in the foregoing analysis since it takes $p$ through its physical values $p = \pm i\omega$ (arguing stability of the solution with time and no time damping) for the branches $\nu_j(\pm i\omega)$ of $C$. The cut shown is a common one needed to make all the $K_\nu$ functions in the integrands of (14) singled-valued [1].

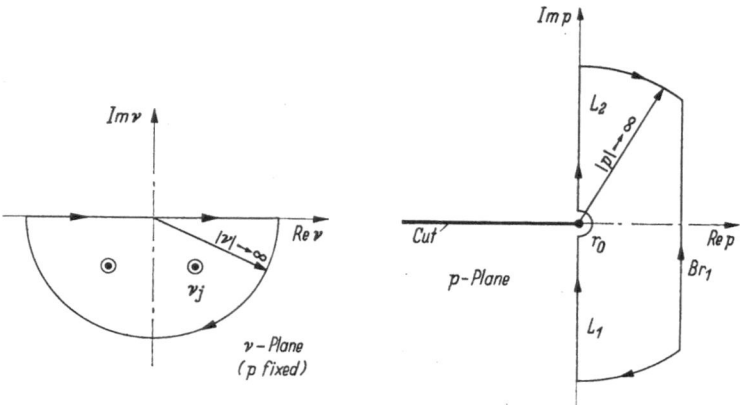

Fig. 2. Integration in the $\nu$-plane.     Fig. 3. Integration in the $p$-plane.

Through the use of the CAUCHY-GOURSAT theorem, and arguments based on the conjugate nature of the roots of $C$ and the integrands in (14), the latter becomes

$$
\left.\begin{array}{c}
u_r(r,\theta,t) \\
u_\theta(r,\theta,t)
\end{array}\right\} = \frac{1}{\pi}\int_0^\infty \operatorname{Re} \sum_{\nu_j(-i\omega)} i\left[ D\Big|_{\substack{\nu=\nu_j \\ p=-i\omega}} e^{-i[\nu_j\theta+\omega t]} + D\Big|_{\substack{\nu=-\bar{\nu}_j \\ p=-i\omega}} e^{i[\bar{\nu}_j\theta-\omega t]} \right] d\omega, \quad (15)
$$

where the roots $\nu_j(-i\omega)$ are found from

$$
C\Big|_{p=-i\omega} = \frac{H^{(1)}_{\nu-2}(\eta)}{H^{(1)}_\nu(\eta)}\cdot\frac{H^{(1)}_{\nu+2}(\zeta)}{H^{(1)}_\nu(\zeta)} + \frac{H^{(1)}_{\nu+2}(\eta)}{H^{(1)}_\nu(\eta)}\cdot\frac{H^{(1)}_{\nu-2}(\zeta)}{H^{(1)}_\nu(\zeta)}
$$
$$
- \left(\frac{c_d^2}{c_s^2}-1\right)\left[\frac{H^{(1)}_{\nu+2}(\zeta)}{H^{(1)}_\nu(\zeta)} + \frac{H^{(1)}_{\nu-2}(\zeta)}{H^{(1)}_\nu(\zeta)}\right] = 0, \quad (16)
$$

and where $D = \left\{\begin{array}{c} \text{in} \\ (14) \end{array}\right\}$, $\eta = \omega a/c_d$, and $\zeta = \omega a/c_s$ (i.e., $u = pa/c_d = i\omega a/c_d$, $v = pa/c_s = i\omega a/c_s$. The symmetries of the present problem

---

[1] The possibility of additional singularities is discussed in [3]. If they exist one would expect additional integrals in the solution but no basic change in its form.

have therefore reduced root finding to those roots of (16), $\nu_j(\eta)$, lying in one quadrant of an eight quadrant $\eta - \nu_j$ (dimensionless frequency-wave number) space. Once these $\nu_j(\eta)$ branches have been determined numerically they would permit (15) to be integrated numerically to give the final results for $u_r$ and $u_\theta$, hence also the stresses.

In analyzing (15) further it is helpful to draw on known properties in wave scattering problems. It may be seen that the terms of (15) correspond to component diffracted and radiated harmonic waves [one for

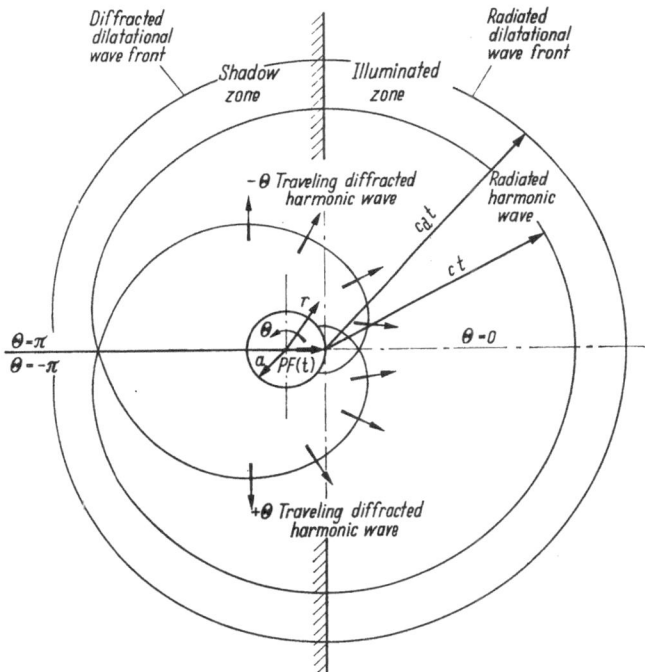

Fig. 4. Wave propagation from a surface line source in a circular cylindrical cavity.

each $(\nu_j, \omega)$ and $(\bar{\nu}_j, \omega)$ pair] travelling in the $\theta$-direction, and outward in $r$, the latter stemming from the $H_\nu^{(1)}(\text{w})$ and $H_\nu^{(1)}(y)$ character of $D$. The first term corresponds to diffracted and radiated waves travelling in the negative $\theta$-direction and the second to those travelling in the positive $\theta$-direction. For the $(-\theta)$ domain the terms interchange their roles. Taking this symmetry into account. Fig. 4 shows the position of two such diffracted and radiated waves, starting from the line source $PF(t)$ travelling with phase velocity $c$, corresponding to a time $t$ when the diffracted waves at the cavity wall have already begun their second trip around the cavity. It is clear that the RIEMANN surface sheets $n = \pm 1$, in addition to $n = 0$, are involved here, hence the corresponding terms in (8).

Fig. 4 points out that the diffracted waves wrap themselves around the cavity continuously over the domain $-\infty < \theta < \infty$, consistent with the governing Eqs. (8), (9), and (10), and the discussion following (8). It is clear that one gets the total effect at a station at time $t$ in the physical plane by projecting all the diffracted waves, which propagate into and on the $-\theta$ and $+\theta$ sheets of the RIEMANN surface, into the physical plane. The number of sheets that come into play [i.e., the number of terms in (8)] in determining an actual disturbance are limited by the length of time that has elapsed since $t = 0$. In Fig. 4 this is evidenced by noting that the dilatational wave fronts (radiated and diffracted) which travel with the speed $c_d$, are outer bounds for the disturbance. Hence, even though each of the terms in (15) correspond to waves spread out on a RIEMANN surface of the doubly infinite number of sheets, the wave front condition leads to cancellation of the negative and positive travelling waves in (15) in advance of the dilatational wave fronts, and the existence of the sum, indicated there only back of these fronts.

It is important to note, as pointed out earlier, that each integral in (15) represents one branch of the frequency equation $\nu_j(\eta)$ in $0 \le \omega < \infty$. This permits a detailed study to be made of the various waves in the problem with an eye on both wave number $\nu$ and frequency $\omega$. Approximation which, in some sense, is a feature of most involved elastic wave problems is particularly aided by this character. The long time approximate solution obtained in the next section is a case in point. This stems from the RAYLEIGH wave root which comes from a large $|\nu|$ and large $\omega$ approximation to one branch of (16).

## (d) Rayleigh Waves and the Long Time Solution

It is pointed out in Ref. [3] that (16) was also the frequency equation VIKTOROV [5] found for the related steady wave propagation problem. He showed the dispersive nature of the waves it governs in his work on steady RAYLEIGH-type waves. The dispersion is not surprising when one considers the curvature of the boundary in this problem. VIKTOROV remarked that with $\eta$ and $\zeta$ real (as they are here) (16) would only have roots corresponding to complex $\nu = ka = k_r a + i k_i a = \nu_r + i \nu_i$ ($k$ being wave number), but offered no proof of his statement. He showed, however, for large $|\nu|$, vanishingly small imaginary part $\nu_i$, and large $\omega$ there was a limiting real root $\nu_0 = k_0 a = \omega a / c_0$ corresponding to the RAYLEIGH surface wave on the free half space ($c_0$ is the RAY-

LEIGH wave speed). This is logical since $\nu_0 = k_0 a \to \infty$ can be interpreted as $k_0 \to \infty$ for fixed $a$, hence this very short wave "does not see the curvature".

The time dependent RAYLEIGH waves are obtained by first picking out of (15) the two terms corresponding to the branch $\nu_j(-i\omega)$ containing the RAYLEIGH wave real pair $(\omega, \nu)$ as a high frequency-short wave limit $(\omega, \nu \gg 1)$. The pertinent branch of $C$ in (16) is defined by this limit. As in [5] the desired approximation of $C$ and other compatible terms in (15) were found by using the appropriate DEBYE asymptotic expansion for the $H_\nu^{(1)}$ functions in (16) and (15). The general expansion needed is one in which both order $\nu$ and argument $\eta$ (or $\zeta$) are large and positive. The present work further imposes that

$$\nu/\eta = c_d/c_0 > \nu/\zeta = c_s/c_0 > 1. \tag{17}$$

The general expansion may be found in ERDELYI et al [6]. (17) means that through the substituion $\nu/\eta = \cosh\alpha$ this expansion for the first order term can be written

$$H_\nu^{(1)}(\eta) \cong -i\sqrt{\frac{2}{\pi\nu\tanh\alpha}}\,e^{-\nu(\tanh\alpha-\alpha)}\{1 + 0(1/\nu)\}, \tag{18}$$

where $\alpha = \tanh^{-1}(q_0/k_0)$, $q_0/k_0 = \sqrt{1 - b^2\varkappa^2}$, $\varkappa^2 = c_0^2/c_s^2$, and $b^2 = c_s^2/c_d^2$. $H_\nu^{(1)}(\zeta)$ is just like (18) except that $\beta$ is substituted for $\alpha$, where $\beta = \tanh^{-1}(s_0/k_0)$, and $s_0/k_0 = \sqrt{1 - \varkappa^2}$.

Making this first order approximation to (16) yields

$$(2/b^2\varkappa^4)\left[(\varkappa^2 - 2)^2 - 4\sqrt{1 - b^2\varkappa^2}\sqrt{1 - \varkappa^2}\right] = 0, \qquad 0 < \varkappa < 1, \tag{19}$$

the well-known equation for the speed of a RAYLEIGH wave on the surface of a free elastic half space. Taking into account the continuity of (18) in $\nu$ and $\omega$, the derivation of (19) proves that the branch of $C$ containing the RAYLEIGH wave pair has the real asymptote

$$\eta = b\varkappa\nu \qquad (\text{or} \quad \nu = k_0 a = \omega a/c_0). \tag{20}$$

From [5] it is clear that the branch approaches this asymptote (for $\omega$, $\nu \gg 1$) through a vanishingly small Im $\nu$.

Taking $F(t)$ as the delta function input $\delta(t)$ the first order approximation of (18) is made to $D\,\big|_{\substack{\nu=\nu_j \\ p=-i\omega}}$ in (15), and through the evenness of $K_\nu$ with respect to $\nu$, together with the use of (20), reduces (15) to

$$\left.\begin{matrix} u_{rR}^\delta(\nu, \theta, t) \\ u_{\theta R}^\delta(r, \theta, t) \end{matrix}\right\} = \frac{2}{\pi}\int_0^\infty \left\{\begin{matrix} \bar{u}_{rR}^\delta(r, \nu, t)\cos\nu\theta \\ \bar{u}_{\theta R}^\delta(r, \nu, t)\sin\nu\theta \end{matrix}\right\}\,d\nu, \tag{21}$$

where $\tilde{u}_{rR}^\delta$ and $\tilde{u}_{\theta R}^s$ are respectively the FOURIER cosine and sine transforms of the RAYLEIGH wave displacements (the superscript $\delta$ corresponding to the delta function input) given by

$$\tilde{u}_{rR}^\delta(r, \nu, t) = \pi S[U_A(r)e^{-A(r)\nu} + U_{\dot B}(r)e^{-B(r)\nu}] \sin D(t)\nu, \qquad (22\,\mathrm{a})$$

$$\tilde{u}_{\theta R}^\delta(r, \nu, t) = -\frac{\pi S}{\varkappa} [V_A(r)e^{-A(r)\nu} + V_B(r)e^{-B(r)\nu}] \sin D(t)\nu, \qquad (22\,\mathrm{b})$$

where

$$S = \frac{Pc_0\varkappa}{\pi\mu a R}, \qquad\qquad D(t) = c_0 t/a,$$

$$U_A(r) = b(2 - \varkappa^2)\left(\frac{a}{b\varkappa r} - \sqrt{\frac{1 + Q_0/k_0}{1 - Q_0/k_0}}\right)\sqrt{\frac{q_0}{Q_0}}, \qquad U_B(r) = \frac{2aq_0}{\varkappa r k_0}\sqrt{\frac{s_0}{S_0}},$$

$$V_A(r) = (2 - \varkappa^2)\frac{a}{r}\sqrt{\frac{q_0}{Q_0}}, \qquad V_B(r) = \frac{2\varkappa q_0}{k_0}\left(\frac{a}{\varkappa r} - \sqrt{\frac{1 + S_0/k_0}{1 - S_0/k_0}}\right)\sqrt{\frac{s_0}{S_0}},$$

$$A(r) = Q_0/k_0 - q_0/k_0 - \tanh^{-1}(Q_0/k_0) + \tanh^{-1}(q_0/k_0),$$

$$B(r) = S_0/k_0 - s_0/k_0 - \tanh^{-1}(S_0/k_0) + \tanh^{-1}(s_0/k_0),$$

$$Q_0(r)/k_0 = \sqrt{1 - b^2\varkappa^2 r^2/a^2}, \qquad S_0(r)/k_0 = \sqrt{1 - \varkappa^2 r^2/a^2},$$

$$R = 2\left\{\frac{8[2 - (1 + b^2)\varkappa^2] + 4(\varkappa^2 - 2)^3 + (\varkappa^2 - 2)^4}{(\varkappa^2 - 2)^2}\right\},$$

in which $r$ has been restricted to the neighborhood of the cavity wall $(a \leq r < a/\varkappa)$ where the major effects occur, rendering $A(r)$ and $B(r)$ both real and $\geq 0$. Both approach zero as $r \to a$.

Eqs. (22) represent simple transforms and their inverses may be found in OBERHETTINGER's tables [7]. The RAYLEIGH wave displacements and circumferential stress at the surface $r = a$, where the maxima occur, are[1]

$$u_{vR}^\delta(a, \theta, t)/u_l = \frac{c_0\varkappa^2\sqrt{1 - b^2\varkappa^2}}{\pi a R}\left[\frac{1}{D - \theta} + \frac{1}{D + \theta}\right], \qquad (23\,\mathrm{a})$$

$$u_{\theta R}^\delta(a, \theta, t)/u_l = -\frac{c_0\varkappa^2(2 - \varkappa^2)}{2aR}[\delta(D - \theta) - \delta(D + \theta)], \qquad (23\,\mathrm{b})$$

$$\sigma_{\theta R}^\delta(a, \theta, t)/\sigma_l = \frac{2c_0\varkappa^2(2 - \varkappa^2)}{aR}[\delta'(D - \theta) + \delta'(D + \theta)], \qquad (23\,\mathrm{c})$$

for $0 < \theta \leq \pi$, and for $-\pi \leq \theta < 0$ after substitution of $(-\theta)$ for $\theta$. In (23) $u_l = P/\mu$, $\sigma_l = P(\lambda + \mu)/a(\lambda + 2\mu)$ and the prime indicates

---

[1] [3] gives the inverses for the entire neighborhood of $r = a$.

differentiation with respect to the argument of the delta function. The expressions (23) satisfy the boundary conditions (5) of the problem.

Substitution of (23) into (8) extends the solution to all of $\theta$. The observer, standing at a station $\theta$ in the physical plane, sees each of the waves in (23) as a periodic phenomenon in $t$ (of period $D = c_0 t/a = 2\pi$). This is in agreement with Fig. 4, particularly the wave front propagation depicted there, since (23) are two-sided wave fronts. It should be emphasized that the closed form solutions (23) are exact solutions for the RAYLEIGH waves in the present problem, valid on the cavity wall ($r = a$), for all $\theta$ according to (8).

It may be noted that these solutions contain pairs of like terms representing waves travelling in both the $+\theta$ and $-\theta$ directions consistent with (15) and Fig. 4. Important is the fact that the form of these waves are independent of the spatial variable $\theta$, i.e., the response is independent of the position $\theta + 2n\pi$. It is clear that in the physical plane $|\theta| < \pi$ (the near field, $\theta$ small) other contributions from (16) must be evaluated for a complete solution (over all time) to the present problem. The RAYLEIGH wave will be shown to be the predominant disturbance there for long time.

The long time solution in the present problem is obtained for $\theta + 2n\pi > 0$ by first imposing $D \gg 1$, $\theta + 2n\pi \gg 1$, and $|D - (\theta + 2n\pi)| < 1$ on the extension (8) of (23), and projecting the result into the physical plane. In the case of (23 b) and (23 c) it may be seen that all the second terms [the $(D + \theta + 2n\pi)$ terms, representing negative $\theta$-travelling waves] can be neglected, and of the positive $\theta$-travelling waves only the ones with $|D - (\theta + 2n\pi)| < 1$ survive. In the case of (23 a), however, the disturbance has a head and tail, as well as the infinite discontinuity at the RAYLEIGH wave arrival time. The contributions of the tails of the negative $\theta$-traveling waves to a disturbance at a station are cancelled by the contributions from the heads of the positive $\theta$-traveling waves, hence the body contributions of the positive $\theta$-traveling waves compose the $\theta + 2m\pi > 0$ solution. One can impose $|D - (\theta + 2n\pi)| < 1$ on the positive $\theta$-traveling waves here also on the basis of the singularity involved. Hence (23) reduce to just the first terms appearing there.

Comparison of this result with the corresponding result for LAMB'S problem of the elastic half space, involving a surface impulsive normal line load (see DE HOOP [8]), shows the two solutions are identical. This agreement has an important side feature. Since other dynamic contributions to the half space problem are known to decay spatially (with coordinate in the propagation direction along the surface) such contributions in the present problem could only come from pairs $(\omega, \nu)$ on the branches of $\nu_j(\eta)$ of (16) which have Im $\nu \neq 0$, according to (15) [note that $\theta$ enters into the solution only through the exponential appearing in (15)]. Hence the RAYLEIGH root is the only real $\nu$ root of (16), proving a statement made by VIKTOROV (made without proof).

The only further consideration necessary for the long time solution to the present problem is the static solution. Since the input is the delta function, the static solution is zero. Therefore, the long time solution (for

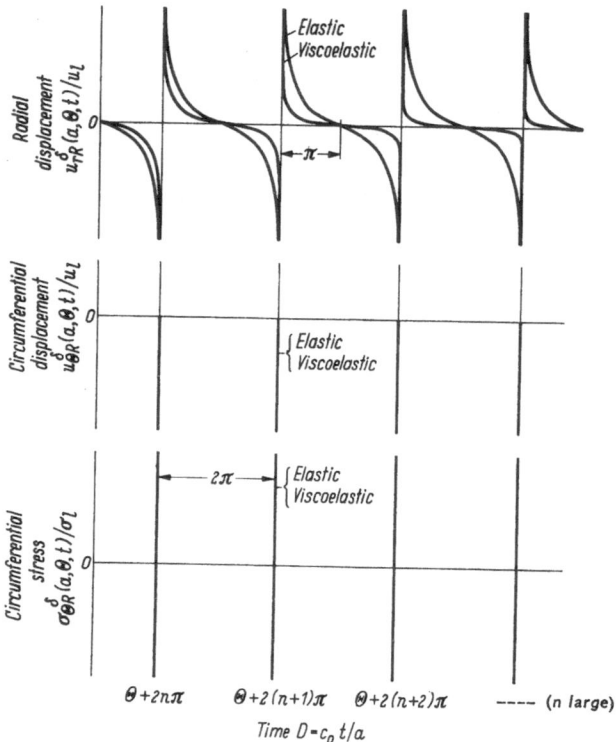

Fig. 5. Response for large time at cavity wall due to the positive $\theta$-traveling RAYLEIGH waves (line load delta-function source).

a particular station in the physical plane) is obtained by projecting the first terms (the positive $\theta$-traveling waves) and second terms (the negative $\theta$-traveling waves, but from the $\theta < 0$ solution) in (23), both extended by (8), into the physical plane. The positive $\theta$-traveling waves part of this solution is illustrated in Fig. 5 as the curves marked "elastic". One can get the total response at a particular station by superposing the set of negative $\theta$-traveling waves on those in Fig. 5. For example, at $\theta = \dfrac{\pi}{2}$ one would have the same figures for $u_{rR}^{\delta}$ and $\sigma_{\theta R}^{\delta}$, except the period would be $\pi$, and $u_{\theta R}^{\delta}$ would change its sign every $\pi$ [note the expected evenness and oddness, respectively, of these terms in (23) with respect to $\theta$]. It should be remarked that the $u_{rR}^{\delta}$ waves in Fig. 5 account for the heads and tails of neighboring waves at a particular arrival time $\theta + 2n\pi$. Because of the cancelling effect these heads and tails have on each other, only a wave or two to each side of the particular arrival time affects the final shape shown.

The infinite discontinuities in (23) and Fig. 5, of course, are directly dependent on the nature of the input function $F(t)$ [here $\delta(t)$]. It is therefore of interest to get the response to the step $H(t)$, which offers a much less severe high frequency input (i.e., $|\tilde{\delta}(i\omega)| = 1$, $|\tilde{H}(i\omega)| \cong 1/\omega$). [3] shows a solution corresponding to (23), and its extension by (8), can be obtained for the step case, through the use of the DUHAMEL integral operating on a series of the first terms on the R.H.S. of (23). This yields the RAYLEIGH positive $\theta$-traveling wave contributions to the long time solution as

$$u_{rR}^{H}(a, \theta, t)/u_l = \frac{\varkappa^2\sqrt{1 - b^2\varkappa^2}}{\pi R} \sum_{n=N_0}^{N} \log |D - (\theta + 2n\pi)|, \quad (24\,\text{a})$$

$$u_{\theta R}^{H}(a, \theta, t)/u_l = -\frac{\varkappa^2(2 - \varkappa^2)}{2R} \sum_{n=N_0}^{N} H[D - (\theta + 2n\pi)], \quad (24\,\text{b})$$

$$\sigma_{\theta R}^{H}(a, \theta, t)/\sigma_l = \frac{2\varkappa^2(2 - \varkappa^2)}{R} \sum_{n=N_0}^{N} \delta[D - (\theta + 2n\pi)], \quad (24\,\text{c})$$

where now $N$ and $N_0$ are large and are determined by the number of periodic waves occuring in a certain domain of large time, and $H[D - (\theta + 2n\pi)]$ is the two-sided step function

$$H[D - (\theta + 2n\pi)] = \begin{cases} -\,{}^1/_2, & D < \theta + 2n\pi, \\ 0, & D = \theta + 2n\pi, \\ {}^1/_2, & D > \theta + 2n\pi. \end{cases}$$

The terms in (24a) are valid in the vicinity of $D - (\theta + 2n\pi) = 0$. The full curve can be obtained by numerically integrating (w.r.t. time away from $D - (\theta + 2n\pi) = 0$) the $u_{rR}^{\delta}$ curve in Fig. 5. Again here the negative $\theta$-traveling waves must be superposed on (24) for the full dynamic contribution to the long time solution in the physical plane. At $\theta = \pi/2$ in the present case the oddness of $u_{\theta R}^{H}/u_\theta$ (w.r.t. $\theta$) leads to a alternating square waves of duration $\pi$, and magnitude $\varkappa^2(2 - \varkappa^2)/2R$, occurring with periodicity $2\pi$. In the present case the static solution has existing constant (w.r.t. time) terms which can be neglected with respect to the terms in (24a, c), hence the latter are again the predominant disturbances in the long time solution. Note that $u_{rR}^H$ and $\sigma_{\theta R}^H$ are still represented by infinite discontinuities. In the case of $u_{\theta R}^H$ the static solution would have to be added to the dynamic for the total long time solution.

## Line Load Problem; Viscoelastic Case

### (a) Statement and Formal Solution of Problem

Solution of the problem in Fig. 1 for a viscoelastic case can be obtained by using a correspondence principle (see BLAND [9]) and a linear viscoelastic model. In the LAPLACE transform of (1) to (6), using (7), the elastic moduli appearing in the transformed stress-strain relations are replaced with corresponding $p$-varying moduli. The latter are transformed linear time derivative operators representing the viscoelastic material assumed. After this substitution the transformed equations, now representing the transformed viscoelastic problem, have the same form as in the elastic case. Hence the transformed viscoelastic solutions for the potentials, displacements, and stresses will have the same form, i.e., (13b, c). One has only to replace the elastic moduli there with the $p$-varying moduli. The additional Fourier transform in the present problem is a separate linear transformation on $\theta$ and does not affect the correspondence principle (which involves the time transform). Hence the formal solution is again (13a).

### (b) Exact Inversion

(14) follows from the same arguments as in the elastic case, where now the roots $\nu_j(p)$ of $C$ represent the viscoelastic spectra. Hence $\left\{\begin{array}{c}\text{in}\\(14)\end{array}\right\}$ represents the Laplace transformed displacement solutions in which $p$-varying moduli (once these are defined for a material) must replace the elastic moduli appearing there. The $p$-varying moduli can be selected on the basis of a model. Consistent with assumptions made in constructing idealized flow theories, it is assumed here that (a) deviator stress-strain relations are based on the behavior of a MAXWELL element (a simple spring and viscous dashpot in series), and (b) the dilatational stress-strain relation is based on elasticity. These assumptions attribute flow to shearing action alone, this having well-known agreement with experiment except for very high pressures. On this basis the transformed

stress-strain relations may be written as

$$P(p)\bar{\sigma}'_r = Q(p)\bar{\varepsilon}'_r, \quad \text{etc.}, \quad R(p)\bar{J_1} = S(p)\bar{I}, \tag{25}$$

where $J_1$ and $I_1$ are the first stress and strain invariants respectively, the primed quantities are deviators, and

$$Q(p)/P(p) = 2\mu p/(p+\gamma), \quad S(p)/R(p) = 3\lambda + 2\mu, \tag{26}$$

where $\gamma = \mu/\mu'$, $\mu'$ being the coefficient of viscosity of the MAXWELL dashpot. Comparing (25) with the corresponding transformed elastic stress-strain relations it is clear that $Q(p)/P(p)$ corresponds to $2\mu$ and $S(p)/R(p)$ to $3\lambda + 2\mu$, or in $\begin{Bmatrix} \text{in} \\ (14) \end{Bmatrix}$ $\mu$ is replaced by $\mu p/(p+\gamma)$. The replacement affects all quantities involving $\mu$, such as the body wave speeds for example,

$$c_d^2 \rightarrow (1/\varrho)[\lambda + 2\mu p/(p+\gamma)], \quad c_s^2 \rightarrow c_s^2 p/(p+\gamma). \tag{27}$$

Inversion of the LAPLACE transform would follow the same technique as that leading to (15). Two additional branch points are introduced through the quantities $k_d$ and $k_s$, i.e.,

$$k_d = \frac{p}{c_d} \sqrt{\frac{p+\gamma}{p+\delta}}, \quad k_s = \frac{\sqrt{p(p+\gamma)}}{c_s}, \tag{28}$$

where $\delta = \lambda\gamma/\lambda + 2\mu$. These are simple, however, and would cause no significant further complexity. Hence, one could through numerical procedures, or approximations such as that accomplished for RAYLEIGH waves in the next section, get information from (15) and (16) for the viscoelastic case for various waves of interest. Fig. 4 and its discussion would therefore also be applicable here.

It should be noted that a similar analysis could be carried out on the basis of other more appropriate models for the particular frequency domains of interest. It also follows that other viscoelastic problems like the present, basically involving dispersive elastic waves, such as those in rods and plates, could be formulated and solved similarly.

### (c) Rayleigh Waves and the Long-Time Solution

From the frequency-wave number relation for the elastic RAYLEIGH wave, and the second of (27), it follows that

$$\nu = \frac{\omega a}{\varkappa c_s} = \frac{\omega a}{c_0} (1 - i\gamma/2\omega), \tag{29}$$

i.e., a complex wave number. Clearly then

$$\lim_{\substack{|v| \to \infty \\ \eta \to \infty}} (v/\eta) \to (c_d/c_0) \left(1 - \frac{i\gamma}{2\omega}\right) = \frac{c_d}{c_0} > 1,$$

$$\lim_{\substack{|v| \to \infty \\ \zeta \to \infty}} (v/\zeta) \to (c_s/c_0) \left(1 - \frac{i\gamma}{2\omega}\right) = \frac{c_s}{c_0} > 1,$$

(30)

i.e., (17), and it follows that (16), through (18), again yields (19), and the elastic RAYLEIGH wave speed. In (29), $-Im(v) = a\gamma/2c_0$ $= a\sqrt{\mu\varrho/2\mu'}\varkappa$ corresponds to an attenuation factor. The asymptote (20) now is displaced from the real plane, parallel to itself along the imaginary axis a distance $-a\gamma/2c_0$. These characteristics are the same as in the related half space harmonic wave case, as given by BLAND ([9], pp. 73—75). (30) also means that the same asymptotic form (18) can be used to approximate the integrand of (15) for the viscoelastic RAY-LEIGH wave root $v_j$. It is evident therefore that (15), corresponding to the viscoelastic case, differs only from (15) by an additional multiplicative attenuation factor $e^{-\gamma a\theta/2c_0}$. It follows that (23) are also multiplied by this factor, i.e., the surface RAYLEIGH waves for the viscoelastic case are

$$\left.\begin{array}{l} u_{rR}^{\delta}(a, \theta, t)/u_l = \\ u_{\theta R}^{\delta}(a, \theta, t)/u_l = \\ \sigma_{\theta R}^{\delta}(a, \theta, t)/\sigma_l = \end{array}\right\} e^{-\gamma\varkappa\theta2/c_0} \text{ [R.H.S. (23)]}$$

(31)

for $0 < \theta \leq \pi$, and for $-\pi \leq \theta < 0$ after substitution of $(-\theta)$ for $-0$, for the delta function input.

Extension of (31) through (8), to large $\theta + 2n\pi$ and $D$, similarly only involves multiplication of the elastic solution by the attenuation factor. This result is plotted in Fig. 5 as the curves marked "viscoelastic". It may be noted that the infinite singularities at $D = \theta + 2n\pi$ are not affected by the added viscosity. However, away from this time there is spatial attenuation as the $u_{rR}$ curve indicates. This is exaggerated in the figure to bring out the effect; the attenuation per circuit is $(1 - e^{-\gamma a\pi/c_0})$ whereas the initial portion of the curve in the figure has been attenuated from the elastic by $1 - e^{-(\gamma a/2c_0)(\theta + 2n\pi)}$, where $n$ is large.

For the step input it follows that another equation like (31) would be formed from (24). Here the step response of $u_{\theta R}$ is of further interest since the viscosity would attenuate its magnitude. It should be pointed out that these results are in agreement with LEE and KANTER's work [10] on the MAXWELL rod. This is not surprising when one considers that there is one dimensional propagation here, $r$ being confined to the near field.

## Plane Pulse Problem; Elastic Case

### (a) Statement of Problem

The problem is sketched in Fig. 6. A plane compressional stress pulse $\sigma_r(x, t) = \sigma_x(x, t) = -\sigma_i J(t + x/c_d)$, travelling along the $x$-axis ($\theta = 0$) in the negative $x$-direction, envelops the cavity wall symmetrically

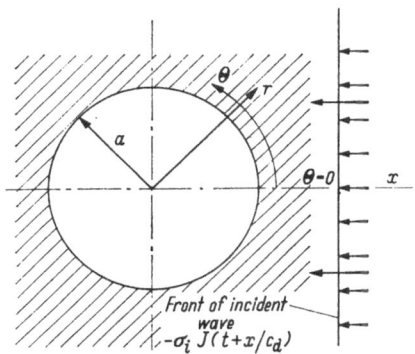

Fig. 6. Problem of plane compressional pulse incident on circular cylindrical cavity.

from $\theta = 0$. The problem is conveniently formulated on the basis of the governing Eqs. (1), (2), (3), and (4) by noting that the incident wave can be written in terms of $\varphi$ by using (2), (3), and (4), i.e.,

$$\sigma_r = \sigma_x = -\sigma_i J(t + x/c_d) = -\varrho \varphi_i F''(t + x/c_d), \qquad (32)$$

where $\sigma_i$ and $\varphi_i$ are magnitude constants (of dimensions lbs/in² and in²/sec² respectively), the primes indicate differentiation with respect to the argument of the input function $F$, and the incident wave is

$$\varphi(x, t) = -\varphi_i F(t + x/c_d). \qquad (33)$$

In the present problem the boundary conditions at $r = a$ are

$$\sigma_r(a, \theta, t) = 0, \qquad \sigma_{r\theta}(a, \theta, t) = 0. \qquad (34)$$

The infinity condition (6) applies here only to the scattered waves and not the incident wave. Statement of initial conditions such as (7) are not required in this problem.

## (b) Formal Solution of Problem

Solution of the present problem follows essentially the same technique used in the line load problem. Here, however, the bilateral LAPLACE transform is used to accommodate the negative travelling incident wave. Hence in (9) and (10) there is need only to replace (9a) with

$$\bar{\varphi}(r, \theta, p) = \int_{-\infty}^{\infty} \varphi(r, \theta, t) e^{-pt} dt. \tag{35}$$

Again (11) are the resultant STURM-LIOUVILLE equations for the system. The solution now is

$$\begin{aligned}
\tilde{\bar{\varphi}} &= \tilde{\bar{\varphi}}_{\text{inc}} + \tilde{\bar{\varphi}}_{\text{sc}} = - \varphi_i f(p) I_{|\nu|}(w) + A_\nu(p, \nu) K_\nu(w), \\
\tilde{\bar{\psi}} &= \tilde{\bar{\psi}}_{\text{sc}} = B_\nu(p, \nu) K_\nu(y),
\end{aligned} \tag{36}$$

where the transformed potentials corresponding to incident and scattered waves have been indicated. $I_{|\nu|}(w)$ is the modified BESSEL function of the first kind. The double transform of $\tilde{\bar{\varphi}}_{\text{inc}}$ in (36) can be verified by employing POISSON's summation formula [3].

Following the same process as that leading to (13), (13a) also gives the formal solution to the present problem, but (13b, c) are replaced by

$$\tilde{\bar{u}}_r(r, \nu, p) = \varphi_i f(p) \left\{ \frac{1}{\Delta(p, \nu)} \left[ k_d Y K'_\nu(w) + \frac{\nu}{r} Z K(y) \right] - k_d I'_{|\nu|}(w) \right\}, \tag{37a}$$

$$\tilde{\bar{u}}_\theta(r, \nu, p) = -i \varphi_i f(p) \left\{ \frac{1}{\Delta(p, \nu)} \left[ \frac{\nu}{r} Y K_\nu(w) + k_s Z K'_\nu(y) \right] - \frac{\nu}{r} I_{|\nu|}(w) \right\}, \tag{37b}$$

where

$$\Delta(p, \nu) = \mu p^2 K_\nu(u) K_\nu(v) C, \quad Y = A I + \frac{2\mu\nu}{a} E II,$$

$$Z = B I + \frac{2\mu\nu}{a} [F + 2(c_d^2 - c_s^2) K_\nu(u)] II,$$

$$I = \left[ \frac{2\mu}{a^2} (\nu^2 - |\nu|) + (\lambda + 2\mu) k_d^2 \right] I_{|\nu|}(u)$$

$$- \frac{2 k_d}{a} [2\lambda(|\nu| + 1) + \mu(4|\nu| + 3)] I_{|\nu|+1}(u),$$

$$II = \frac{(|\nu| - 1)}{a} I_{|\nu|}(u) - k_d I_{|\nu|+1}(u),$$

$$E = c_d^2 [K_{\nu+2}(v) - K_{\nu-2}(v)], \qquad F = c_s^2 [K_{\nu+2}(u) + K_{\nu-2}(u)].$$

and $A$, $B$, and $C$ are given in (13).

### (c) Exact Inversion

It may be noted with $C$ in $\Delta(p, \nu)$ in (37) the same frequency equation governs, so that a solution such as (15), with roots given by (16), would result for the first terms in (37). The second terms there represent the incident wave and are easily inverted through the known pair, (33), and the first term in (36). A sketch, similar to that in Fig. 4 is shown in Fig. 7. Here no diffraction starts until the incident wave front reaches the vertical line, $\theta = \pm\pi/2$, through the cavity center, i.e., diffracted waves have their origin at $\theta = \pm\pi/2$, $r = a$, propagating into the

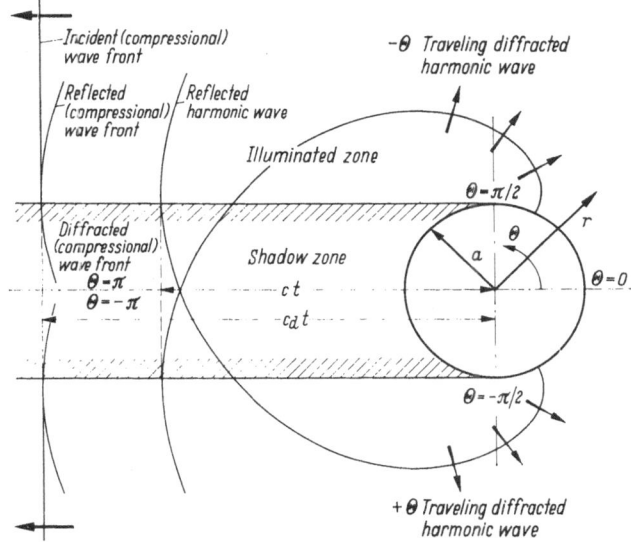

Fig. 7. Scattering of a plane pulse by a circular cylindrical cavity.

shadow-zone. This occurs when time $t = 0$ (note the incident wave front reaches the cavity at $\theta = 0$ at time $t = -a/c_d$). The reflected wave, only partially shown in the figure, surrounds the cavity, except in the shadow zone where it does not occur. To the right of the cavity, the reflected wave has reached its extreme, a distance $ct + a$, from $r = a$, $\theta = 0$, along the line $\theta = 0$ (Note also the reflected (compressional) wave front would be there too, a distance $c_d t + a$). The analog of (15) here indicates that a system of such diffracted and reflected harmonic waves are generated by the incident wave as it impinges on the cavity boundary. It should be noted that the reflected waves in the present problem (Fig. 7) are analogous to the radiated waves in the line load problem (Fig. 4). Both, of course, are also $\theta$-travelling waves, being continuations of the diffracted waves in the shadow zone.

### (d) Rayleigh Waves and the Long Time Solution

The RAYLEIGH waves were obtained from the first two terms in (37) in the same way as they were from (13 b, c) in the line load problem [3]. Analogously $F(t + x/c_d)$ is taken as a linear function, making $\sigma_r$ the delta function, which gives

$$\tilde{u}_{rR}^{\delta}(r, v, t) = \sqrt{\pi/2}\, M\, [Q_F(r)\,\mathrm{e}^{-F(r)v} + Q_G(r)\,\mathrm{e}^{-G(r)v}]\,\frac{\sin E(t)v}{\sqrt{v}}, \quad (38\text{a})$$

$$\tilde{u}_{\theta R}^{\delta}(r, v, t) = -\sqrt{\pi/2}\, M\, [R_F(r)\,\mathrm{e}^{-F(r)v} + R_G(r)\,\mathrm{e}^{-G(r)v}]\,\frac{\sin E(t)v}{\sqrt{v}}, \quad (38\text{b})$$

where

$$M = -\frac{\sigma_i b^2 \varkappa^2}{\pi \varrho c_0 \sqrt{q_0/k_0}\, R}, \qquad\qquad E(t) = D(t) + \pi/2,$$

$$Q_F(r) = \frac{N\, U_A(r)}{b\,(2 - \varkappa^2)}, \qquad\qquad Q_G(r) = \frac{\varkappa k_0 P\, U_B(r)}{q_0},$$

$$R_F(r) = \frac{N\, V_A(r)}{b \varkappa\,(2 - \varkappa^2)}, \qquad\qquad R_G(r) = \frac{k_0 P\, V_B(r)}{q_0},$$

$$F(r) = A(r) + A', \qquad\qquad G(r) = B(r) + A',$$

$$N = (2 - \varkappa^2)k + 4s_0 l/b\varkappa k_0,$$

$$P = q_0 k/b\varkappa k_0 + (2 - \varkappa^2)l/b^2\varkappa^2,$$

$$k = [2 - \varkappa^2 + 4(1 - q_0/k_0)/b^2],$$

$$l = 2 - q_0/k_0, \qquad\qquad A' = \tanh^{-1}(q_0/k_0) - q_0 k_0,$$

where $\tilde{u}_{rR}^{\delta}$ and $\tilde{u}_{\theta R}^{\delta}$ are FOURIER cosine and sine transforms, respectively.

The inverses are again simple and may be found in the tables [7]. The long time solution here is obtained in the same manner as in the line load problem. Here the spatially non-decaying, positive $\theta$-traveling periodic RAYLEIGH circumferential stress wave on the surface $r = a$ is[1]

$$a\,\sigma_{\theta R}^{\delta}(a, 0, t)/c_0\sigma_i = -0.15 \left[\frac{(E - \theta)\,[\sqrt{(A')^2 + (E - \theta)^2} + 2A']}{\sqrt{\sqrt{(A')^2 + (E - \theta)^2} + A'\,[(A')^2 + (E - \theta)^2]^{3/2}}}\right] \tag{39}$$

an odd function with respect to $(E - \theta)$. [3] shows a curve for the long time solution and since the input is the delta function this is the total

---

[1] [3] gives complete RAYLEIGH wave solutions for displacements and stress (for $\lambda \pm \mu$) for the region near the cavity. The complete long time solutions for the surface displacements, velocities and accelerations are also obtained there.

long time solution. It should be noted that $E - \theta$ instead of $D - \theta$ occurs here, pointing out that (39) only represents a RAYLEIGH wave generated by the incident pulse interaction with the cavity wall at $\theta = \pi/2$, i.e., $|E - \theta| < \varepsilon$, for $t$, $\theta > 0$ can only be satisfied for $\theta > \pi/2$[1]. As might be expected the plane wave character (distributed loading) of the input results in a non singular RAYLEIGH wave. (39) is again extended by (8) and the resultant curve is analogous to that of $u_{rR}^{\delta}$ in (23a), which was also odd with respect to the RAYLEIGH wave arrival time. [3] shows this response for the $+\theta$-traveling waves for long time. Since the input was a delta function, there is no static stress to be considered.

The RAYLEIGH wave response $\sigma_{\theta R}^{H}$ to the step input in $\sigma_r$, was obtained by first applying the Duhamel integral, and numerical integration, to

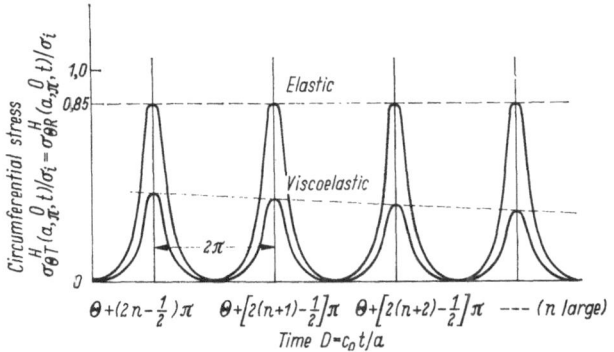

Fig. 8. Long time total circumferential stress response to the plane step wave source at the cavity wall for the poles $\theta = 0, \pi$.

the long time solution obtained for $\sigma_{\theta R}^{\delta}$ for the positive $\theta$-traveling waves ([3] also shows this result). Accounting for the negative $\theta$-traveling waves and the static stress $\sigma_{\theta S}$ for the step input, given by $\sigma_{\theta S}(a, \theta)/\sigma_i = -(4/3(\ )1 - \cos 2\theta)$ for $\lambda = \mu$, gives the total long time stress $\sigma_{\theta T}^{H}$. $\sigma_{\theta S}$ is even in $\theta$, varying from zero at $\theta = 0$ to $-(8/3)$ at $\pi/2$, and back to zero at $\theta = \pi$. Hence it is a compressional stress. $\sigma_{\theta R}^{H}$ is tension everywhere in $\theta$. Hence the latter's effects will be most severe at $\theta = 0$, $\pi$ where this stress is the only component in $\sigma_{\theta T}^{H}$. Further, the evenness of $\sigma_{\theta R}^{H}$ w.r.t. $\theta$ leads to double its value at the poles 0, $\pi$, as a periodic phenomenon, i.e., the sum of positive $\theta$- and negative $\theta$-traveling waves. Fig. 8 is a plot of this response as a function of time (marked "elastic").

---

[1] The same restriction is exhibited in FRIEDLANDER's work (see pp. 719–720, in [4]).

18*

## Plane Pulse Problem; Viscoelastic Case

The analysis used for the analogous line load case carried over to the present case such that solutions are again the attenuation factor $e^{-\gamma a\theta/2c_0}$ times the elastic solutions. The plot of $\sigma_{\theta T}^H = \sigma_{\theta R}^H$ at the cavity surface, at the poles 0, $\pi$ (for $\lambda = \mu$), appears in Fig. 8 marked "viscoelastic". The same attenuation characteristics as in Fig. 5 are involved here. It should be noted that as the viscosity coefficient $\mu'$ becomes infinite, $\gamma \to 0$ and all viscoelastic solutions reduce to the elastic case.

### Bibliography

[1] SOLDATE, A. M., and J. F. HOOK: National Engr. Sci. Co. Tech. Doc. Rep. No. AFSWC-TDR-62-30 for Kirtland Air Force Base, N.M., (Mar. 1962), Appendix III.

[2] LLOYD, J. R., and J. MIKLOWITZ: Proc. 4th U.S. Nat. Congr. Appl. Mech. Vol. 1, 1962, pp. 255—268.

[3] MIKLOWITZ, J.: National Engr. Sci. Co. Tech. Doc. Rep., No. RTD—TDR 63—3052 (AFWL, Kirtland Air Force Base, N. Mex.) Nov. 1963.

[4] FRIEDLANDER, F. G.: Comm. Pure Appl. Math. 7, 705—732 (1954). See also Sound Pulses, Cambridge U. 1958.

[5] VIKTOROV, I. A.: Sov. Phys. Accoust., Amer. Inst. Phys. Transl. 4, 131 (1958).

[6] ERDELYI, A., et al.: Higher Transcendental Functions, Vol. 2, New York: McGraw-Hill Book Co. 1953.

[7] OBERHETTINGER, F.: Tabellen zur Fourier Transformation, Berlin/Göttingen/Heidelberg: Springer-Verlag 1957.

[8] DE HOOP, A. T.: Second Annual Rep., Seismic Scattering Proj., Inst. Geophys. UCLA 1957.

[9] BLAND, D. R.: Theory of Linear Viscoelasticity, London: Pergamon 1960, pp. 95—96.

[10] LEE, E. H., and I. KANTER: J. Appl. Phys. 24, 1115 (1953).

# Some Dynamic Characteristics of Rocks

By **Werner Goldsmith** and **Carl F. Austin**

University of California, Berkeley, Cal.,
and U.S. Naval Ordnance Test Station, China Lake, Cal., U.S.A.

## Abstract

The literature in the field of dynamic behavior of rock materials is relatively sparse and incomplete. Details and even raw data on pulse propagation and internal structural response of rock to the passage of a transient of more than negligible amplitude is virtually non-existent, particularly under controlled conditions. The present investigation consists of a study of the transmission and decay of pulses produced both by impact and explosive loading in a rock classified as a diorite. The samples were chosen and prepared with sufficient care and tested under identical conditions so that reproducibility of results comparable to that expected in metallic systems was obtained, indicating that the concept of non-reproducibility in geologic materials is not well founded.

Ballistically-suspended HOPKINSON bars of diorite of 3/4-inch diameter and approximately 22 inches long were subjected to longitudinal impact by spherical and flat-ended projectiles of hardened steel at an initial velocity of about 3300 in/sec. Strain gages attaches to the specimens at various stations permitted a recording of the shape and velocity of propagation of the resultant wave. Similar experiments were performed on an aluminum alloy bar of identical size in order to assess the magnitude of the dispersion resulting from the three-dimensional character of the rod. The nature of the transformation of the pulse during passage permits an assessment of the validity of various models proposed for geologic substances in the field of seismology. A further detailed study of the internal response of the samples to such pulse passage was accomplished by means of microscopy and static tests, which correlated very well with the dynamic results. The average propagation velocity of a pulse generated by a contact explosion in a rectangular block of diorite was also determined.

## 1. Introduction

In recent years, the propagation of waves in rocks and soils has assumed increasing technological importance, and the subject has thus received corresponding attention. While previously, virtually the exclusive domain of the geophysicist, this topic is also currently being examined by engineers and physicists. Since the equations of motion and boundary and initial conditions are identical to those employed in problems involving more familiar materials, the difference in the analytical representation of the physical situation and results deduced therefrom must be attributed to the differences in the material behavior of the medium. Here, however, geological raw materials exhibit characteristics which are considerably at variance from those of the more conventional transmitting media. The present paper will be concerned with some aspects of the response of a particular type of rock to dynamic loads, and the difference between this response and that of a common aluminum alloy subjected to identical loading conditions. The purpose of this study is a preliminary effort to gain an understanding of the nature of the dissipative process present in the transmission of waves in such a substance with the eventual objective of establishing a relation describing the material behavior which is of relatively simple form, yet sufficiently accurate to provide usable results in the prediction of wave phenomena.

A popular concept, commonly expressed in the literature, is the idea that geologic materials such as rock are extremely inhomogeneous, highly directional in behavior, and suffer from problems of variable grain size and chemical composition. This notion is well founded when random samples are chosen from random locations, for a study of a single rock will yield widely divergent results if the criterion of the rock name is the sole basis for the selection of specimens. Thus a single rock name encompasses materials that can vary widely in composition, grain size, structure, alteration, the presence of inclusions, and related effects that cause significant differences in mechanical properties. On the other hand, careful selection of a single rock type, as in the present investigation, with attention to details of physical structure in all respects will give accurately reproducible data, regardless of the locality from which the samples are collected, indicating that the concept of the non-reproducibility of properties of geologic materials is a myth. In other words, when the same precautions are observed in the selection of natural specimens as in the composition, manufacturing procedure, processing and heat-treatment of a specified type of steel, aluminum alloy or plastic, the data derived from a number of samples subjected to identical experimental conditions are equally uniform.

In addition to the sample selection procedure *per se*, significant sources of variation in collected data for rocks commonly result from poor preparation techniques, which selectively damage the crystal structure or chemical composition of portions of sample suites, and from the poor choice of test methods. As an example of the latter, a pulse length shorter than the grain size in a given test material will lead to highly variable results and the need for interpretation based on statistical methods. In view of this, the methods of sample selection, preparation, and testing in a project must be carefully delineated in any meaningful discussion of both static and dynamic properties of such easily damaged materials as rocks or other similar brittle solids. However, when properly executed, such procedures permit interpretations of behavior and behavior changes to be drawn from relatively few experiments involving both macroscopic measurements of pulse metamorphosis and travel time and microscopic examination of the internal grain structure response of the materials to pulse passage, accomplished by careful thin section techniques.

The present investigation is concerned with the determination of various properties of a specific type of rock, a diorite. Specimens were subjected to standard static compressive tests and to impulsive loading by two types of projectiles. The changes in wave form and the velocity of pulse propagation between measuring stations were observed, some of the results being compared to data obtained under identical conditions in a 2024-T4 aluminum bar. Microscopic examinations were conducted of the crystal structure of both virgin and shocked rock specimens. Finally, the average velocity of shock propagation of a high-amplitude pulse produced by a contact explosion was determined. Corresponding properties of the rock and the aluminum alloy are presented to indicate the nature of the difference of behavior of the two materials under these loading conditions. The results of the pulse propagation tests were compared to the predicted performance of various analytical models proposed for geologic substances.

## 2. Summary of Previous Investigations

The propagation of seismic waves and interpretation of seismograms have been analyzed most frequently on the basis of a homogeneous, isotropic, albeit layered, elastic medium [1, 2]. In the case of underground explosions, the investigation of the transient generated at "moderate" pressures has been pursued on the same basis in a continuous medium [3—10], extending even to an explanation of fracture processes [9—14]. The partial success of such an approach to the failure phenomenon is attributable to the high ratio of compressive to tensile strength of rocks which permits an incident compressive pulse small in comparison with the breaking strength to produce spalling upon reflection from a free surface as a tensile wave. This concept has been utilized in a recent investigation of the dynamic tensile strengths of basalt and sandstone determined from measurements in a HOPKINSON bar and the use of a one-dimensional theory of elastic wave propagation [15].

The propagation of pulses at high pressures has been universally described on the basis of hydrodynamic theory involving the RANKINE-HUGONIOT shock front conditions which assume a negligible rigidity of the medium and a state of thermodynamic equilibrium. Experiments seem to confirm that both of these conditions are fairly well satisfied for a sufficiently high shock pressure and short rise time. It has been suggested [16] that an elastic analysis of the propagation phenomenon would be appropriate in the region beyond the fracture zone, that is, at pressures below the crushing strength of the material. However, the scattered investigations of the transition of the pulse from one regime to the other have been inconclusive.

Consideration has also been given to the nature [17, 18] and mathematical representation of dissipative mechanisms in the analysis of pulse propagation in the earth, primarily in an attempt to explain discrepancies appearing in seismographic records interpreted on a purely elastic basis. This effort has proceeded in two directions: (a) Laboratory determinations of the attenuation of waves in rock specimens, ascertained from the measured decay of ultrasonic pulses [19—21] and the energy dissipated per cycle in the transverse vibrations of beams [22], or field measurements of the propagation characteristics of waves initiated by explosive sources, and (b) an analysis of several theoretical models based upon incorporation of various simple frictional processes. The latter approach is complicated by conflicting experimental evidence. Attenuation factors have been determined experimentally which vary as different powers of the frequency of harmonic components present; alternatively, in a number of cases, the dissipation function appears to be independent of frequency [20, 23, 24]. In particular, the ratio of energy loss per cycle to maximum elastic energy stored, $\Delta E/E$, in transverse vibration tests was found to be independent of frequency in the range from 40 to 120 cycles per second for specimens of sandstone, oolitic and shelly limestone, granite, dolerite and diorite specimens, the latter exhibiting the lowest of the values at 5 per cent [22].

A corollary controversy in geophysical circles has developed around the question as to whether this attenuation is accompanied by dispersion. It has even been suggested that internal friction for consolidated rock appears to have a negligible effect at frequencies up to 100 cycles per second [1]—in contradiction to the results cited above — and would thus not significantly affect seismic phenomena, although it might be of considerable importance in the propagation of explosively- or impact-generated waves. Only Pierre shale of Eastern Colorado appears to conform experimentally to the attenuation and dispersion requirements of a simple, visco-elastic solid, namely the KELVIN-VOIGT model [25—27].

This solid has been employed as the most common anelastic model for the earth, and its specification leads to the one-dimensional differential equation of motion

$$\left[1 + \frac{\eta}{E} \frac{\partial}{\partial t}\right] \frac{\partial^2 u}{\partial x^2} = \frac{1}{c_0^2} \frac{\partial^2 u}{\partial t^2}, \tag{1}$$

where $u$ is the displacement in the direction of propagation $x$, $t$ is time, $\eta$ is the STOKES' viscosity term, $E$ is YOUNG's modulus, and

$$c_0 = (E/\varrho)^{1/2} \tag{2}$$

is the "rod" wave velocity, with $\varrho$ as the density of the material. Numerous theoretical investigations applying to seismic phenomena have been conducted with the aid of Eq. (1) [1, 24—35]. It has been shown that the propagation of a plane harmonic compression wave through such a medium, characterized by frequency $\omega$ and wave number $k$, can be described by the component

$$u = u_0 e^{-\alpha x} e^{i(\omega t - k x)} \tag{3}$$

provided the frequency-dependent phase velocity $c \equiv \omega/k$ and attenuation coefficient $\alpha$ are determined by the relations

$$\frac{c}{c_0} = \left[-\left(\frac{1 + \sqrt{1 + \xi^2}}{\xi}\right)^2 + 2\left(1 + \sqrt{1 + \xi^2}\right) + 1\right]^{1/2}, \quad \text{with} \quad \xi \equiv \frac{\omega \eta}{E}, \tag{4}$$

and

$$\alpha = \frac{\omega}{\xi c}\left[-1 + \sqrt{1 + \xi^2}\right]. \tag{5}$$

The decay constant can also be expressed in terms of the logarithmic decrement $\delta$ and hence as a function of $\Delta E/E$ by

$$\alpha = \frac{\delta \omega}{2 \pi c} \quad \text{and} \quad \frac{\Delta E}{E} = 1 - e^{-2\delta}. \tag{6}$$

Among many well-known solutions of Eq. (3), the case of a step pulse of displacement at the origin has been determined in the vicinity of the principal part of the disturbance [31].

In order to obtain a better fit with a large body of experimental data, a solid friction type of mechanism has been attributed to rocklike materials, so that the wave propagation equation for rods is given by [23, 24]

$$\left[1 + \frac{1}{|\omega|Q_0} \frac{\partial}{\partial t}\right] \frac{\partial^2 u}{\partial x^2} = \frac{1}{c_0^2} \frac{\partial^2 u}{\partial t^2}, \tag{7}$$

where $Q_0$ is assumed to be a dimensionless constant which is connected to the energy loss per cycle through the relation [36]

$$\frac{1}{Q_0} = \frac{1}{2\pi} \frac{c}{c_0} (1 - e^{-2\delta}). \tag{8}$$

The absolute value of the angular frequency, $|\omega|$, is included in Eq. (7) to assure absorption of both positive and negative frequency terms of the FOURIER spectrum of the pulse. The dissipative quantity in this relation corresponds to the well-known structural damping factor frequently used in vibration analysis. A harmonic component $e^{i(\omega t - kx)}$ will propagate in such a medium with a frequency-independent phase velocity $c$ given by

$$c = \frac{c_0}{\gamma} \left[1 + \frac{1}{Q_0^2}\right]^{1/2}, \qquad \text{where} \qquad \gamma \equiv \left[\frac{1}{2}\left(1 + \sqrt{1 + Q_0^{-2}}\right)\right]^{1/2}, \tag{9}$$

and a frequency-dependent absorption coefficient $\alpha$ specified as

$$\alpha = \frac{\Phi\omega}{c} \left[1 + \frac{1}{Q_0^2}\right]^{1/2}, \qquad \text{with} \qquad \Phi \equiv \left[\frac{1}{2}\left(-1 + \sqrt{1 + Q_0^{-2}}\right)\right]^{1/2}. \tag{10}$$

An initial impulse of displacement at $x = 0$, represented by $I$, leads to displacement

$$u = \frac{I\Phi c_0 x}{\pi} \left[\frac{(1 + Q_0^{-2})^{1/2}}{\tau^2 c_0^2 (1 + Q_0^{-2} + x^2 \Phi^2)}\right] \qquad \text{with} \qquad \tau \equiv t - \frac{\gamma x}{c_0[1 + Q_0^{-2}]^{1/2}}. \tag{11}$$

with a peak value $u_{\max}$ of

$$u_{\max} = \frac{I c_0 [1 + Q_0^{-2}]^{1/2}}{\pi \Phi x}. \tag{12}$$

A similar impulse of stress $S$ results in the displacement

$$u(x, \tau) = \frac{S}{\pi\varrho} \left[\frac{(x/(2Q_0)) - \Phi c_0 \tau}{x^2 \Phi^2 + \tau^2 c_0^2 (1 + Q_0^{-2})}\right]. \tag{13}$$

With the aid of LIGHTHILL's method of characteristics, the variation of peak strain with distance for an impressed exponentially decaying strain pulse at the origin, given by $\varepsilon_0 e^{-\beta t}$ for $t \geq 0$, can be derived to first order in $1/Q_0$ as

$$\varepsilon = \varepsilon_0 e^{-\pi\beta x/Q_0}. \tag{14}$$

A closed-form solution has also been obtained using the method of superposition for a pulse $\psi(0, t)$ impressed at $x = 0$ in a medium, whose attenuation coefficient can be expressed as [37]

$$\alpha(\omega) = b \, | \, \omega \, |^{1/\cdot} \tag{15}$$

where $b$ is constant. In this case, the disturbance propagated from the exponentially decaying source $\psi(0, t) = A_0 e^{-\beta t}$ for $t \geq 0$ can be represented as

$$\psi(x, t) = A_0 e^{-\beta(t - (x/c_0))} Re \left| e^{ibx\sqrt{2\beta}} Erfc \left( \frac{bx}{\sqrt{2(t - (x/c_0))}} + i \sqrt{\beta\left(t - \frac{x}{c_0}\right)} \right) \right|. \tag{16}$$

A similar method has been employed to permit the retention of a linear form of the wave equation, and hence the validity of the superposition principle, by a suitable choice of frequency dependence of the absorption coefficients, based on a first-power law of attenuation [36]. However, it is not claimed that such a procedure will shed light on the actual mechanism of propagation in geologic materials.

Available information concerning the mechanical behavior of rocks can be classified as data specifying either elastic, attenuating, or hydrodynamic properties. Although such sources are not abundant, the use and interpretation of results in these categories must be approached with caution including due regard for the problems of reproducibility. Furthermore, much of these data have been obtained under static, quasi-static, or ultrasonic conditions in the laboratory, with the notable exception of References [15] and [22], the latter reporting a slight increase in the value of YOUNG's modulus of 2.5 per cent over the frequency range from 40 to 120 cycles per second for seven rocks, including diorite. However, no variation was found in this parameter over the range from 140 to 4500 cycles per second in another study [38]. Most dynamic determinations have been performed in situ using seismic exploration techniques, where knowledge of the character of the medium is virtually absent.

Some data concerning the elastic constants and density of rocks may be found in References [39—43], while an exhaustive investigation of the mechanical properties of randomly selected rocks has been undertaken [44—46] using standardized testing and sampling techniques on laboratory specimens. Longitudinal bar velocities were obtained by calculation from the observed fundamental mode of longitudinal vibration and the measured length and density of the specimen. Results of this type have also been compared to field measurements of the arrival times of pulses generated by shallow underground explosions [47]. Furthermore, a host of experiments have been performed to determine the velocity of propagation in rocks by ultrasonic techniques [21, 48—50].

Although some earlier experiments have been performed to ascertain the static compressibility of rocks and minerals [51, 52]—where the compressibility of most igneous rocks, including diorite, was found to be essentially independent of pressure — the principal results concerning the high-pressure behavior of rocks have only been obtained very recently, employing the detonation of a high explosive in intimate contact with the specimen and the measurement of both shock and particle velocity. From these data, the stress and density extant in the material can be computed by means of the hydrodynamic theory [53]. Information has been obtained in this manner for gabbro, dunite, basalt, granite, marble, rock salt, limestone, tuff, greywhacke, shale, and concrete [16, 54, 55]. Polymorphic transitions similar to those found in metals under similar circumstances have been observed in gabbro [54]; the effect of such changes on wave propagation phenomena is discussed in Reference [53]. Otherwise no effort seems to have been expended in correlating the microstructural behavior with the passage of pulses.

### 3. Selection and Preparation of Test Materials

The selection of a diorite, an igneous plutonic rock, for the examination of the behavioral characteristics of a rock was made on a two-fold basis. The grain structure of the chosen material is typical of the medium- to coarse-grained crystalline or hypidiomorphic granular rocks of the world. Thus any useful data obtained concerning the properties of a diorite should be reasonably applicable to other common plutonic rocks with similar grain structure, such as granites, adamellites, and the syenites. A plutonic rock was chosen for these initial studies because such rocks are more uniform three-dimensionally than either sediments or metamorphics, both of which are usually planar in structure. The use of a massive plutonic rock obviates many of the problems of reproducibility and validity of sampling commonly encountered in studies of the layered or planar geologic materials.

The specific choice of diorite as a representative plutonic rock was based on the chemistry of the constituent minerals and on the overall internal structure as well as on the form of available field occurrences. The first feature is optimal since the rock lacks potash feldspar, a mineral that weathers rapidly to clays upon exposure in outcrop and that is also readily attacked by deuteric fluids during the cooling history of the rock. Selection of a diorite free of potash feldspar implies that small, easily-handled joint blocks from field occurrence could serve as specimens and still be free of the weakening effects of chemical alteration due to weathering. The use of small joint blocks from the outcrop was required to prevent pretest damage to the grain structure by sledging or blasting, a serious problem in many commercially quarried rock samples. The material selected was further checked for possible internal grain boundary damage as the result of geologic processes by means of petrographic techniques. No measurable grain-boundary crushing or deformation was noted, although some constituent minerals (interstitial quartz) exhibited considerable strain. To investigate the possibility of non-uniform behavior of the test diorite as a result of this or other unnoted strains, a set of cores were drilled in three mutually perpendicular directions. Specimens from these cores did not exhibit any directional properties.

The test diorite, like many plutonic rocks, contains scattered inhomogeneities in grain size, though not in mineralogy. These areas of inhomogeneity are fine-grained diorite that forms small xenoliths. The presence of xenoliths in the test specimens was minimized by the small size plus visual inspection and rejection of suspect samples. On the basis of internal structure, mineralogy, uniformity of composition, lack of preferred orientation of either minerals or internal stress, and the presence of blocks of practical size for handling, the diorite chosen was considered a valid and worthwhile sample. This material was found in Wilson Canyon at the U.S. Naval Ordnance Test Station, China Lake, California, and forms a surface exposure measurable in terms of miles.

The preparation of the specimens to be employed in the static and dynamic tests was performed in such a fashion as to minimize possible damage to the structure of the sample. The material collected at the field outcrop consisted of naturally occurring joint blocks. Only boulders of proper size were utilized in order to prevent shock damage from sledging or blasting. The use of these small samples from the surface of the ground does admit of the possibility of general grain bond weakening through the process of temperature cycling, as the sample site air temperature often fluctuates in excess of 50 °F daily. After transport to the laboratory, all material conceivably damaged by weathering was removed and the ends of the sample blocks were cut parallel by trimming with a large diamond saw to a $12'' \times 12'' \times 24''$ size.

The specimens were obtained by boring holes through the long dimension of the block by means of an annular drill consisting of a diamond set masonry bit mounted on a thin-walled extension tube without core barrel, which was hand fed at 900 rpm. The resulting rods were approximately 24 inches long with a diameter of 0.76-inch, round to within 0.001-inch as drilled. The vibration from drilling may have caused some internal damage to the grain structure during the drilling process. Thermal damage to the specimen during this operation was mitigated by the combination of a large flow of cooling water and a slow feed rate of the drill press.

Specimen breakage occurred as a result of the drill encountering small fractures that had not been detected during the selection of the sample block. A few rods also broke during handling after termination of drilling due to the completion of fractures already present in the rock. The finished bars were trimmed square on the ends with a carborundum cut-off saw, which induces less vibration than a diamond wheel, and were then ground flat by hand with a 400 grit. All subsequent transport was accomplished with the greatest care to eliminate the possibility of damage to individual test rods by rough handling.

## 4. Composition and General Physical Properties of Diorite Specimens

Diorite is defined as an intrusive rock composed of 5 to 50 per cent dark minerals containing a plagioclase with a composition of oligoclase to andesine, exhibiting a total feldspar content containing less than

Fig. 1. Normal grain structure in fresh diorite specimens employed for wave propagation tests (crossed Nicols illumination).

5 per cent potash feldspar, and with a total quartz and feldspar content of less than 5 per cent quartz. The samples of diorite used to form the test specimens contain 20 per cent dark minerals, andesine plagioclase, a total quartz and feldspar content comprised of 3 to 5 per cent quartz

and a total feldspar content containing less than 3 per cent potash feld-spar. The grain size is typical for intrusive rocks. Scattered mineral grains will occasionally reach 5 to 6 millimeters in length, but the average length throughout the rock is closer to $1^1/_2$ to 2 millimeters. Diorite differs somewhat in its constituent composition from the related intrusive ma-terials gabbro, quartz, diorite, and granodiorite, but is identical in mine-ralogy to the extrusive rock andesite which, however, is much more fine-grained as a result of its much more rapid cooling history.

Structurally, the fabric of diorite consists of an interlocking mass of dark minerals and plagioclase feldspar as shown in Fig. 1, which re-presents a virgin specimen. The dark minerals are predominantly sub-hedral to nearly anhedral hornblende, a double chain silicate, and an-hedral magnetite, an oxide with a spinel structure. The plagioclase feld-spar, andesine, is a three-dimensional network structure, showing pro-minent polysynthetic twinning plus growth zoning and is present in subhedral grains. The quartz present appears as anhedral interstitial fillings. The grain boundaries of all constitutent minerals are sharp and free of deformational textures [56].

Table 1 presents physical properties determined for the test specimens of diorite and also indicates typical values obtained for similar rocks by other investigators.

## 5. Equipment, Instrumentation and Procedure

The apparatus employed for the HOPKINSON bar experiments has previously been described in detail [57, 58]. Briefly, a ballistically-suspended rod of diorite or 2024-T4 aluminum alloy of 3/4-inch diameter with a length ranging from 21 to 24 inches was subjected to the longi-tudinal impact of a $^1/_2$-inch diameter projectile fired by means of an air gun at a nearly constant velocity of 3300 in/sec. The hard-steel bullets were either spheres or flat-nosed cylinders of $^1/_2$-inch length, counter-bored at the free end to maintain a constant projectile weight of 8.35 grams. Strain pulses in the bars were detected by means of strain gages of the FAP-12-12 foil or SR-4 wire resistance type, mounted in pairs at opposite ends of a diameter at various stations of the bar, and connected so as to eliminate any antisymmetric component of the tran-sient. The signals generated were transmitted to and photographed on the screen of commercial oscilloscopes having a bandpass ranging from d.c. to more than 1 megacycle. Each record was individually calibrated.

The satisfactory employment of strain gages for non-metallic ma-terials had been previously determined [59]. Since these gages were of

Table 1. *Physical Properties of some Common Hypidiomorphic Granular Rocks*

| | Specific gravity | Strength, lb/in² | | Porosity, per cent | Velocity, ft/sec | | Static Young's modulus, lb/in² × 10⁶ | Poisson's ratio |
|---|---|---|---|---|---|---|---|---|
| | | Compressive | Tensile | | Longitudinal | Transverse | | |
| Diorite of present investigation | 2.83 | 32,000 | 800 | — | 12,000[1] 23,500[2] | — | 5.06 | — |
| Diorite (45) (Utah) | 2.74 | 48,300 | — | 0.25 | 18,200[3] | — | 12.2 | 0.25 |
| Diorite (46) (Michigan) | 3.01 | 39,800 | — | — | 19,700[3] | — | 15.47 | 0.29 |
| Granite (46) (Colorado) | 2.67 | 25,500 | 590 | 0.5 | 10,400[3] | — | 3.94 | −0.19 (!) |
| Granite (56), range of values | 2.56— 2.74 | 10,000— 60,000 | 410— 560 | 0.23— 1.75 | 13,000— 18,700 | 6,900— 10,800 | 5.03— 9.57 | 0.20— 0.26 |

[1] Surface measurement of propagation of longitudinal pulse propagation due to impact.
[2] Average velocity of explosively-induced shock propagation.
[3] Calculated from observed fundamental mode of longitudinal vibration of specimen.

sufficient size to cover several grain boundaries, their application in the present sequence of tests seemed to be warranted.

In view of a common triggering circuit for the oscilloscopes, the velocity of propagation of the pulses could be determined accurately from the arrival time of the first disturbance on each record. The interruption by the projectile of two light beams focussed on miniature photocells located near the slotted muzzle end of the gun barrel permitted a recording of the time of passage of the bullet on a counter with a microsecond scale, providing an accurate monitor of the magnitude of the impact velocity. The second photocell initiated the triggering signal through an adjustable delaying circuit.

Specimens subjected to the collision of spherical projectiles were suspended with their impact surface located some distance from the muzzle end in order to record the rebound velocity of the bullet by means of a stroboscopic camera. While this procedure proved to be successful in the case of the aluminum bars, the rebound trajectory of the projectile could not be observed for the diorite specimens by this technique, attributable either to the field of view being obscured by the fragments invariably emitted from the sample upon impact or to the negligible rebound velocity of the bullet, or both. Since these data were to be employed solely as a check of the magnitude of the impulse produced in the bar, it was not considered worthwhile to complicate the experimental procedure by employing a high-speed framing camera for the determination of this parameter. Measurement of the rebound velocity was not even considered in the tests involving the flat-nosed bullets where the specimens were placed as close as possible to the muzzle of the gun in an attempt to achieve a reasonably plane impact condition. This sequence of experiments was conducted with the object of achieving a faster rise time of the pulse, and hence a different harmonic spectrum, than could be obtained with spherical projectiles.

The aluminum alloy bars, whose properties had been previously investigated [58], were used repeatedly in the experiments, a small portion of the rod near the impact end being cut off and the new terminal face being reground after each shot. No variations were observed either in macroscopic behavior or microstructural pattern of the material as a result of repetitive shocking provided the immediate region around the small permanent crater formed by the collision was so eliminated. The impact of spherical projectiles on diorite at the standard velocity of 3300 in/sec resulted in complete disintegration of the collision end of the rod up to a length of 3/4 inches from the point of contact, leaving a highly irregular fracture surface on the remnant of the bar; however, the distal end of the rod remained intact. In the case of flat-ended projectiles, the same demolition pattern occurred at the impact end, but

the bar also fractured at an additional position as the result of the action
of the reflected tensile wave, indicating a more severe loading condition
under these circumstances.

Most of the tests involved the use of two strain gage observation
stations spaced 10.68 inches apart, the gages being located from the ends
of the virgin bar at nearly equidistant positions ranging from 5 to
6 inches. The data from a typical run are presented in Fig. 2. When
a once-shocked sample of the rock was to be used for a second experiment, the bar was simply inverted in position; but for additional tests
on that particular specimen, the same cut-off and grinding procedure
as employed in the original sample preparation was followed. In two

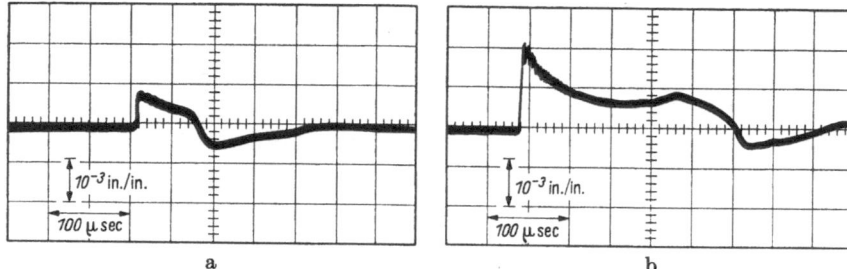

Fig. 2a and b. Strain gage data obtained during the impact of a $^1/_4$-inch diameter steel sphere at a
velocity of 3294 in/sec. on a 0.76-inch diameter bar of diorite. Strain gages (a) and (b) were located
5.19 and 15.87 inches from the impact point, respectively.

successive impacts of the spherical projectile on the same specimen,
five observation stations were employed; here, the bar for the second
test was refaced at the impact end and placed in the same position
as the fresh specimen.

The suspected damage produced in the grain structure by the passage
of the pulses was subjected to examination by a comparison of both
static compressive stress-strain data and microscopic inspection of virgin
and shocked specimens. Extreme care was again exercised to minimize
additional grain damage during preparation. Compression samples were
cut with a carborundum saw to the desired length and then wax-mounted
on an aluminum V-block for the additional removal of a 1/16″ length
from each end by grinding, finishing with a 0.7 micron grit to ensure
the absence of saw-damaged material. The specimens were then removed with trichloroethylene heated to 80 °C, a temperature too low
to produce significant grain bond weakening according to thermal shock
studies performed on this diorite, and subsequently air cooled.

All samples were ground parallel to 0.0001 inch across a 0.76 inch
diameter. Two lengths of test section were employed, 1.241 inches and
1.005 inches, yielding length to diameter ratios of 1.63 and 1.32, respec-

tively. These values are intermediate to those recommended in the literature [44, 60]. Since variations in the compressive strength of granite with moisture content had been shown to be small [44], no special effort was exercised to control this factor other than to assure identical moisture content of the specimens at the time of testing. The samples were loaded without a cap in a standard commercial testing machine at the rate of 100 lb/sec, representing a value midway in the range of 100 to 400 lb/in²/sec, where published information has indicated no change in results with loading rate for the same rock type [44]. The swivel head of the testing machine was not locked in position as suggested [60], but was adjusted to be parallel to the samples.

The examination of the internal structure of the diorite consisted of the grinding of standard petrographic thin sections, 0.03 mm thick, followed by microscopic inspection for shattered grain boundaries and fractures. Care had to be exercised in this interpretation to ensure that breakage caused by grinding and mounting processes is not attributed to stress-wave induced damage.

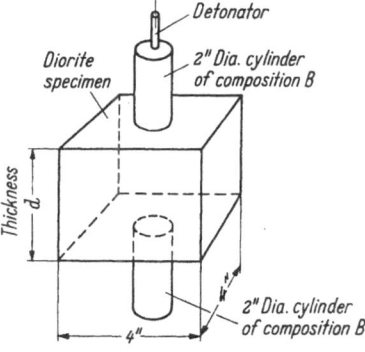

Fig. 3.
Sketch of a donor-receptor arrangement.

The contact explosion experiments were conducted by means of the donor-receptor arrangement sketched in Fig. 3, the thickness $d$ of the specimen being varied from $^1/_2$ to 3.5 inches. The explosive consisted of composition $B$ whose threshhold detonation pressure of 300,000 lb/in² served as the basis for the computation of the attenuation of the pulse at the center of the specimen with variations in thickness. Shock velocity measurements were carried out by means of a high-speed camera.

## 6. Results and Discussion

The results of the dynamic investigations are summarized in Table 2 for both the diorite and aluminium specimens. The pulse data for diorite are reported in terms of the measured peak strain, while comparable information for aluminum has been converted to force, using a dynamic YOUNG's modulus equal to the static value of $10^7$ lb/in². Fig. 4 and Fig. 5 show the respective effects of a spherical and a flat-faced projectile striking virgin bars of the rock at comparable velocities; although the amplitudes in the second case are considerably higher, the general pulse

Table 2. *Summary of Dynamic Tests on 3/4-Inch Diameter Rods of Diorite and Aluminum*

| Specimen material | Run no. | Type of Bullet* | Number of previous shocks | Distance of first gage G(1) to impact point, inches | Initial projectile velocity, in/sec | Strain amplitude, $10^{-3}$ in/in of initial pulse at station | | | | | Propagation velocity, $10^3$ in/sec between stations | | | |
|---|---|---|---|---|---|---|---|---|---|---|---|---|---|---|
| | | | | | | G(1) | G(2) | G(3) | G(4) | G(5) | G(1)–G(2) | G(2)–G(3) | G(3)–G(4) | G(4)–G(5) |
| Diorite | 1 | S | 0 | 5.5 | 3327 | 1.82 | — | 0.56 | 0.56 | 0.18 | | 144.5 | 144.8 | 144.4[1] |
| | 2 | S | 1 | 4.34 | 3253 | 2.26 | 1.06 | 0.68 | 0.59 | 0.20 | 103.0 | 123.5 | 136.0 | 175.0[2] |
| | 3 | S | 0 | 5.19 | 3294 | 1.69 | 0.65 | — | — | — | 144.2 | — | — | — |
| | 4 | S | 0 | 5.13 | 3291 | 1.54 | 0.39 | — | — | — | 142.4 | — | — | — |
| | 5 | S | 0 | 5.54 | 3294 | 1.59 | 0.99 | — | — | — | 144.8 | — | — | — |
| | 6 | S | 1 | 6 | 3289 | 1.55 | 1.25 | — | — | — | 142.5 | — | — | — |
| | 7 | S | 2 | 4.05 | 3275 | 2.01 | 0.68 | — | — | — | 143.0 | — | — | — |
| | 8 | S | 3 | 3.88 | 3181 | 1.85 | 0.76 | — | — | — | 104.0 | — | — | — |
| | 9 | S | 4 | 2.5 | 3291 | 1.77 | 1.02 | — | — | — | 106.8 | — | — | — |
| | 10 | S | 5 | 3.56 | 3302 | 2.26 | 0.60 | — | — | — | 127.0 | — | — | — |
| | 11 | F | 0 | 5.5 | 3211 | 2.44 | 1.16 | — | — | — | 144.8 | — | — | — |
| | 12 | F | 1 | 5.61 | 3201 | 1.52 | 0.82 | — | — | — | 129.6 | — | — | — |
| | 13 | F | 1 | 5.62 | 3215 | 2.70 | 1.00 | — | — | — | 131.3 | — | — | — |
| | 14 | F | 4 | 2.53 | 3164 | 3.24 | 1.28 | — | — | — | 142.5 | — | — | — |
| | 15 | F | 5 | 3.12 | 3167 | 3.72 | 1.52 | — | — | — | 104.0 | — | — | — |
| 2024–T4 Aluminum alloy | 16 | S | — | 6.05 | 3278 | 2.27 | 2.21 | — | — | — | 205.4 | — | — | — |
| | 17 | F | — | 5.5 | 3213 | 3.51 | 3.26 | — | — | — | 205.4 | — | — | — |

\* $S$ = 1/2-inch diameter steel sphere, weight = 8.35 grams

$F$ = 1/2-inch diameter flat-nosed cylinder, 1/2-inch long, hollowed out to weigh 8.35 grams.

[1] Overall velocity $c$ = 144,500 in/sec.    [2] Overall velocity $c$ = 131,500 in/sec.

Gage stations for runs 1 and 2 are 3.5 inches apart. 5 observation stations were employed for these runs, but the record from station 2 (G(2)) was lost in run 1.

Gage stations for runs 3–17 are 10.68 inches apart. 2 observation stations were employed for these runs.

shapes produced by the two types of bullets are similar. The strain-time curves shown here are typical; their appearance consists of a rapid rise followed by a decay until the effect of the pulse reflected from the free end manifests its influence. The reflected transient generally tends to

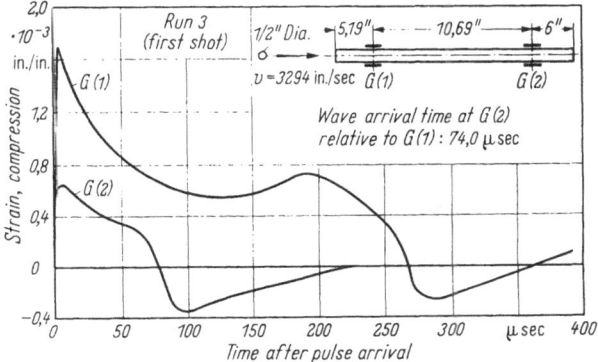

Fig. 4. Strain-time curves at two stations of a diorite bar of 0.76 inches diameter resulting from the impact of a $^1/_2$-inch diameter spherical projectile composed of hard steel at a velocity of 3294 in/sec.

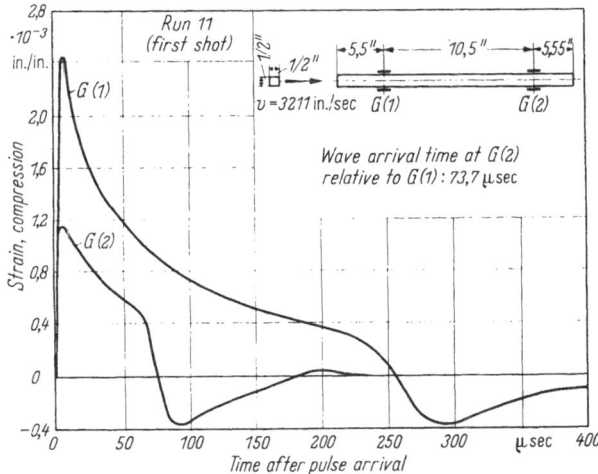

Fig. 5. Strain-time curves at two stations of a diorite bar of 0.76 inches diameter resulting from the impact of a $^1/_2$-inch diameter flat-nosed projectile of hard-steel at a velocity of 3211 in/sec.

produce a strain of opposite phase, but is sometimes preceded by a small reflected compressive precursor. In some instances, the gages closest to the impact point appeared to have been subjected to a permanent strain before the arrival of the reflected wave.

Fig. 6 and Fig. 7 present similar results for four and five gage stations, respectively, in a virgin and shocked diorite bar subjected to collision by a spherical projectile. A remarkable feature of these diagrams is the fact

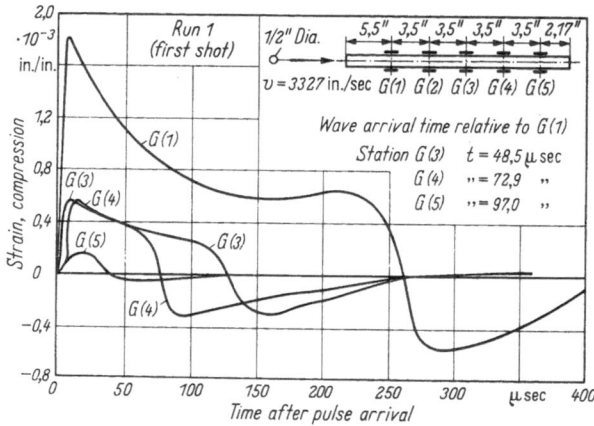

Fig. 6. Strain-time curves at four stations of a virgin diorite bar of 0.76 inches diameter resulting rom the impact of a spherical hard-steel projectile of ¹/₂-inch diameter at a velocity of 3327 in/sec.

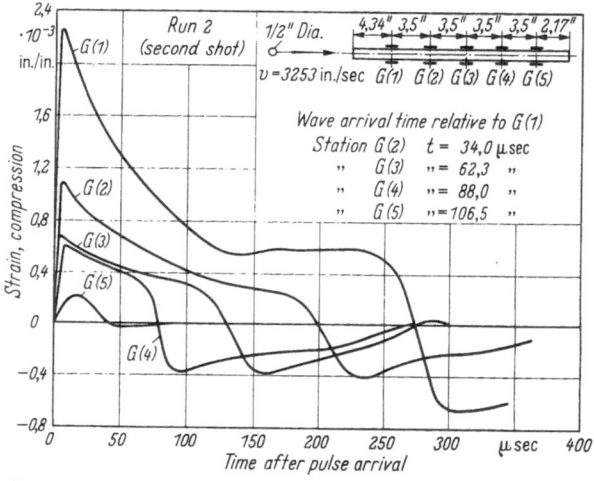

Fig. 7. Strain-time curves at five stations of a one-shocked diorite bar of 0.76 inches diameter resulting from the impact of a spherical hard-steel projectile of ¹/₂-inch diameter at a velocity of 3253 in/sec.

that the amplitude of the reflected component increases with distance during its traverse towards the impact end. It is hypothesized that this phenomenon might be due to the greater deformation response of the increasingly more damaged material to a given pressure pulse or possibly due to the presence of stored energy in the grains which is released upon arrival of the reflected wave. The initial shape of the pulse cannot be predetermined, being controlled by the material properties and the fragmentation process near the impact point as well as by the geometry and initial velocity of the bullet. A logarithmic plot of the portion of the curves representing the incident wave indicated the existence of exponential decay with a virtually constant attenuation factor of $0.93 \times 10^4 \sec^{-1} \pm 7$ per cent. An examination of the remaining data indicated a similar trend, independent of the bullet geometry, although a somewhat larger scatter in the magnitude of the attenuation coefficient was obtained. In any event, no dispersion of the initial pulse was observed within the limits of experimental error.

On the other hand, this transient was subjected to drastic attenuation in the traverse of the bar. An examination of the peak amplitudes recorded at stations _1_ to _4_ in Fig. 6 and Fig. 7 indicated that this decay was not linearly exponential, but varied more nearly as $x^{-0.55}$, where $x$ is the distance traversed in inches. The results from the station nearest the free end were discarded for this purpose, since the pulse did not fully develop at this position before the arrival of a reflected wave. Any frequency dependence of the attenuation factor should be canceled out by the conditions of pulse propagation. Since the strain-time curves at all stations exhibit the same exponential decay, their FOURIER spectra and ratio of energy levels of their respective components will be identical, and the attenuation parameter should consequently be identical at all positions. The absence of exponential attenuation was also indirectly supported by the auxiliary data listed in Table 2 which, under the hypothesis of linear exponential decay, exhibited attenuation factors ranging from $-0.02$ to $-0.13$. According to Eq. (10), an average value of $\alpha$, say $-0.1$, would require a single equivalent frequency of the pulse spectrum of 3,600,000 rad/sec to correlate with the value of $\Delta E/E$ cited by Reference [22] for diorite. This frequency is not only well above the recording capability of the equipment, but does not seem to correspond to the dominant frequency of the pulse which, in view of the magnitude of the decay constant, would appear to be more of the order of 10 kilocycles. The exponent of $-0.55$ does not provide the best fit for the remainder of the data and is thus subject to adjustment pending further investigation, but the power law is a better representation of the complete set of results than the exponential decay.

In order to assess the effects of dispersion resulting from the three-dimensional nature of the bar, some auxiliary experiments were conducted under identical conditions on 3/4-inch diameter bars of 2024-T4 aluminum alloy, a material which had been previously subjected to extensive investigations involving pulse propagation [57, 58].

The results of these "calibration" tests are shown in Fig. 8 and Fig. 9; the transient produced by the spherical projectile was transmitted between the measuring stations virtually intact, with a loss of less than 5 per cent of the impulse. A larger dispersive effect was observed in the case of the cylindrical bullet, although the general pulse form was still preserved. The reduction in amplitude even in the second instance can not begin to account for the attenuation observed in the diorite bars, which must conse-

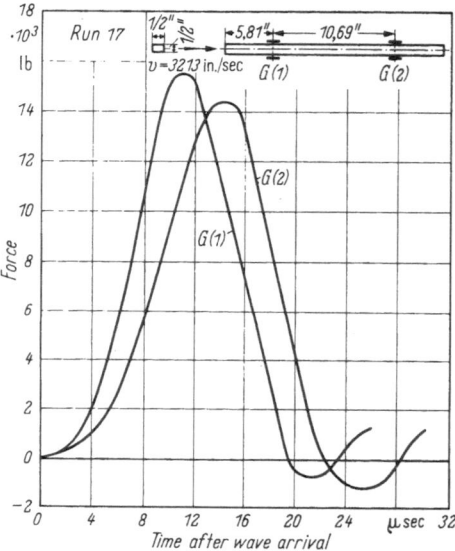

Fig. 8. Force-time curves at two stations in a ³/₄-inch diameter 2024-T 4 aluminum alloy bar resulting from the impact of a ¹/₂-inch diameter hard-steel spherical projectile at a velocity of 3278 in/sec.

Observed change of momentum of the bullet:      0.177 lb-sec;
Impulse recorded by gage 1:      0.174 lb-sec;
Impulse recorded by gage 2:      0.168 lb-sec.

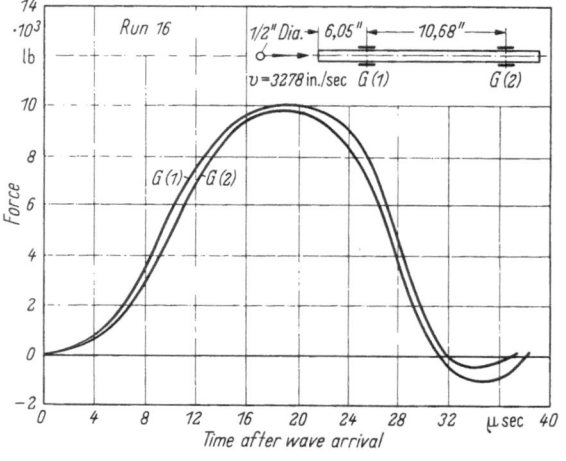

Fig. 9. Force-time curves at two stations in a ³/₄-inch diameter 2024-T 4 aluminum alloy bar resulting from the impact of a ¹/₂-inch diameter hard-steel cylindrical projectile with a flat nose at a velocity of 3213 in/sec.

Initial bullet momentum:      0.153 lb-sec;
Positive impulse recorded by gage 1:      0.150 lb-sec;
Positive impulse recorded by gage 2:      0.146 lb-sec.

quently be attributed to an internal frictional mechanism such as relative sliding along the grain boundaries.

In view of the experimental evidence, some conclusions can be drawn concerning the applicability of the various proposed models discussed in Section 2. The KELVIN-VOIGT solid or other types of visco-elastic models must be rejected as suitable representations for the diorite, since all of these models will exhibit dispersive characteristics in contrast to the observed results. The same objection applies to the use of the hypothetical substance defined by Eqs. (15) and (16). The solid friction mechanism expressed by Eq. (7) is far more acceptable in view of the absence of dispersive features, but, pending additional study, the exponential decay and frequency dependence stipulated by Eq. (10) and the analogous result predicted by Eq. (14) may need to be adjusted. A proper model for the diorite must incorporate the features of absence of dispersion and an attenuation most likely based on some power of the ratio of travel distance to bar diameter.

The velocities of propagation of the pulse presented in Table 2 are uniformly consistent for the virgin bars at a value of 144,000 in/sec $\pm$ 1.5 per cent. This magnitude is somewhat less than that reported for other diorites, which were determined by the resonant frequency method — although in the range of some granites — and considerably smaller than values determined from field tests. However, the discrepancy between field and laboratory tests is not unique [47], since ratios of more than 2 to 1 have been reported for substances very similar to the diorite employed in the present investigation. This difference is presumably due to the absence of a controlled medium and the effect of confinement in the field. On the basis of the quoted measurement, the computed value of the "dynamic" YOUNG's modulus for the diorite is $5.5 \times 10^6$ lb/in². A consistent trend can be observed from the data of Table 2 in that successive shocks in the rock generally result in lower velocities of pulse propagation, indicating a reduction in the modulus with number of collisions, as the porosity and hence density of the material remained essentially invariant. An examination of the data from Run 2, a specimen previously subjected to an impact, reveals that the velocity of propagation increased in sections successively further removed from the original collision end, indicating that the reduction in modulus and thus strength also depends upon location relative to the point of prior contact. This feature is responsible for the observed variations in strain amplitude exhibited by specimens previously subjected to the same number of shocks.

Corroborating evidence of the existence of significant shock damage were obtained from static compression tests and microscopic examination of both virgin and shocked specimens. Table 3 presents a compilation of the compression test results. Three unshocked samples, E-1, E-2, and E-3 — two being cut from the same rod, the third being selected at random — yielded essentially identical ultimate strengths of 31,800 lb/in² $\pm$ 2 per cent. Specimens C-1, D-1 and D-2 were removed

from a single rod which had been subjected to an impact at each end
by a spherical projectile travelling at 3300 in/sec. Samples D-1 and D-2
were cut from the strain-gage stations of the rod and exhibited essentially

Table 3. *Summary of Compression Data*

| Sample no. | Length, in. | Diameter, in. | Area, in² | Load to failure, lb | Breaking strength, lb/in² | Number of previous shocks | Velocity of impact, in/sec (¼-inch diameter steel sphere) |
|---|---|---|---|---|---|---|---|
| A—1 | 1.0067 | 0.7659 | 0.461 | 12,500 | 27,115 | 2 | 3,285 3,300 |
| A—2—2 | 1.0055 | 0.7660 | 0.461 | 13,000 | 28,200 | 1 | 3,285 |
| B—1 | 1.2400 | 0.7613 | 0.455 | 13,000 | 28,571 | 1 | 3,300 |
| C—1 | 1.2411 | 0.7633 | 0.458 | 8,000 | 17,467 | 2 | 3,300 |
| D—1 | 1.2427 | 0.7626 | 0.457 | 9,450 | 20,678 | 2 | 3,300 |
| D—2 | 1.2396 | 0.7616 | 0.456 | 9,600 | 21,053 | 2 | 3,300 |
| E—1 | 1,2392 | 0.7604 | 0.454 | 14,700 | 32,379 | 0 | ... |
| E—2 | 1.2410 | 0.7653 | 0.460 | 14,400 | 31,304 | 0 | ... |
| E—3 | 1.2410 | 0.7662 | 0.461 | 14,550 | 31,562 | 0 | ... |

identical strengths of 20,678 lb/in² and 21,053 lb/in², while the third
sample, which displayed a considerably lower value of 17,467 lb/in², was
cut from a section immediately adjacent to an impact end.

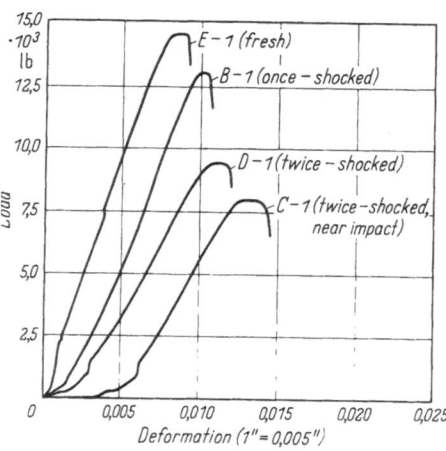

Fig. 10. Comparison of load-deformation curves
for diorite subjected to different conditions of
repeated impact.

Additional rods were tested to
further verify these results. Two
singly-shocked rods, samples B-1
and A-2-2, also yielded nearly
identical values of 28,571 lb/in²
and 28,200 lb/in², respectively,
somewhat lower than that of the
virgin material. The second of
these specimens was subjected to
a second impact after which a test
section A-1 yielded a strength of
27,115 lb/in². All of these changes
in ultimate strength are in the
direction of lowered sample
strength with increasing numbers
of impacts and also with increasing proximity to the impact point,
resulting in a distinct weakening
of the material. The validity of the change in strength between samples
A-1 and A-2 is considered to be particularly strong since the same rod
was employed for both specimens. The individual rods were found to
perform quite uniformly in the compression tests.

Fig. 10 presents load-deformation curves for samples E-1, D-1, C-1, and B-1, showing the variations in response as the history of shocking is varied. The tangent modulus of the unshocked specimen corresponds to a value of $5.06 \times 10^6$ lb/in², approximately 10 per cent lower than the corresponding dynamic value. Although metals apparently exhibit the same elastic modulus under both static and dynamic conditions at moderate pressures, the increase in modulus observed for the diorite might well correspond to the observed increase in strength of metallic substances in the plastic range under dynamic conditions. Specimens D-1 and C-1 yielded values of the tangent modulus of $3.32 \times 10^6$ and $3.06 \times 10^6$ lb/in², respectively, representing a reduction of almost 40 per cent of this parameter from that of the virgin specimen. The significant decrease in both ultimate strength and tangent modulus of the shocked material emphasizes the serious nature of the damage produced in the grain structure and bonding elements. This effect also accounts for the lower velocity of pulse propagation in shocked and refaced diorite bars. The concept was further confirmed by microscopic examination. Fig. 11 is a typical microphotograph of a thin section located $1^1/_2$ inches from the point of impact, which clearly exhibits a number of microfractures not present in the virgin material (Fig. 1). Any grain damage beyond this region was too subtle to show in the specimens utilized. This fracturing process is one of the principal mechanisms of energy dissipation resulting

Fig. 11. Photomicrograph of portion of one-shocked diorite rod adjacent to the point of impact and showing development of microfractures.

in pulse attenuation. All these results clearly indicate the presence of a new material after each shock, and attempted re-use of each sample by refacing the impact ends must be approached in the light of this situation.

The average velocity of propagation of explosively-induced shocks in diorite for specimen thicknesses in the range from 2 to 3 inches was found to be 282,000 in/sec, corresponding to an incident pressure in the sample of about $5 \times 10^6$ lb/in². The velocity of propagation in 2024-T4 aluminum alloy under the same conditions of incident pressure was found to be 296,000 in/sec [61]. The ratio of propagation velocities at this high pressure to that at the moderate pressures encountered in the pulse propagation produced by impact loading of a projectile for diorite is hence equal to 1.96, while that for aluminum is only 1.44, indicating about the same change in density of both materials. The total attenuation of the shock wave pressure in a diorite specimen 2.78 inches thick between the centers of the two interfaces was found to be $3.98 \times 10^6$ lb/in², although the greater portion of this attenuation must be attributed to lateral expansion of the wave.

## 7. Conclusions

1. Repeatable and consistent data can be obtained in specimens of rock provided sufficient care is exercised in the selection and preparation of samples. The results obtained concerning pulse form, velocity of propagation, and grain bond damage in diorite rods of $3/_4$-inch diameter due to the impact of $1/_2$-inch diameter steel projectiles at a velocity of about 3300 in/sec exhibited little scatter and could be readily interpreted.

2. The pulse form produced in the rock is not directly controllable and its shape is not significantly influenced by projectile geometry. The pulse propagates without dispersion and with an attenuation proportional so some power of the ratio of distance traversed to bar diameter. A solid-friction type of mechanism accounts for the first feature, but may require some modification to properly represent the observed attenuation phenomenon.

3. The ultimate strength of the diorite and its static Young's modulus decrease in direct proportion to the number of impacts previously experienced by the specimen. The damage to the internal structure is a function of this parameter as well as to the proximity to the impact point. The velocity of propagation, and hence the dynamic Young's modulus are also lower in a shocked specimen of the rock.

4. Under identical loading conditions, the rock selected exhibits drastically different characteristics from an aluminum alloy used as a reference with respect to pulse form and manner of propagation, change in mechanical properties, and microscopic damage.

# 8. Acknowledgment

The authors are grateful for the technical assistance of Mr. T. W. LIU of the University of California at Berkeley, and Mr. L. COSNER of the Michelson Laboratory of the U.S. Naval Ordnance Test Station, China Lake, Cal. The work was principally performed under the sponsorship of the U.S. Naval Ordnance Test Station, China Lake and was also in part supported by the National Science Foundation.

## Bibliography

[1] EWING, W. M., W. S. JARDETZKY and F. PRESS: Elastic Waves in Layered Media, New York: McGraw-Hill Book Co. 1957.

[2] BREKHOVSIKH, L. A.: Waves in Layered Media, New York: Academic Press 1960.

[3] SHARPE, J. A.: The Production of Elastic Waves by Explosion Pressure, Geophysics 7, 144, 311 (1942).

[4] BLAKE, F. G., Jr.: Spherical Wave Propagation in Solid Media, J. Acoust. Soc. Amer. 24, 211 (1952).

[5] SELBERG, H. L.: Transient Compression Waves from Spherical and Cylindrical Cavities, Arkiv för Fysik 5, 97 (1952).

[6] OBERT, L., and W. I. DUVALL: Generation and Propagation of Strain Waves in Rock, Part 1, U.S. Bureau of Mines Report of Investigations 4683, 1950.

[7] DUVALL, W. I., and B. PETKOF: Spherical Propagation of Explosion-generated Strain Pulses in Rock, U.S. Bureau of Mines Report of Investigations 5483, 1959.

[8] PEARSE, G. E.: Rock Blasting, Mine and Quarry Engineering 21, 25 (1955).

[9] HINO, K.: Fragmentation of Rock through Blasting, Quarterly of the Colorado School of Mines 51, 191 (1956).

[10] HINO, K.: Velocity of Rock Fragments and Shape of Shock Wave, J. of Industrial Explosives (Japan) 17, 236 (1956).

[11] RINEHART, J. S., and J. PEARSON: Behavior of Metals under Impulsive Loads, American Society for Metals, Cleveland, Ohio, 1954.

[12] RINEHART, J. S.: On Fractures Caused by Explosions and Impacts, Quarterly of the Colorado School of Mines 55, No. 4 (1960).

[13] BROBERG, K. B.: Shock Waves in Elastic and Elastic-Plastic Media, Kungl. Tekniska Högskolan (Stockholm), Avhandling No. 11, 1956.

[14] DUVALL, W. I., and T. C. ATCHISON: Rock Breakage by Explosives, U.S. Bureau of Mines Report of Investigations 5356, 1957.

[15] BACON, L. O.: A Method of Determining Dynamic Tensile Strength of Rock at Minimum Loading, U.S. Bureau of Mines Report of Investigations 6067, 1962.

[16] GRINE, D. R.: Finite Amplitude Stress Waves in Rocks, Stanford Research Institute, Poulter Laboratories Technical Report 012—59, 1959.

[17] KNOPOFF, L., and G. J. F. McDONALD: Models for Acoustic Loss in Solids, J. Geophysical Research 65, 2191 (1960).

[18] ZENER, C.: Elasticity and Anelasticity of Metals, University of Chicago Press 1948.

[19] COLLINS, F., and C. C. LEE: Seismic Wave Attenuation Characteristics from Pulse Experiments, Geophysics 21, 16 (1956).

[20] PESELNICK, L., and I. ZIETZ: Internal Friction of Fine-grained Limestones at Ultrasonic Frequencies, Geophysics 24, 285 (1959).

[21] AUBERGER, M., and J. S. RINEHART: Ultrasonic Velocity and Attenuation of Longitudinal Waves in Rocks, J. Geophysical Research 66, 191 (1961).

[22] BRUCKSHAW, J. McG., and P. C. MAHANTA: The Variation of the Elastic Constants of Rocks with Frequency, Petroleum 17, 14 (1954).

[23] KNOPOFF, L.: The Seismic Pulse in Materials Possessing Solid Friction, I: Plane Waves, Bulletin of the Seismological Society of America 46, 175 (1956).

[24] KNOPOFF, L., and G. J. F. MACDONALD: Attenuation of Small Amplitude Stress Waves in Solids, Reviews of Modern Physics 30, 1178 (1958).

[25] RICKER, N.: Form and Nature of Seismic Waves and the Structure of Seismograms, Geophysics 5, 348 (1940).

[26] RICKER, N.: The Form and Law of Propagation of Seismic Wavelets, Proceedings of the Third World Petroleum Congress, Sec. I, 1951, p. 514.

[27] McDONAL, F. J., F. A. ANGONA, R. L. MILLS, R. L. SENGBUSH, R. G. VAN NOSTRAND and J. E. WHITE: Attenuation of Shear and Compressional Waves in Pierre Shale, Geophysics 23, 421 (1958).

[28] RICKER, N.: Further Developments in the Wavelet Theory of Seismogram Structure, Bulletin of the Seismological Society of America 33, 197 (1943).

[29] RICKER, N.: Wavelet Functions and their Polynomials, Geophysics 9, 314 (1944).

[30] RICKER, N.: The Computation of Output Disturbances from Amplifiers for True Wavelet Inputs, Geophysics 10, 207 (1945).

[31] JEFFREYS, H.: Damping in Bodily Seismic Waves, Monthly Notices of the Royal Astronomical Society, Geophysical Supplement 2, 318 (1931).

[32] RICKER, N., and W. A. SORGE: The Primary Seismic Disturbance in Shale, Bulletin of the Seismological Society of America 41, 181 (1951).

[33] VAN MELLE, F. A.: Note on The Primary Seismic Disturbance in Shale, Bulletin of the Seismological Society of America 44, 123 (1954).

[34] SEZAWA, K.: On the Decay of Waves in Viscoelastic Solid Bodies, Bulletin of the Earthquake Research Institute, Tokyo University, Vol. 3, 1927, p. 43.

[35] KNOPOFF, L.: The Attenuation of Compression Waves in Lossy Media, Bulletin of the Seismological Society of America 46, 47 (1956).

[36] FUTTERMAN, W. I.: Dispersive Body Waves, J. Geophysical Research 67, 5279 (1962).

[37] LAMB, G. L.: The Attenuation of Waves in a Dispersive Medium, J. Geophysical Research 67, 5273 (1962).

[38] BIRCH, F., and D. BANCROFT: Elasticity and Internal Friction in a Long Column of Granite, Bulletin of the Seismological Society of America 28, 243 (1938).

[39] REICH, H.: Geologische Unterlagen der angewandten Geophysik, Handbuch der Experimentalphysik, Vol. 25, pt. 3, Leipzig: Akademische Verlagsgesellschaft 1930, p. 1.

[40] DOBRIN, M. B.: Introduction to Geophysical Prospecting, 2nd ed., New York: McGraw-Hill 1960.

[41] BIRCH, F., J. F. SCHAIRER and H. C. SPICER (Editors): Handbook of Physical Constants, Geological Society of America Special Paper 36, 1942.

[42] PROTODYAKONOV, M. M.: Methods of Studying the Strength of Rocks, used in the U.S.S.R., International Symposium on Mining Research, Vol. 2, New York: Pergamon Press 1962, p. 649.

[43] GRIFFITH, J. H.: Physical Properties of Typical American Rocks, Iowa State College of Agriculture, Engineering Experiment Station Bulletin 131, 1937.

[44] OBERT, L., S. L. WINDES and W. I. DUVALL: Standardized Tests for Determining the Physical Properties of Mine Rock, U.S. Bureau of Mines Report of Investigations 3891, 1946.

[45] WINDES, S. L.: Physical Properties of Mine Rock, Parts I and II, U.S. Bureau of Mines Report of Investigations 4459 and 4727, 1949 and 1950.

[46] BLAIR, B. E.: Physical Properties of Mine Rock, Parts III and IV, U.S. Bureau of Mines Report of Investigations 5130 and 5244, 1955 and 1956.

[47] NICHOLLS, H. R.: In Situ Determination of the Dynamic Elastic Constants of Rock, International Symposium on Mining Research, ed. by G. B. CLARK, New York: Pergamon Press 1962, p. 727.

[48] HUGHES, D. S., and H. J. JONES: Variation of Elastic Moduli of Igneous Rocks with Pressure and Temperature, Bulletin of the Geological Society of America 61, 843 (1950).

[49] HUGHES, D. S., and C. MAURETTE: Variation of Elastic Wave Velocities in Basic Igneous Rocks with Pressure and Temperature, Geophysics 22, 23 (1957). (See also Institut Français du Pétrole et Annales des Combustibles Liquides, Revue 12, 730 (1957).)

[50] BIRCH, F.: The Velocity of Compressional Waves in Rocks to 10 Kilobars, J. Geophysical Research 65, 1083 (1960); 66, 2199 (1961).

[51] BRIDGMAN, P. W.: The Physics of High Pressure, London: Bell & Sons 1949.

[52] ADAMS, L. H., and E. D. WILLIAMSON: On the Compressibility of Minerals and Rocks at High Pressures, J. of the Franklin Institute 195, 474 (1923).

[53] RICE, M. H., R. G. McQUEEN and J. M. WALSH: Compression of Solids by Strong Shock Waves, Solid State Physics, Vol. 6, New York: Academic Press 1958, p. 1.

[54] HUGHES, D. S., and R. G. McQUEEN: Density of Basic Rocks at Very High Pressures, Transactions of the American Geophysical Union 39, 959 (1958).

[55] LOMBARD, D. B.: The Hugoniot Equation of State of Rocks, Proceedings of the 4th Symposium on Rock Mechanics, 1961. Bulletin of the Mineral Industries Experiment Station, Mining Engineering Series, College of Mineral Industries, Pennsylvania State University. [See also University of California Lawrence Radiation Laboratory (Livermore) Report No. 6311.]

[56] AUSTIN, C. F., and J. K. PRINGLE: Rocks that occur as Brittle Solid Test Materials at the U.S. Naval Ordnance Test Station, China Lake, California, NAVWEPS Report 7928, NOTS TP 2955, July 1962.

[57] CUNNINGHAM, D. M., and W. GOLDSMITH: Short-time Impulses produced by Longitudinal Impact, Proceedings of the Society for Experimental Stress Analysis 16, No. 2, 153 (1959).

[58] GOLDSMITH, W., and P. T. LYMAN Jr.: The Penetration of Hard-Steel Spheres into Plane Metal Surfaces, J. Appl. Mech. 27, 717 (1960).

[59] GOLDSMITH, W., and G. W. NORRIS Jr.: Stresses in Curved Beams due to Transverse Impact, Proceedings of the Third U.S. National Congress of Applied Mechanics, 1958, p. 153.

[60] GROSVENOR, N. E.: Specimen Proportion Key to Better Compressive Strength Tests, Mining Engineering 15, 31 (1963).

[61] GORANSON, R. W., D. BANCROFT, B. L. BURTON, T. BLECHAR, E. E. HOUSTON, E. F. GITTINGS and S. A. LANDEEN: Dynamic Determination of the Compressibility of Metals, J. Appl. Phys. 26, 1472 (1955).

# On Some Simplifications in the Description of the Motion of a Soft Soil

## By S. S. Grigorian

### Institute of Mechanics of Moscow State University, U.S.S.R.

The system of equations given in earlier papers [1, 2] to describe the motion of a soft soil is very complicated. But when dealing with various special problems, considerable simplifications of this system may be possible. According to the character of the problem under consideration, it is therefore desirable to determine what simplifications are possible and what simplified system of equations should be used. A full classification of all possible problems for the aforementioned system of equations is given below together with the derivation of simplified systems for each class of problem.

1. The basic equations of soft soils given in [1, 2] are as follows

$$\varrho \left( \frac{\partial v_i}{\partial t} + v_j \frac{\partial v_i}{\partial x_j} \right) = \varrho F_i^e - \frac{\partial p}{\partial x_i} + \frac{\partial S_{ij}}{\partial x_j}, \qquad \frac{\partial \varrho}{\partial t} + \frac{\partial (\varrho v_i)}{\partial x_i} = 0,$$

$$p = f(\theta, \theta_*), \qquad \theta = 1 - \frac{\varrho_0}{\varrho}, \qquad \theta_* = 1 - \frac{\varrho_0}{\varrho_*},$$

$$\frac{\partial \theta_*}{\partial t} + v_i \frac{\partial \theta_*}{\partial x_i} = \left( \frac{\partial \theta}{\partial t} + v_i \frac{\partial \theta}{\partial x_i} \right) e\,(\theta - \theta_*)\, e \left( \frac{\partial \theta}{\partial t} + v_i \frac{\partial \theta}{\partial x_i} \right),$$

$$G \left( \frac{\partial v_i}{\partial x_j} + \frac{\partial v_j}{\partial x_i} - \frac{2}{3} \frac{\partial v_k}{\partial x_k} \delta_{ij} \right) = \frac{\breve{d} S_{ij}}{dt} + \lambda S_{ij},$$

$$\frac{\breve{d} S_{ij}}{dt} = \frac{\partial S_{ij}}{\partial t} + v_k \frac{\partial S_{ij}}{\partial x_k} - \Omega_{ik} S_{jk} - \Omega_{jk} S_{ik}, \tag{1}$$

$$2\Omega_{ij} = \frac{\partial v_i}{\partial x_j} - \frac{\partial v_j}{\partial x_i}, \qquad \lambda = \frac{2GW - F'(p)\,(\partial p/\partial t + v_i\,\partial p/\partial x_i)}{2F(p)}$$

$$\times\, e\,[J_2 - F(p)]\, e \left[ 2GW - F'(p) \left( \frac{\partial p}{\partial t} + v_i \frac{\partial p}{\partial x_i} \right) \right],$$

$$2W \equiv S_{ij} \left( \frac{\partial v_i}{\partial x_j} + \frac{\partial v_j}{\partial x_i} \right), \qquad 2J_2 \equiv S_{ij} S_{ij}.$$

Here, $x_i$ are rectangular Cartesian coordinates and $t$ the time, $\varrho$ is the density, $p$ the pressure, $v_i$ the velocity, $F_i^e$ the body force per unit mass,

$\delta_{ij}$ the unit tensor, $S_{ij}$ the stress deviator tensor, $\Omega_{ij}$ the vorticity tensor and $\tilde{d}/dt$ denotes the JAUMAN's material derivative.

These equations contain three functions that characterize the soil and must be determined experimentally:

$$p = f\,(\theta, \theta_*), \quad J_2 = F(p), \quad G = G(\theta_*). \tag{2}$$

A method of determining these functions by means of static and dynamic experiments has been proposed in [1, 2], and experimental results have been presented in [3, 4].

The fundamental problem of soil mechanics arising for the system (1) is posed as follows. The medium occupies a given domain $\Omega$ and is initially at rest. The surface tractions $\overline{P}_n$ are given on part of the boundary of this domain as functions of position and time, and the displacements $\overline{U}$ on the remainder of the boundary.

Since the boundary functions $\overline{P}_n$, $\overline{U}$ can always be specified in the universal form

$$\overline{P}_n = \sigma_0\,\overline{\Pi}_n\left(\frac{x_i}{l}, \frac{t}{t_0}\right), \quad \overline{U} = V_0 t_0\,\overline{U}\left(\frac{x_i}{l}, \frac{t}{t_0}\right), \tag{3}$$

where $\overline{\Pi}_n$, $\overline{U}$ are dimensionless functions of dimensionless arguments, the boundary conditions introduce the parameters $\sigma_0$, $V_0$, $t_0$, $l$ into the mathematical formulation of the problem. These parameters are a characteristic stress, velocity, time, and length. When the other conditions are the same, different classes of problems will be characterized by the difference in the numerical values of these parameters. However, to be able to classify problems in this manner, one must enlarge this group of parameters by parameters contained in the system of Eqs. (1). To obtain the latter we remember that expressions (2) can always be represented in the form

$$p = K f(\theta, \theta_*), \quad J_2 = \sigma_*^2 F(p/\sigma_*), \quad G = G_0 g(\theta_*), \tag{4}$$

where

$$f(\theta, \theta_*) \to \theta \quad \text{for} \quad \theta_* \to \min \theta_*, \quad \theta \to 0,$$

$$F(p/\sigma_*) \to 1 \quad \text{for} \quad p/\sigma_* \to 0,$$

$$g(\theta_*) \to 1 \quad \text{for} \quad \theta_* \to \min \theta_*.$$

We now introduce the notation

$$S_\infty = \lim_{p/\sigma_* \to \infty} \sqrt{\sigma_*^2 F(p/\sigma_*)} = \sigma_* \sqrt{F(\infty)}. \tag{5}$$

The following estimates are a result of experiment:

$$K \sim G_0, \qquad \sigma_* \ll G_0, \tag{6}$$

and it can also be supposed that $S_\infty$ has the same order as $G_0$ or is of smaller order:

$$S_\infty \lesssim G_0, \tag{7}$$

Thus, in addition to the previous group of parameters, we have $K$, $G_0$, $\sigma_*$, $S_\infty$, $\varrho_0$, so that the full system of representative parameters of the problem finally consists of

$$\varrho_0 \; V_0, \; t_0, \; l, \; K, \; G_0, \; \sigma_*, \; S_\infty, \; \varrho_0. \tag{8}$$

It should be noted that when the boundary conditions are formulated purely in stresses or purely in displacements, the system (8) will include only $\sigma_0$ or only $V_0$, while $\sigma_0$ and $V_0$ will be connected by some relation.

Our classification will finally be reduced to the derivation of some estimates and inequalities for dimensionless combinations constructed from the parameters (8). The fulfillment of these estimates and inequalities will lead to corresponding simplifications in the system (1).

2. We begin our considerations with a purely elastic case in which the quantities $\sigma_0 < \sigma_*$ and $u_0 \sim V_0 t_0$ are very small. In Eq. (1), $\lambda \equiv 0$, and the relations between the deviators of the stress and strain rate tensors becomes

$$G \left( \frac{\partial v_i}{\partial x_j} + \frac{\partial v_j}{\partial x_i} - \frac{2}{3} \frac{\partial v_k}{\partial x_k} \delta_{ij} \right) = \frac{\tilde{d} S_{ij}}{dt}. \tag{9}$$

We now have the following formulas

$$u_i = V_0 V_i \left( \frac{x_k}{l}, \; \frac{t}{t_0} \right), \qquad S_{ij} = \sigma_0 \Sigma_{ij} \left( \frac{x_k}{l}, \; \frac{t}{t_0} \right), \tag{10}$$

the nondimensional functions $V_i$, $\Sigma_{ij}$ and their derivatives with respect to the dimensionless arguments being of order unity. With this in mind, we can estimate the order of all terms in Eq. (9), and reduce it to

$$G_0 \frac{V_0}{l} \sim \sigma_0 \left( \frac{1}{t_0} + \frac{V_0}{l} \right). \tag{11}$$

Since in this case $\sigma_0 \sim \sigma_* \ll G_0$, Eq. (11) furnishes

$$\frac{V_0 t_0}{l} \sim \frac{\sigma_0}{G_0} \ll 1. \tag{12}$$

This indicates that the displacements $u_0 \sim V_0 t_0$ and the strains $u_0/l$ are small, and the convective terms, which are comparatively small, should be neglected in taking the material derivative:

$$\frac{\tilde{d}}{dt} \sim \frac{d}{dt} \sim \frac{\partial}{\partial t}. \tag{13}$$

Thus, the system (1) reduces to

$$\varrho_0 \frac{\partial v_i}{\partial t} = \varrho_0 F_i^e - \frac{\partial p}{\partial x_i} + \frac{\partial S_{ij}}{\partial x_j}, \qquad \frac{\partial \theta}{\partial t} + \frac{\partial v_i}{\partial x_i} = 0,$$

$$p = K\theta, \qquad G_0 \left( \frac{\partial v_i}{\partial x_j} + \frac{\partial v_j}{\partial x_i} - \frac{2}{3} \frac{\partial v_k}{\partial x_k} \delta_{ij} \right) = \frac{\partial S_{ij}}{\partial t}. \tag{14}$$

The second and fourth Eqs. (14) can be integrated, giving HOOKE's law,

$$S_{ij} = G_0 \left( \frac{\partial u_i}{\partial x_j} + \frac{\partial u_j}{\partial x_i} + \frac{2}{3} \theta \delta_{ij} \right), \qquad \theta = -\frac{\partial u_k}{\partial x_k}, \qquad p = K\theta, \tag{15}$$

where $u_i = \int\limits_0^t v_i \, dt$ is a displacement.

Thus in the case under consideration the system (1) reduces to the equations of the linear theory of elasticity.

For Eqs. (14) it is interesting to consider problems of two kinds, essentially dynamical and quasi-statical ones. In the former case, all terms in the momentum equation have the same order, that is,

$$\varrho_0 \frac{V_0}{t_0} \sim \frac{\sigma_0}{l}. \tag{16}$$

The comparison (16) to (12) gives us

$$\frac{l}{t_0} \sim \sqrt{\frac{G_0}{\varrho_0}} \sim C, \qquad \sigma_0 \sim \varrho_0 V_0 C, \qquad V_0 \sim \theta C,$$

since $\sigma_0 \sim K\theta \sim G_0\theta$.

In these estimates $C$ is a characteristic velocity of the elastic wave propagation. All these formulas are well-known relations for elastic waves.

When the motion is quasi-static, we have

$$\varrho_0 \frac{V_0}{t_0} \ll \frac{\sigma_0}{l} \quad \text{or} \quad \frac{l}{t_0} \ll \sqrt{\frac{G_0}{\varrho_0}} \sim C. \tag{17}$$

This condition determines a time scale $t_0$ for the boundary functions (3) at which it is admissible to neglect the elastic wave propagation. If we introduce a "wave" time scale $t_w \sim l/C$, we can rewrite condition (17) in the form

$$t_0 \gg t_w. \tag{18}$$

20*

3. When motions are accompanied by small elastic-plastic strains, $\sigma_0 \sim \sigma_*$, $u_0 \sim V_0 t_0 \ll l$, but $\lambda \neq 0$. The estimates (13) obviously remain unchanged. Estimation of the terms in the law of elastic-plastic flow

$$G \left( \frac{\partial v_i}{\partial x_j} + \frac{\partial v_j}{\partial x_i} - \frac{2}{3} \frac{\partial v_k}{\partial x_k} \delta_{ij} \right) = \frac{\tilde{d} S_{ij}}{dt} + \lambda S_{ij}$$

gives

$$G_0 \frac{V_0}{l} \sim \frac{\sigma_0}{t_0} + \left( \frac{G_0 V_0}{\sigma_0 l} + \frac{1}{t_0} \right) \sigma_0. \tag{19}$$

On the right hand side of (19) all terms must have the same order since the elastic and plastic parts of the strain have the same order in the case under consideration. This condition is satisfied if

$$\frac{G_0 V_0 t_0}{\sigma_0 l} \sim 1. \tag{20}$$

At the same time, condition (20) leads to the fulfillment of (19) and, moreover, this condition becomes identical to (12).

The conditions characterizing the dynamic or the quasistatic case as well as the resulting estimates for $\sigma_0$ and $V_0$ remain the same as before, and the system (1) takes the form

$$\varrho_0 \frac{\partial v_i}{\partial t} = \varrho_0 F_i^e - \frac{\partial p}{\partial x_i} + \frac{\partial S_{ij}}{\partial x_j}, \quad \frac{\partial \theta}{\partial t} + \frac{\partial v_i}{\partial x_i} = 0,$$

$$p = f(\theta, \theta_*), \quad \frac{\partial \theta_*}{\partial t} = \frac{\partial \theta}{\partial t} e (\theta - \theta_*) e \left( \frac{\partial \theta}{\partial t} \right),$$

$$G_0 \left( \frac{\partial v_i}{\partial x_j} + \frac{\partial v_j}{\partial x_i} - \frac{2}{3} \frac{\partial v_k}{\partial x_k} \delta_{ij} \right) = \frac{\partial S_{ij}}{\partial t} + \lambda S_{ij}, \tag{21}$$

$$\lambda = \frac{2 G_0 W - F'(p) \partial p / \partial t}{2 F(p)} e [J_2 - F(p)]$$

$$\times e \left[ 2 G_0 W - F'(p) \frac{\partial p}{\partial t} \right].$$

4. We next consider motions with large elastic-plastic deformations but moderate stresses, that is we suppose that $V_0 t_0 \sim l$ and $\sigma_* \ll \sigma_0 \ll G_0$. This case is of practical interest, since, for example, for sand we have $\sigma_* \sim 0.5 \text{ kg/cm}^2$, $G_0 \sim 10^3 \text{ kg/cm}^2$ and are consequently dealing with stresses of the order $\sigma_0 \sim 10 \text{ kg/cm}^2$ to $100 \text{ kg/cm}^2$. In this case

$$\frac{G_0 V_0 t_0}{\sigma_0 l} \gg 1, \tag{22}$$

and it follows from the flow law (19), that the elastic components may be neglected. The equation of continuity then shows that

$$\frac{\theta}{t_0} + \frac{V_0}{l} \sim 0.$$

It follows from the experimental data that at $p \sim \sigma_0 \sim 10 \text{ kg/cm}^2$ to $100 \text{ kg/cm}^2$ we have $\theta \sim 0.01$. Consequently the last relation shows, that in the continuity equation all the terms containing derivatives of the density should be neglected. Finally the estimation of terms in the momentum equation for a dynamical problem gives

$$\varrho_0 \left( \frac{V_0}{t_0} + \frac{V_0^2}{l} \right) \sim \frac{\sigma_0}{l} \quad \text{or} \quad \sigma_0 \sim \varrho_0 V_0^2, \tag{23}$$

and this equation therefore remains unchanged except for the replaceiment of $\varrho$ by $\varrho_0$. We thus find from (1) that

$$\varrho_0 \left( \frac{\partial v_i}{\partial t} + v_j \frac{\partial v_i}{\partial x_j} \right) = \varrho_0 F_i^e - \frac{\partial p}{\partial x_i} + \frac{\partial S_{ij}}{\partial x_j}, \quad \frac{\partial v_i}{\partial x_i} = 0,$$

$$\frac{\partial v_i}{\partial x_j} + \frac{\partial v_j}{\partial x_i} = \frac{W}{F(p)} e[J - F(p)] e(W) S_{ij}. \tag{24}$$

It should be noted, that after determination of the distribution of $p$ from (24), the distribution of $\varrho$ can be found separately by using the relation $p = f(\theta, \theta_*)$ and the expression for $\partial p/\partial t + v_i \, \partial p/\partial x_i$.

In the dynamic case, we have

$$\frac{l}{t_0} \sim \sqrt{\frac{\sigma_0}{\varrho_0}} \sim \sqrt{\frac{\sigma_0}{G_0}} C \sim \sqrt{\theta} \, C \quad \text{or} \quad t_0 \sim \frac{1}{\sqrt{\theta}} t_w. \tag{25}$$

Now it should be kept in mind, that all the conclusions drawn above apply only to regions far from the wave fronts. In regions close to these fronts there arises a situation identical with that in the preceding section. Indeed, the conditions that must be satisfied on the wave fronts furnish the following estimates for the regions close to these fronts:

$$t_0 \sim t_w, \quad V_0 \sim \theta C, \quad \sigma_0 \sim \varrho_0 \theta C^2.$$

Thus the motion in the regions close to the front is described by the system (21). It should furthermore be noted, that when in regions close to the fronts there exist large gradients and the wave traverses distances that are much larger than the length of the wave (i.e. when the wave is "short"), one must take the convective terms in the material derivatives into account [5]. Nevertheless, in these cases the fundamental equations can also be considerably simplified in the way explained in [5] which permits a relatively simple and complete analysis. However these considerations will not be presented here.

In the case under consideration there are therefore two kinds of essentially dynamical motions of the medium: in the region close to the front, where the motion has a clearly expressed wave character, and in the region far from the front, where wave effects are negligible and the motion is described as one of an incompressible medium. It may be mentioned, that a similar situation arises in the description of the motions of fluids of very small compressibility (water, but not gases). For example, after an underwater explosion the domain, in which the water is drawn into motion, quickly becomes divided into two regions with essentially different characters of motion: a region close to the front and a region surrounding the gas bubble. In the former the compressibility of the water plays a role and the motion is propagated as a wave, while in the neighbourhood of the bubble the motion can be very well described as that of an incompressible ideal fluid [6]. Obviously, in the case of fluid motion as well as in the present case this division results from the same cause, the small compressibility of the medium.

Although the simplified systems of equations describing the motion in these two regions are different, they may be combined into one system, which will automatically reduce to the system appropriate for each region. This combined system is

$$\varrho_0 \left( \frac{\partial v_i}{\partial t} + v_j \frac{\partial v_i}{\partial x_j} \right) = \varrho_0 F_i^e - \frac{\partial p}{\partial x_i} + \frac{\partial S_{ij}}{\partial x_j}, \quad \frac{\partial \theta}{\partial t} + \frac{\partial v_i}{\partial x_i} = 0,$$

$$p = f(\theta, \theta_*), \quad \frac{\partial \theta_*}{\partial t} = \frac{\partial \theta}{\partial t} e(\theta - \theta_*) e\left( \frac{\partial \theta}{\partial t} \right),$$

$$G \left( \frac{\partial v_i}{\partial x_j} + \frac{\partial v_j}{\partial x_i} - \frac{2}{3} \frac{\partial v_k}{\partial x_k} \delta_{ij} \right) = \frac{\partial S_{ij}}{\partial t} + \lambda S_{ij}, \tag{26}$$

$$\lambda = \frac{2GW - F'(p)\,\partial p/\partial t}{2F(p)} \, e[J_2 - F(p)] \, e[2GW - F'(p)\,\partial p/\partial t].$$

This system differs from that for the region close to the front only by the presence of the convective terms in the material derivative of the velocity. But in this region these terms are small so that the system does not differ from (21). On the other hand, in the region far from the wave front, the system (26) practically coincides with the system (24), since in this region all time derivatives, except those in the momentum equation, are negligibly small.

For quasi-static motion we obviously have

$$t_0 \gg \frac{1}{\sqrt{\theta}} \, t_w \gg t_w, \tag{27}$$

(see (25)), i.e. we have a stronger condition than in the former cases, where it was sufficient to satisfy condition (18). Note, that if $\sigma_0 \sim 10 \text{ kg/cm}^2$ to $100 \text{ kg/cm}^2$, $\theta \sim 0.01$, $l \sim 10$ m, and $C \sim 100$ m/sec, we shall have $t_w / \sqrt{\theta} \sim 1$ sec, so that the condition for motion (27) requires $t_0 \gg 1$ sec. Thus, for the usual problems on foundations and building supports this condition will always be satisfied, and the equations for dynamic motion need only be used for problems connected with blast and impact.

When (27) is satisfied, Eqs. (24) furnish the fundamental equations

$$-\frac{\partial p}{\partial x_i} + \frac{\partial S_{ij}}{\partial x_j} + \varrho_0 F_i^e = 0, \quad \frac{\partial v_i}{\partial x_i} = 0,$$

$$\frac{\partial v_i}{\partial x_j} + \frac{\partial v_j}{\partial x_i} = \frac{W}{F(p)} \, e[J_2 - F(p)] e(W) S_{ij}. \tag{28}$$

These are the equations of quasi-static rigid-plastic flow of soil with developed plastic deformations, or the equations of "limiting equilibrium" of the medium. In the particular case of a plane problem, these equations yield the equations of plane limiting equilibrium of the statics of granular media [7]. Indeed, as a result of the flow law we have $S_{zz} = 0$. This reduces the yield condition to

$$\frac{1}{2} [(\sigma_{xx} + p)^2 + (\sigma_{yy} + p)^2 + 2\sigma_{xy}^2] = F(p)$$

and gives

$$\sigma_{zz} = -p, \quad p = -(\sigma_{xx} + \sigma_{yy})/2,$$

thus reducing the yield condition to the final form

$$\frac{1}{4} [(\sigma_{xx} - \sigma_{yy})^2 + 4\sigma_{xy}^2] = F\left[-\frac{1}{2}(\sigma_{xx} + \sigma_{yy})\right].$$

In the particular case $F(p) = (kp + b)^2$ (see [3, 4]) we have

$$\frac{1}{4} [(\sigma_{xx} - \sigma_{yy})^2 + 4\sigma_{xy}^2] = \frac{k^2}{4}\left[-(\sigma_{xx} + \sigma_{yy}) + 2\frac{b}{k}\right]^2.$$

In [7] the condition of the limiting equilibrium is given in the form

$$\frac{1}{4} [(\sigma_{xx} - \sigma_{yy})^2 + 4\sigma_{xy}^2] = \frac{\sin^2\varrho}{4}[-(\sigma_{xx} + \sigma_{yy}) + 2H]^2,$$

where $\varrho$ and $H$ are the angle of internal friction and the cohesion, respectively. But for some notations, these conditions are therefore identical. When adding the equations of equilibrium

$$\frac{\partial \sigma_{xx}}{\partial x} + \frac{\partial \sigma_{xy}}{\partial y} + \varrho_0 F_x^e = 0, \quad \frac{\partial \sigma_{xy}}{\partial x} + \frac{\partial \sigma_{yy}}{\partial y} + \varrho_0 F_y^e = 0$$

we obtain a system of three equations for $\sigma_{xx}$, $\sigma_{xy}$, $\sigma_{yy}$. This system is the usual basis for the statics of granular media [7]. It should be noted, that even in the plane case these equations are not sufficient for the complete statement and correct solution of problems in statics of granular media (quite aside from the fact that the traditional statics of granular media does not offer any equations for the description of three-dimensional problems): the plane problem of limiting equilibrium cannot always be formulated in terms of stresses alone, and even when a problem can be formulated in this manner, it cannot be solved correctly without reference to the velocity field. This situation is familiar from the theory of plasticity where the necessity of constructing velocity fields in the solution of problems of plasticity has only recently been recognized. To designate a solution which includes the discussion of the velocity field the special term "full solution" has been introduced. After this revision of the theory, some solutions obtained earlier were found to be incorrect. Similarly incorrect solutions are found in the statics of granular media.

The system (28) permits us, first, to analyze not only a plane but an arbitrary three-dimensional problem of limiting equilibrium, and, secondly, not only problems formulated in stresses, but problems with mixed boundary conditions. It should be noted, of course, that there are problems where the main, and often the unique cause of deformation is the compressibility of the medium (uniaxial compression, for example). In these cases one must use the full equations (1) in the quasistatic case.

5. When the elastic-plastic deformations as well as the stresses are large, we have $V_0 t_0 \sim l$, $\sigma_0 \sim S_\infty$. If we have in addition $S_\infty \ll G_0$, the problem coincides with the preceding one. If, on the other hand, $S_\infty \sim G_0$ no simplification is possible in system (1), as the value of $\theta$ is no longer small: $\theta \sim 1$. The condition for quasi-static motion retains the form (27). For dynamic motion, there is no division of the domain of motion into two regions of different types of motion, since $\theta \sim 1$, and the motion in the entire domain is described by the complete system (1).

6. Finally, when the motion is accompanied by very large stresses $\sigma_0 \gg S_\infty$, we have $S_{ij} \sim S \ll \sigma_0 \sim p$, and it will be possible to neglect the shear stresses throughout. As a result, the system (1) will reduce to the equations of motion of an ideal compressible fluid with possibly irreversibility of the volume deformation:

$$\varrho \left( \frac{\partial v_i}{\partial t} + v_j \frac{\partial v_i}{\partial x_j} \right) = \varrho F_i^e - \frac{\partial p}{\partial x_i}, \quad p = f(\theta, \theta_*),$$

$$\theta = 1 - \frac{\varrho_0}{\varrho}, \quad \theta_* = 1 - \frac{\varrho_0}{\varrho_*}, \quad \frac{\partial \varrho}{\partial t} + \frac{\partial (\varrho v_i)}{\partial x_i} = 0, \qquad (29)$$

$$\frac{\partial \theta_*}{\partial t} + v_i \frac{\partial \theta_*}{\partial x_i} = \left( \frac{\partial \theta}{\partial t} + v_i \frac{\partial \theta}{\partial x_i} \right) e(\theta - \theta_*) e \left( \frac{\partial \theta}{\partial t} + v_i \frac{\partial \theta}{\partial x_i} \right).$$

The discussion of quasi-static motions is then not interesting because these motions are trivial.

This concludes the consideration of all possible types of motions of soft soils. Obviously a similar analysis can be carried out for other rigid media (for example, plastic metals), and the results will be completely analogous to those obtained above.

### References

[1] GRIGORIAN, S. S.: Doklady AN USSR 124, No. 2 (1959).
[2] GRIGORIAN, S. S.: Prikl. Mat. Mekh. 24, No. 6 (1960).
[3] ALEXEENKO, V. D., S. S. GRIGORIAN, A. F. NOVGORODOV and G. V. RYKOV: Doklady AN USSR 133, No. 6 (1960).
[4] ALEXEENKO, V. D., S. S. GRIGORIAN, L. I. KOSHELEV, A. F. NOVGORODOV and G. V. RYKOV: Prikl. Mekh. Tekh. Fiz. No. 2 (1963).
[5] CHRISTIANOVICH, S. A.: Prikl. Mat. Mekh. 20, No. 5 (1956).
[6] COLE, R.: Underwater explosions (Russian translation), Moscow 1950.
[7] SOKOLOVSKII, V. V.: Statics of granular medium, Moscow 1960.

# One-Dimensional Inelastic Wave Propagation in Soils: Experimental and Theoretical Investigations

By Werner Heierli

Eidgenössische Technische Hochschule, Zürich, Switzerland

## Summary

Plane waves in one-dimensional inelastic bodies of soil are investigated. The dynamic stress-strain diagram of two types of soil, sand and gravel, is determined in an oedometer operated dynamically. A theory is developed for the calculation of plane waves in any kind of non-linear, inelastic material with heterogeneities and any type of reflective surfaces. Examples show the great influence of inelasticity on the propagation and attenuation of stress waves. Wave propagation experiments were conducted, and the results are compared to the theoretical findings.

## 1. Introduction

Inelasticity is one of the outstanding characteristics of soils, and one that has long been used in engineering. Before there was any considerable interest in inelastic wave propagation, soils were compacted for the construction of highways and dams. Compaction is possible only if a material is inelastic. In foundation engineering, use is made of the inelasticity of soils when heavy buildings have to be built on thick compressible strata: a deep excavation is made so that the subsoil is unloaded. When the building is erected, the soil is reloaded, and the settlements are small because the compressibility for reloading is much lower than for virgin loading. This fact is a direct consequence of the inelasticity of soils.

Phenomena related to inelastic wave propagation were of considerable importance in the Second World War. It was soon realized that intense pressure waves from explosions were attenuated after traveling through a layer of soil. Thus the ruins of destroyed structures served to protect people in basements. Since rocks behave inelastically under high stresses in a similar way to soils, underground fortifications in rocks afforded protection through inelastic wave attenuation. Presently, wave propagation through soils and rocks is the subject of extensive investigations in connection with underground protective construction against nuclear blasts.

The general case of two- or three-dimensional wave propagation in inelastic materials is an extremely involved problem, both from a theoretical standpoint and with regard to the experimental determination of the basic material properties. In this paper, a theoretical and experimental investigation of one-dimensional wave propagation is presented. A wave is called one-dimensional so long as all quantities can be described as functions of only one space coordinate and the time. In this sense, plane, cylindrically symmetrical and spherically symmetrical waves are one-dimensional.

This paper describes the determination of the non-linear inelastic stress-strain diagram, the development of a mathematical method for calculating plane waves under very general assumptions and experiments for checking the theoretical calculations. The investigations form part of the doctoral dissertation of the author [1] at Eidgenössische Technische Hochschule (ETH), Zürich, Switzerland.

## 2. The Dynamic Stress-Strain Diagram

The basic material property involved in the investigation of plane waves through soils is the complete stress-strain diagram. Since different soils can exhibit very different stress-strain characteristics, experimental determinations have to be made with each type of soil. An experimental setup, therefore, was developed in which the dynamic stress-strain diagram could be determined.

Since the properties of soils depend upon the specific stress conditions and also on the rate of loading, the soil in the stress-strain test ($p$-$\varepsilon$-test) has to be subjected to a state of stress as similar as possible to the state occurring during the actual applications, e.g. during nuclear blasts. In a plane wave, the soil cannot expand perpendicularly to the direction of propagation, and therefore, lateral expansion has to be prevented in the test setup. Under such conditions, a specimen of soil cannot fail. The rate of loading should be approximately the same in the test as in the application.

Fig. 1 shows a vertical section through the device called a *dynamic oedometer*. The soil is contained in a rigid steel container. At the top and at the bottom, pressure gages are placed that measure the average dynamic pressure. The upper gage is hit during the test by a loading system consisting of a falling weight, an anvil with appropriate cushioning and a plunger that transmits the blow onto the pressure gage. The deformation of the soil specimen is monitored by two wire potentiometers at diametrically opposite positions. The two readings are added electrically so that the influence of any tilting of the top gage is eliminated. The top gage is held down by three springs (one of which is shown in section) in order to avoid a separation of the gage from the soil during the rebound (unloading) phase of the test.

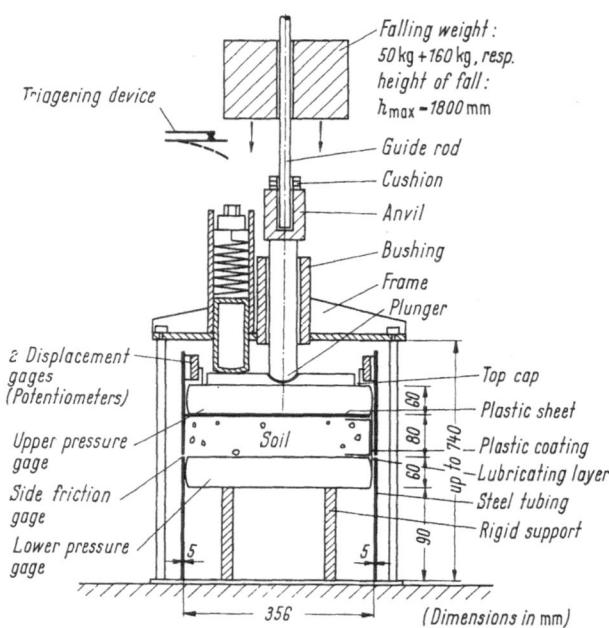

Fig. 1. Dynamic oedometer.

It is required that the state of stress in the dynamic oedometer should be the same as in a plane wave. This condition can only be satisfied if the friction of the soil along the casing of the setup (side friction) is eliminated. It would be very difficult to evaluate the amount of side friction theoretically, and therefore, a direct dynamic measurement is made in the test. The casing is interrupted at the level of the lower pressure gage and sustained by four supports on which strain gages are mounted (side friction gage). These strain gages are connected in a Wheatstone bridge in such a way that the output is proportional to the total induced side friction. The side friction measured in this way proved to be quite small (Fig. 2, lower right hand side). The main reason for this is the fact the height-diameter ratio of the soil specimen was less than $1/4$.

A determination of the dynamic stress-strain diagram is only possible if the stresses in the specimen are the same at all points of a cross-section of the specimen at any time. In a static test, this condition is fulfilled if there is no side friction. In a dynamic test, however, it is also necessary that there should be no wave propagation in the test setup. If there is wave propagation, the stress-strain diagram cannot be constructed

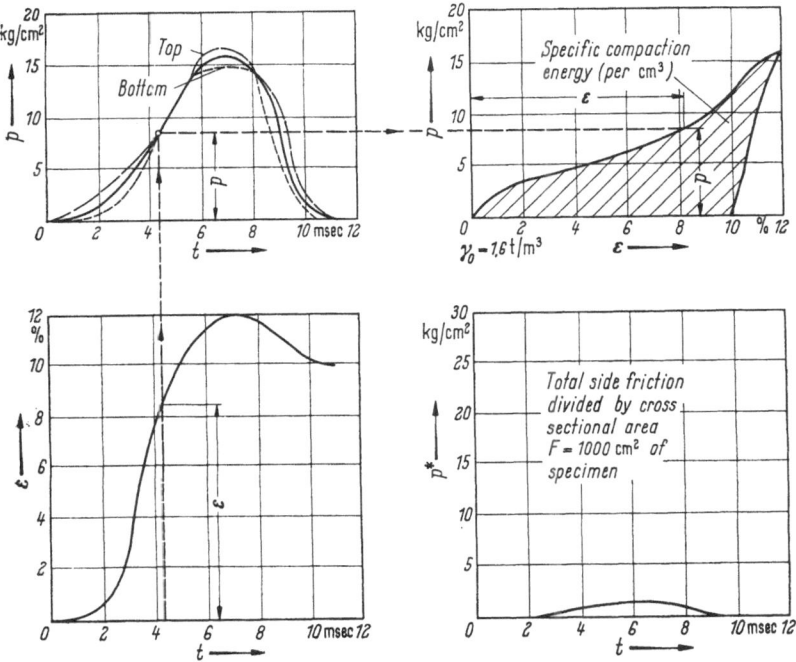

Fig. 2. Construction of dynamic $p$-$\varepsilon$ diagram.

because the state of stress and of deformation is not the same throughout the specimen. A simple wave calculation shows that, in a dynamic test, the stresses at the top and at the bottom of a specimen are approximately the same if the length of the specimen is only a small fraction of the wave length used. The pressure measurements at the top and bottom gage in Fig. 2 show that this requirement is fulfilled. The deviations between the two measurements are negligible. It should be mentioned that, for very fast tests, this wave legth condition can become difficult to realize, especially for relatively coarse grained soils.

Fig. 2 shows the stress-strain diagram that was constructed from the pressure and deformation measurements which are both functions of time. The elimination of the parameter time, of course, means that physical data about the soil are lost since the stress-strain diagram is time-dependent. However, it was found that if the pressure-time curves imposed are only approximately the same in two tests, the resulting stress-strain curves do not differ appreciably from each other. In other words, a stress-strain diagram holds for a certain range of pressure-time curves. This fact is borne out by Figs. 3 and 4 which show a plot of many determinations of $p$-$\varepsilon$ diagrams with different initial densities and slightly different rates of loading. Fig. 3 is for a rather uniform sand and Fig. 4 is for a gravel with sand. The grain size distributions are given in Fig. 5. For comparison the static stress-strain curves are shown in each diagram. They differ very appreciably from the dynamic curves at low densities (low $\varepsilon$'s) but not so markedly at high densities. This general trend has also been observed in other soil dynamics investigations.

Figs. 3 and 4 show the general character of dynamic stress-strain curves. The loading curves are first concave upwards, but, over a wider range of pressures, exhibit a tendency to be concave downward. This is not surprising since, the more the soil is compacted, the greater is its resistance to further compaction. The inelasticity is most pronounced at low densities, whereas at larger densities, more of the deformation is recovered on unloading. The unloading curves are for the most part approximately parallel straight lines. The subsequent reloading curves differ considerably from the foregoing unloading curves in the smaller density range, but less at large densities.

The difference between the static and the dynamic stress-strain diagrams gives rise to the interesting question of where this time sensitivity of the soil stems from. It was conjectured that the behavior of the air in the pores could cause part of this unexplained discrepancy. An adiabatic calculation of the pore air stresses, however, showed that the air does not noticeably contribute to the stress-strain behavior except at the very end of unloading.

For the dynamic oedometer tests as well as for field tests, dynamic soil pressure gages and accelerometers were developed. The design, the dynamic calculation of these gages as well as the tests and calibrations performed are described in detail in [1].

For references to dynamic soil properties, the reader is referred to [1] and [2].

# 3. Theory

At the beginning of this paper, the problem was restricted to the consideration of one-dimensional waves. The present section deals with plane waves. For cylindrically or spherically symmetrical waves, an extension of the method is planned. Apart from this rather stringent condition that the wave should be one-dimensional, no further assumptions are made. The stress-strain properties of the soil can be of a very general nature; the nonlinear inelastic $p$-$\varepsilon$ diagram can have any shape whatsoever, including reloading curves that differ from the foregoing

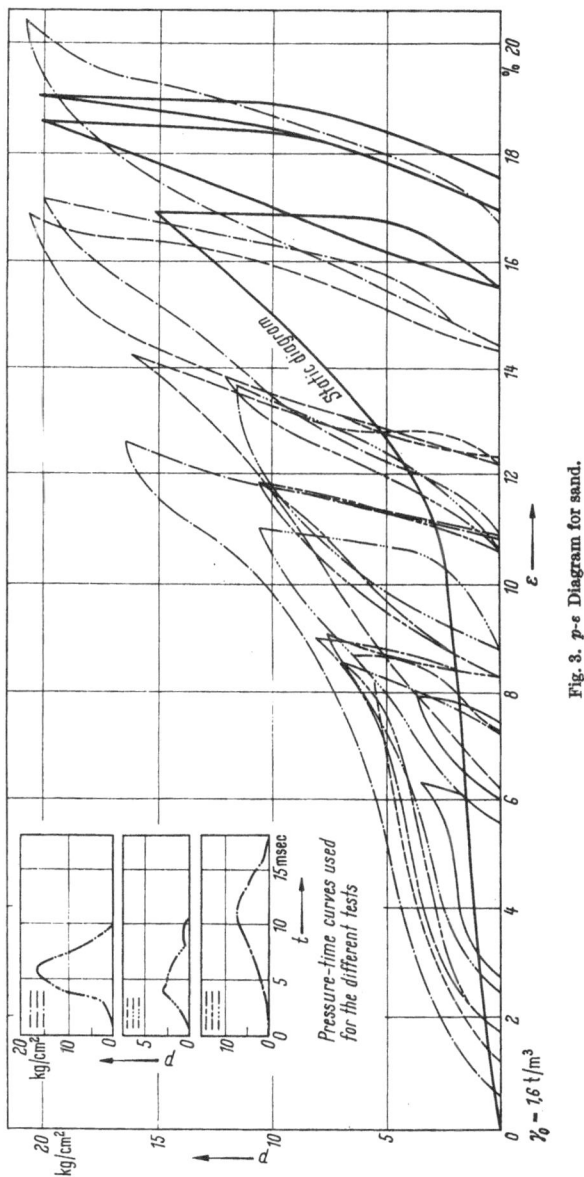

Fig. 3. $p$-$\varepsilon$ Diagram for sand.

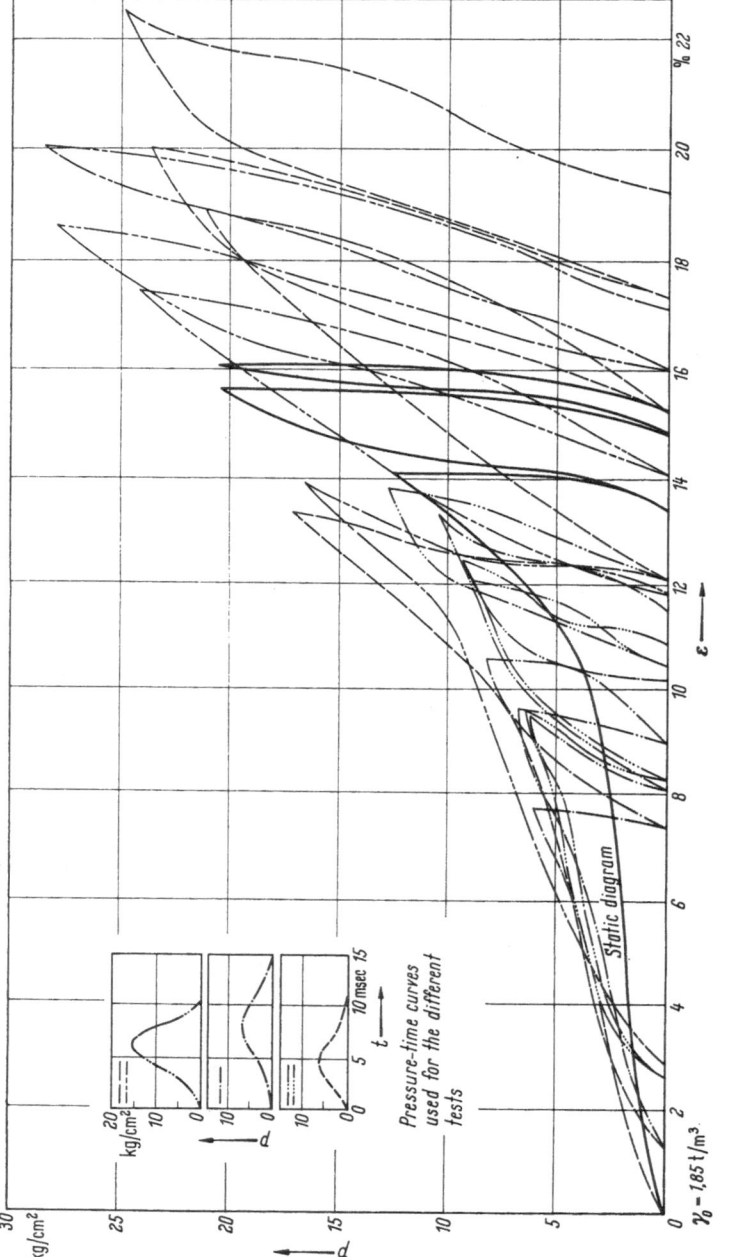

Fig. 4. *p-ε* Diagram for gravel with sand.

unloading curves (as those shown in Figs. 3 and 4). The boundaries may be free (pressure specified) or reflective surfaces (rigid, elastic, rigid-plastic, nonlinearly yielding, etc.), and the input curve, e.g. the pressure-time curve at the input end, may have any form (with nonzero rise-time, with non-monotonic decay, etc.). The material may contain heterogeneities so long as the problem remains one-dimensional.

The problem of calculating plastic waves has been treated by TH. v. KÁRMÁN [4] for cases involving loading. For problems with unloading, E. H. LEE [5] gives an extension of the method of characteristics. An approach different from the method of characteristics can be obtained if the input pressure (or velocity) vs. time curve is approximated by a number of steps[1] (Fig. 6). At each step, the stress or particle velocity changes abruptly, between the discontinuities, they remain constant. For

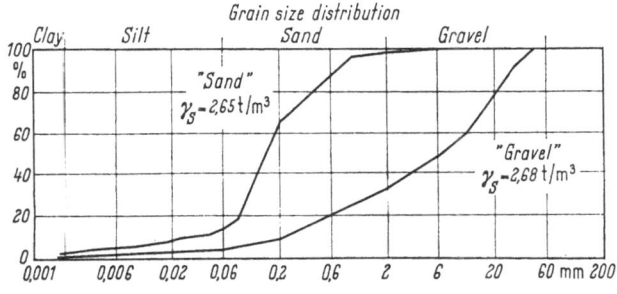

Fig. 5. Grain size distributions.

the stress increase during one step, a constant modulus is chosen corresponding to the secant of the stress-strain curve. The subdivisions of the pressure input curve (the steps) have to be small enough to give a sufficiently good approximation to the actual input and to ensure that the polygon of secants corresponding to the steps in the $p$-$\varepsilon$ diagram does not noticeably deviate from the actual $p$-$\varepsilon$ diagram.

As a first stage of the derivation, the propagation of a single step of loading is shown (Fig. 6): $p_i$ is the stress ahead of the step, $p_{i+1}$ the stress behind the step; $u_i$ and $u_{i+1}$ are the corresponding particle velocities. With a time element $\Delta t$ and a space element $\Delta x$, the equation of continuity can be written in the form

$$\Delta \varepsilon_{i+1}(\Delta x + u_i \Delta t) = \Delta u_{i+1} \Delta t, \tag{1}$$

where $\Delta \varepsilon_{i+1}$ denotes the strain increment[2].

$$\varrho_i \Delta u_{i+1} \Delta x = \Delta p_{i+1} \Delta t, \tag{2}$$

---

[1] This method will subsequently be referred to as the method of impulses.

[2] Stresses are regarded as positive when compressive and strains are regarded as positive when the volume decreases.

where $\varrho_i$ is the mass density at time $t_i$ and at the location $x_i$ to $x_{i+1}$. For relatively soft materials, $\varrho_i$ may undergo appreciable changes during the propagation of the wave, and therefore, the density is not assumed to be constant with time. Likewise, in Eq. (1) it was not assumed that the deformations are small, and therefore, the term $u_i \Delta t$ in the parenthesis had to be added.

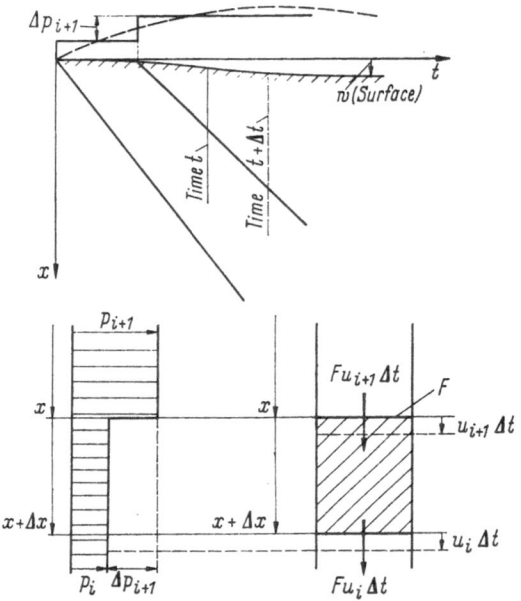

Fig. 6. Derivation of method of impulses.

The stress-strain properties of the material are expressed by

$$\Delta p_{i+1} = T_{i+1} \Delta \varepsilon_{i+1}, \tag{3}$$

where $T_{i+1}$ is the secant modulus belonging to the stress increase $\Delta p_{i+1}$. Introducing the relative wave propagation velocity

$$c_{i+1} = \Delta x / \Delta t \tag{4}$$

(relative to the particle moving with a velocity $u_i$) and the absolute propagation velocity

$$c'_{i+1} = (\Delta x / \Delta t) + u_i, \tag{5}$$

we evaluate $c_{i+1}$ by using the three basic equations

$$c_{i+1} = \frac{1}{2} \left[ -u_i \pm \left( u_i^2 + \frac{4 T_{i+1}}{\varrho_i} \right)^{1/2} \right]. \tag{6}$$

For materials having a relatively high modulus $T_{i+1}$, the simpler, well-known relations

$$c_{i+1} = c'_{i+1} = (T_{i+1}/\varrho_i)^{1/2}$$   (6a)

are found.

From NEWTON's second law (2) the change in particle velocity can be evaluated as

$$\Delta u_{i+1} = \frac{\Delta p_{i+1}}{\varrho_i c_{i+1}} = \frac{2\Delta p_{i+1}}{\varrho_i[-u_i \pm (u_i^2 + 4T_{i+1}/\varrho_i)^{1/2}]}.$$   (7)

The plus or a minus signs appearing in Eqs. (6) and (7) stand for waves traveling in the positive and in the negative $x$-directions, respectively (incident and reflected waves). For materials having a relatively high modulus $T_{i+1}$, the simpler relation

$$\Delta u_{i+1} = \pm \frac{\Delta p_{i+1}}{(T_{i+1}\varrho_i)^{1/2}}$$   (7a)

is found.

The derivation above gives the changes in stress and in particle velocity for a loading wave. So long as no heterogeneity is met, the wave travels at a constant velocity $c'_{i+1}$ which is shown as a straight line in the

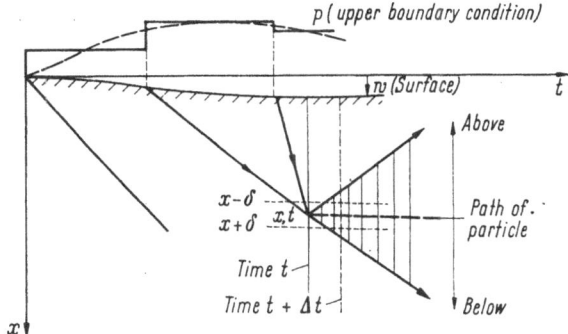

Fig. 7. Unloading, $x$-$t$ plane.

$x$-$t$-diagram. Across this wave propagation line, stress and particle velocity change abruptly. As soon as unloading occurs, however, a different phenomenon is observed (Fig. 7). Since the material is in-elastic, the modulus $T_{i+1}$ for unloading is larger than the loading modulus, and therefore, the unloading wave travels faster than the loading wave. Eventually, the former will reach the latter, and an interference of the two waves will occur.

Fig. 8 shows the stresses at time $t$ (solid lines) and at time $t + \Delta t$ (dashed lines). At time $t$, the stresses and particle velocities above and below the particle path (Fig. 7) are different from each other. With the boundary conditions shown in Fig. 7, the particle velocities above will be larger than below because of the inelasticity of the material. Therefore, the material above will impact on the material below, and waves will emanate from the meeting point both in the positive and in the negative $x$-direction.

These refracted and reflected waves can be evaluated in the following manner. First, a stress increment $\Delta p_{i+1}$ is assumed above and below the particle path. These increments must be such that, after the reflection, the total stress is the same above and below (see dashed lines, Fig. 8). With these assumed stress increments, the corresponding particle velo-

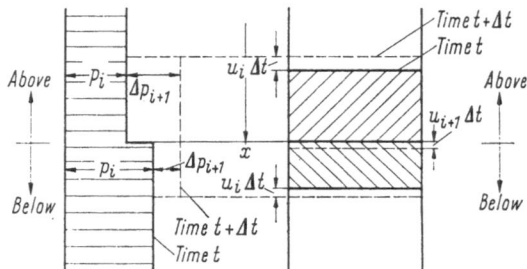

Fig. 8. Unloading, stresses and particle velocities vs. depth.

cities are evaluated above and below the particle path [Eqs. (7), (7a)]. Added to the velocities $u_i$ before the passage of the wave, they give the new total velocities $u_{i+1}$ above and below, and if the guess was correct, they are equal. If they do not agree, they can be determined in an iterative manner. It is evident that, in each iteration, new moduli have to be determined from the $p$-$\varepsilon$ diagram because, in general, the modulus changes with the guessed amounts of stress increase.

After the whole $x$-$t$ plane has been evaluated with a network of wave propagation lines, all results can be obtained readily because in each field the stress and the velocity are known. In the next section, two examples are calculated and shown in detail. Charts for rapid evaluation of simple cases and applications to underground protective construction problems are given in [3]. A more detailed description and instruction for use of the method is presented in [1].

## 4. Experimental Verification of Theoretical Method

The foregoing section gives a method for calculating theoretically the propagation of plane waves into very generally inelastic materials. In the section about the stress-strain diagram, a possible method is shown of determining the dynamic inelastic properties of the material. It is of

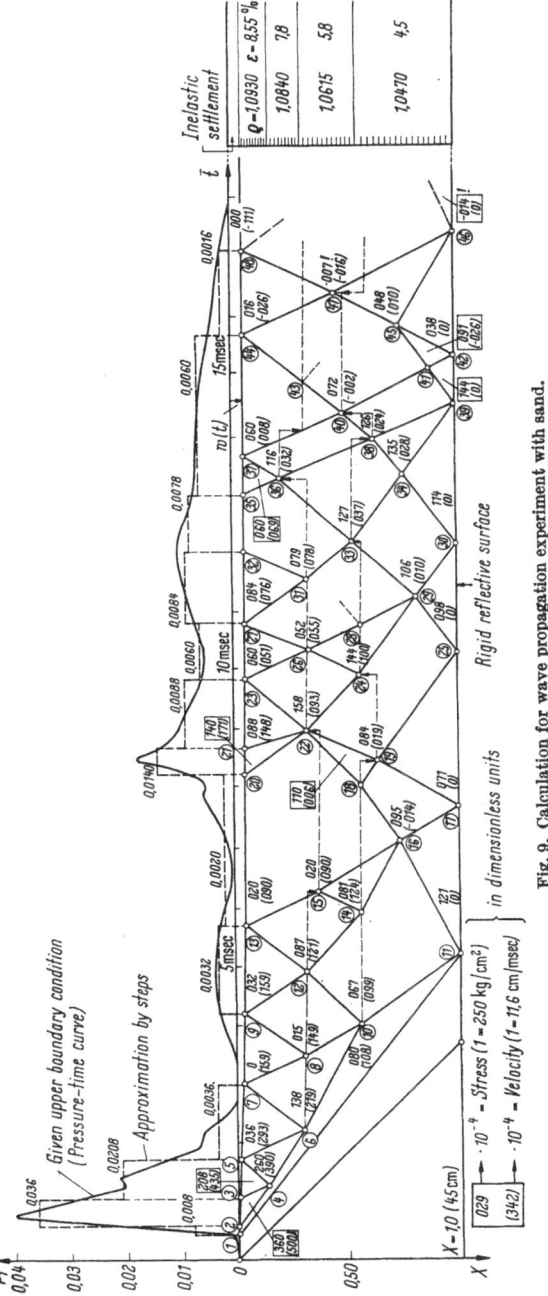

Fig. 9. Calculation for wave propagation experiment with sand.

particular interest now to observe inelastic waves in an experiment, to calculate the measured phenomena by means of the theory and to compare the results.

Plane waves can be created by uniformly loading a large area compared to the depth of soil involved, or by packing the soil in a rigid frictionless container and by loading it with a piston. For the present investigation, the latter way was chosen, except for some field tests described in [1]. It is evident that the requirements for building a wave propagation setup are somewhat similar to those established for the dynamic oedometer, especially in so far as the side friction is concerned. There is, however, one important difference: in the dynamic oedometer, the aim was to have the same stresses and particle velocities throughout the sample at any given time (quasi-static state of stress), and therefore,

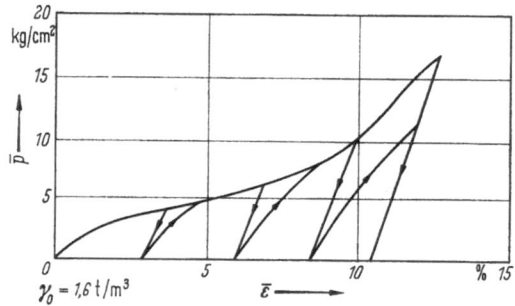

Fig. 10. Stress-strain diagram for Fig. 9 (sand).

the length of the sample was chosen small compared to the wavelength (the product of the loading time and the relevent velocity of propagation). In the wave propagation setup, however, the length of soil should be large compared to the wavelength, otherwise, no significant wave phenomena can occur, and no inelastic effects can be observed. The drawback in this length requirement is that it is difficult to keep the side-friction down to a tolerable amount with a long sample (see dynamic oedometer, second section). Nevertheless, the same basic instrumentation (dynamic oedometer) could be used in the wave propagation tests as in the $p-\varepsilon$ tests (Fig. 1).

The duration of loading was reduced, the soil was tested only in its loosest state to give maximum attenuation, and the sample length was increased to 45 cm (height-diameter ratio 1.25). The side friction was considerably greater than in the $p$-$\varepsilon$ tests and amounted to about one third of the attenuated pressure at the reaction end. The influence of the side friction is, of course, to reduce the pressure at the lower (reaction) end of the sample and also the displacement at the top.

Fig. 9 shows the theoretical calculation for a wave experiment with sand. The input is the step-wise approximated pressure-time curve at the top of the sample. The $p$-$\varepsilon$ diagram, taken from Fig. 3, is shown in Fig. 10. The method of impulses yields the theoretical values for the displacement at the top and for the reflected pressure at the bottom.

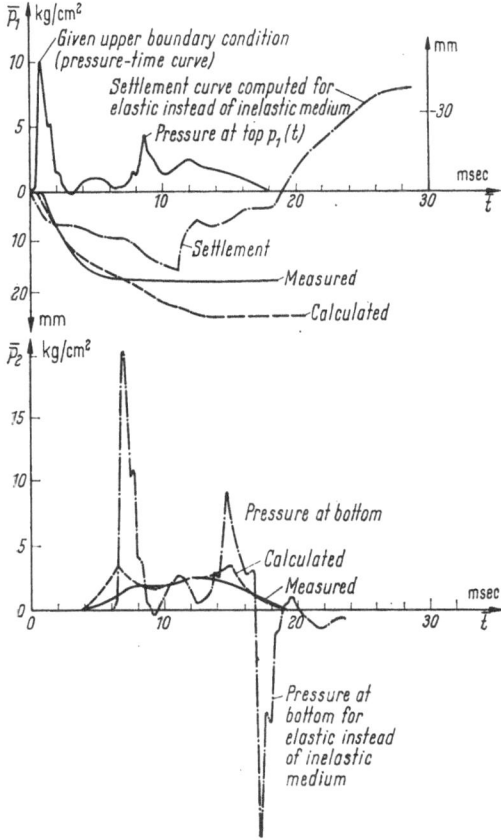

Fig. 11. Comparison theory-experiment for Fig. 9.

Fig. 11 gives the comparison between theory and experiment. The measurements are in reasonable agreement with the results of the calculations. The calculated values are somewhat higher than the theoretical ones, both with regard to the displacement and the pressure. This trend can be explained by the influence of the side friction. The results of elastic calculations are given in the same figure as dash-dotted lines for comparison. They totally disagree with the measurements.

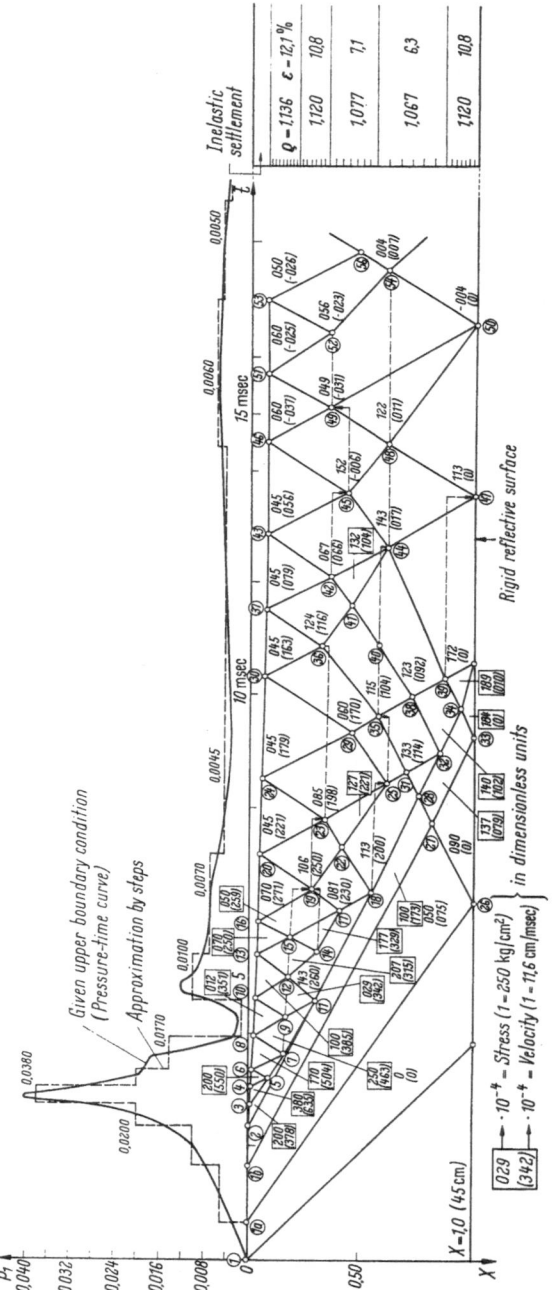

Fig. 12. Calculation for wave propagation experiment with gravel.

A similar comparison between theory and experiment is given in Figs. 12 and 14 with the $p-\varepsilon$-diagram in Fig. 13 (taken from Fig. 4). The same conclusions apply to this experiment with gravel; the only

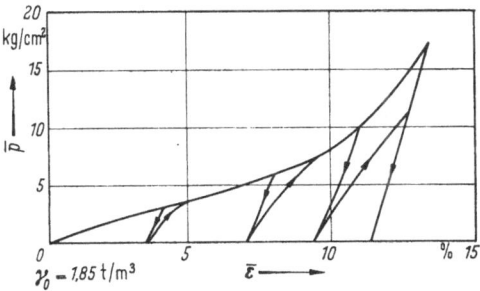

Fig. 13. Stress-strain diagram for Fig. 12 (gravel).

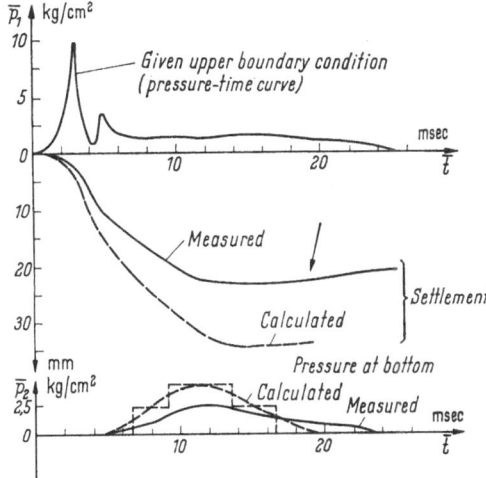

Fig. 14. Comparison theory-experiment for Fig. 12.

difference is that the gravel exhibited more side friction than the sand. Some additional experiments were conducted under a large plate loaded dynamically. For shallow depths, good agreement could be reached between the one-dimensional inelastic theory and the measurements [1].

## 5. Conclusions

The following conclusions can be drawn from the investigations presented:

1. The stress-strain diagrams of sand and gravel were determined dynamically. The side friction problem could be overcome by using a

small height-diameter ratio. The dynamic effects in the $p$-$\varepsilon$ diagram decreased with increasing density.

2. A theoretical method was developed by which the propagation of a plane wave into any kind of nonlinear inelastic body with any type of boundaries can be calculated. For this calculation, the input is approximated by steps which are smoothed out again in the results. A comparison with another method that exists for special inelastic $p$-$\varepsilon$ diagrams and boundary conditions [7] showed that even a relatively coarse subdivision into steps yields good results [3].

3. Wave propagation experiments with loose sand and gravel showed reasonable agreement between theory and experiment. The discrepancies observed could be further diminished by a smaller height-diameter ratio (less side-friction).

Applications of these investigations are possible in the fields of soil and rock dynamics in modern protective construction, in dynamic compaction theory as well as in impact problems with complex material properties and boundary conditions.

## 6. Acknowledgments

The author is indebted to Professors G. SCHNITTER and H. ZIEGLER of ETH, Zürich, for their many suggestions and assistance. Thanks are due to Mr. CH. SCHAERER and a number of his collaborators at the Laboratories for Hydraulics and Soil Mechanics at ETH. Professor R. V. WHITMAN of Massachusetts Institute of Technology contributed to some theoretical developments through his valuable criticism.

### References

[1] HEIERLI, W.: Die Dynamik eindimensionaler Bodenkörper im nichtlinearen, nichtelastischen Bereich, Dissertation (1961) ETH, Zürich, Switzerland.
[2] WHITMAN, R. V.: Testing soils with transient loads, ASTM Spec. Techn. Publ. 232, 242 (1957).
[3] HEIERLI, W.: Proc. Soil Mech. and Foundation Div., ASCE SM 6, 33 (1962).
[4] v. KÁRMÁN, TH.: NDRC Report No. A-29, OSRD No. 365 (1942).
[5] LEE, E. H.: Q. Appl. Math. 10, 335 (1953).
[6] COLLATZ, L.: Numerische Behandlung von Differentialgleichungen, 2. Aufl., Berlin/Göttingen/Heidelberg: Springer 1955.
[7] SALVADORI, M. G., R. SKALAK and P. WEIDLINGER: Proc. ASCE, Eng. Mech. Div. EM 3, No. 2829 (1961).

# Propagation des surfaces de discontinuité dans un milieu élastoplastique

## Par Jean Mandel

Ecole Polytechnique, Paris, France

L'auteur a étudié précédemment [1] la propagation des surfaces de discontinuité dans un milieu élastoplastique pour lequel existe un potentiel plastique identique au critère d'écoulement. Il examine maintenant le cas où cette hypothèse n'est pas vérifiée. Il se limite aux surfaces de discontinuité d'ordre deux (pas de discontinuité de vitesse).

## 1. Introduction

Dans un milieu élastoplastique (sans viscosité) on peut avoir quatre types de surfaces de discontinuité suivant la nature de la transformation (élastique $E$ ou plastique $P$) des deux côtés de la surface (la flèche indiquant le sens de la transition subie par un élément de matière):

$$E \to E \text{ ondes élastiques}, \quad P \to E \text{ frontière de décharge},$$
$$P \to P \text{ ondes plastiques}, \quad E \to P \text{ frontière de charge}.$$

Nous rappelons les résultats obtenus [1] dans le cas où il existe un potentiel plastique identique à la fonction de charge.

### (a) Ondes ordinaires plastiques

Pour chaque direction de la normale à la surface d'onde il existe trois célérités possibles et les vecteurs $[\gamma]$ (discontinuité de l'accélération) correspondants sont orthogonaux. Les célérités des trois ondes plastiques sont séparées par celles des ondes élastiques et leur sont au plus égales.

### (b) Frontières de décharge

Leurs célérités sont comprises dans les intervalles entre les célérités des ondes plastiques et des ondes élastiques de même numéro d'ordre.

### (c) Frontières de charge

Leurs célérités sont extérieures aux intervalles précédents.

Comme dans l'étude précitée nous utiliserons les relations de compatibilité d'Hadamard, appliquées aux équations dynamiques[1]:

$$\frac{\partial \sigma_{ij}}{\partial x_j} - \varrho \gamma_i = 0 \qquad (1.1)$$

($\sigma_{ij}$ désignant les coordonnées du tenseur des contraintes, $\gamma_i$ celles de l'accélération, $\varrho$ la masse volumique), et aux relations entre les vitesses de déformation et les vitesses de contraintes. Mais nous supposons maintenant que la déformation plastique élémentaire est définie par:

$$d\varepsilon_{ij}^p = g\,\varphi_{ij}\left(\frac{\partial f}{\partial \sigma_{hk}}\,d\sigma_{hk}\right), \qquad (1.2)$$

$$g(\sigma_{pq}) > 0 \qquad \text{pour} \qquad \frac{\partial f}{\partial \sigma_{hk}}\,d\sigma_{hk} > 0, \qquad \text{avec} \qquad g(\sigma_{pq}) > 0,$$

la fonction de charge $f(\sigma_{hk})$ pouvant dépendre de l'état d'écrouissage et $\varphi_{ij}$ désignant un tenseur symétrique qui ne coincide pas avec $\partial f/\partial \sigma_{ij}$ $= f_{ij}$. Les relations de PRANDTL-REUSS s'écrivent alors (en confondant le tenseur vitesse de contraintes avec la dérivée matérielle du tenseur $\sigma$ par rapport à des axes fixes, ce qui est justifiable lorsque les coefficients d'élasticité sont assez grands vis à vis des contraintes):

$$v_{ij} = L_{ij,hk}\dot{\sigma}_{hk} + g\,\varphi_{ij}f_{hk}\dot{\sigma}_{hk}, \qquad (1.3)$$

$v_i$, désignant les vitesses de déformation et $L_{ij,hk}$ les compliances élastiques, quantités pourvues de symétries des coefficients d'élasticité. En particulier

$$L_{ij,hk} = L_{hk,ij},$$

mais cette symétrie n'existe plus pour le produit $\varphi_{ij}f_{hk}$ lorsque $\varphi_{ij} \neq f_{ij}$.

## 2. Ondes Plastiques — Cas général

Résolvons les équations (1.3) sous la forme

$$\dot{\sigma}_{ij} = \mu_{ij,hk}v_{hk}. \qquad (2.1)$$

En désignant par $v_i$ les coordonnées de la vitesse:

$$2\,v_{hk} = \frac{\partial v_h}{\partial x_k} + \frac{\partial v_k}{\partial x_h}$$

---

[1] On adopte dans toute la suite la convention de l'indice muet.

et en raison de la symétrie de $\mu$ en $h$ et $k$ on peut écrire:

$$\dot{\sigma}_{ij} = \mu_{ij,hk} \frac{\partial v_k}{\partial x_h}. \tag{2.2}$$

Si $\alpha_i$ désignent les cosinus directeurs de la normale à l'onde dans la direction de sa propagation et $\Omega$ sa célérité[1] les relations de compatibilité donnent:

$$\left[\frac{\partial \sigma_{ij}}{\partial x_j}\right] = -\frac{\alpha_j}{\Omega}[\dot{\sigma}_{ij}] = -\frac{\alpha_j}{\Omega}\mu_{ij,hk}\left[\frac{\partial v_k}{\partial x_h}\right], \tag{2.3}$$

$$\left[\frac{\partial v_k}{\partial x_h}\right] = -\frac{\alpha_h}{\Omega}[\gamma_k], \tag{2.4}$$

d'où

$$\left[\frac{\partial \sigma_{ij}}{\partial x_j}\right] = \frac{\alpha_j \alpha_h}{\Omega^2}\mu_{ij,hk}[\gamma_k]. \tag{2.5}$$

Or ceci, d'après (1.1) doit être égal à $\varrho[\gamma_i]$. D'où l'équation

$$B_{ik}[\gamma_k] = \varrho\Omega^2[\gamma_i] \tag{2.6}$$

en posant

$$B_{ik} = \mu_{ij,hk}\alpha_j\alpha_h. \tag{2.7}$$

D'après (2.6), $\varrho\Omega^2$ est une valeur propre de la matrice $B$ et $[\gamma]$ est le vecteur propre correspondant. Mais la matrice $B$ n'étant pas symétrique lorsque $\varphi_{ij} \neq f_{ij}$, il n'y a plus orthogonalité des vecteurs propres correspondant à deux célérités différentes. De plus, en l'absence de renseignements sur les tenseurs $f_{ij}$ et $\varphi_{ij}$, il n'est pas possible de préciser le nombre des célérités plastiques et leurs positions par rapport aux célérités élastiques. C'est pourquoi dans la suite nous nous limiterons au cas d'un milieu isotrope. Nous entendons par là un milieu isotrope en l'absence de contraintes et à écrouissage isotrope. Les tenseurs $f$ et $\varphi$ ont alors mêmes directions principales (celles du tenseur des contraintes).

### 3. Cas d'un milieu isotrope

En posant:

$$\Psi = gf_{hk}\dot{\sigma}_{hk} \tag{3.1}$$

(vitesse de charge) on peut inverser les relations (1.3) sous la forme:

$$\dot{\sigma}_{ij} = \lambda(\dot{\theta} - \Phi\Psi)\delta_{ij} + 2\mu(v_{ij} - \Psi\varphi_{ij})$$

---

[1] Par rapport à la matière dans son état actuel.

$\lambda$, $\mu$ désignant les coefficients de Lamé, $\Phi = \sum_i \varphi_{ii}$, $\theta = \sum_i v_{ii}$, $\delta_{ij}$ le symbole de Kronecker. En formant la combinaison $f_{ij}\dot{\sigma}_{ij}$ on en tire:

$$\Psi = \frac{\lambda F \delta_{hk} + 2\,\mu f_{hk}}{g^{-1} + \lambda F \Phi + 2\mu f_{\sigma\tau}\varphi_{\sigma\tau}}\, v_{hk} \qquad (3.2)$$

d'où:

$$\dot{\sigma}_{ij} = \lambda \dot{\theta} \delta_{ij} + 2\mu v_{ij} - r(\lambda F \delta_{hk} + 2\mu f_{hk})(\lambda \Phi \delta_{ij} + 2\mu \varphi_{ij})v_{hk} \qquad (3.3)$$

en posant:

$$\frac{1}{r} = \frac{1}{g} + \lambda F \Phi + 2\mu f_{\sigma\tau}\varphi_{\sigma\tau}. \qquad (3.4)$$

Autrement dit:

$$\mu_{ij,hk} = \lambda \delta_{ij}\delta_{hk} + \mu(\delta_{ih}\delta_{jk} + \delta_{jh}\delta_{ik})$$
$$- r(\lambda \Phi \delta_{ij} + 2\mu \varphi_{ij})(\lambda F \delta_{hk} + 2\mu f_{hk}), \qquad (3.5)$$

$$B_{ik} = (\lambda + \mu)\alpha_i \alpha_k + \mu \delta_{ik} - r(\lambda \Phi \alpha_i + 2\mu \varphi_{ij}\alpha_j)(\lambda F \alpha_k + 2\mu f_{hk}\alpha_h). \qquad (3.6)$$

En prenant la normale à l'onde au point étudié comme direction $0x_1$ ($\alpha_1 = 1$, $\alpha_2 = \alpha_3 = 0$), la matrice $B$ prend la forme

$$\begin{bmatrix} \lambda + 2\mu - rUV & -2\mu r U f_{12} & -2\mu r U f_{13} \\ -2\mu r\varphi_{12} V & \mu - 4\mu^2 r \varphi_{12} f_{12} & -4\mu^2 r\varphi_{12}f_{13} \\ -2\mu r\varphi_{13} V & -4\mu^2 r\Phi_{13}f_{12} & \mu - 4\mu^2 r\,\varphi_{13}f_{13} \end{bmatrix}$$

en posant:
$$U = \lambda \Phi + 2\mu \varphi_{11}$$
$$V = \lambda F + 2\mu f_{11}$$

Orientons l'axe $0x_2$ suivant la projection sur le plan tangent à l'onde du vecteur $f(f_{11}, f_{12}, f_{13})$ de telle sorte que $f_{13} = 0$. On voit alors que l'une des racines de l'équation en $S$ est: $S = \mu$ et que le vecteur propre correspondant est perpendiculaire au plan formé par le vecteur $f$ et la normale à l'onde ($[\gamma_1] = [\gamma_2] = 0$). On a une onde transversale. Comme d'ailleurs, d'après (2.4), seules les composantes $v_{13} = v_{31}$ du tenseur des vitesses de déformations sont discontinues, on a par (3.2) où $\delta_{13}$ et $f_{13}$ sont nuls: $[\Psi] = 0$. Il s'agit donc d'une onde *neutre* (pas de discontinuité de la vitesse de charge) ce qui explique qu'elle ait même célérité que les ondes élastiques transversales.

Orientons maintenant l'axe $0x_2$ suivant la projection sur le plan tangent à l'onde du vecteur $\varphi$ ($\varphi_{11}, \varphi_{12}, \varphi_{13}$) de telle sorte que $\varphi_{13} = 0$. Nous voyons alors que les vecteurs propres correspondant aux racines autres que $S = \mu$ sont dans le plan formé par la normale à l'onde et le vecteur $\varphi$ (car $[\gamma_3] = 0$). Les racines en question sont celles de l'équation du second degré:

$$G(S) \equiv (\lambda + 2\mu - S)(\mu - S) - r\,h(S) = 0 \qquad (3.7)$$

où

$$h(S) \equiv (\lambda \Phi + 2\mu\varphi_{11})(\lambda F + 2\mu f_{11})(\mu - S) + 4\mu^2 \varphi_{12} f_{12}(\lambda + 2\mu - S).$$

Si $0x_2$ est orienté arbitrairement dans le plan tangent à l'onde la quantité $\varphi_{12} f_{12}$ est à remplacer par $\boldsymbol{\varphi} \cdot \boldsymbol{f} - \varphi_{11} f_{11} = \varphi_{12} f_{12} + \varphi_{13} f_{13}$, d'où:

$$h(S) \equiv (\lambda \Phi + 2\mu\varphi_{11})(\lambda F + 2\mu f_{11})(\mu - S)$$
$$+ 4\mu^2 (\varphi_{12} f_{12} + \varphi_{13} f_{13})(\lambda + 2\mu - S). \qquad (3.8)$$

La discussion des racines de $G(S)$ et de leur position par rapport à $\mu$, $\lambda + 2\mu$ et $0$ nécessite le calcul de $h(\mu)$, $h(\lambda + 2\mu)$ et $G(0)$. Si l'on désigne par $\alpha_1$, $\alpha_2$, $\alpha_3$ les cosinus directeurs de la normale à l'onde par rapport aux directions des contraintes principales $\sigma_1$, $\sigma_2$, $\sigma_3$ et par $f_1$, $f_2$, $f_3$, $\varphi_1$, $\varphi_2$, $\varphi_3$ les valeurs principales des tenseurs $f$ et $\varphi$, on a:

$$f_{11} = f_1 \alpha_1^2 + f_2 \alpha_2^2 + f_3 \alpha_3^2 \quad \text{(expression analogue pour } \varphi_{11}),$$

$$f_{12}\varphi_{12} + f_{13}\varphi_{13} = f_1\varphi_1\alpha_1^2 + f_2\varphi_2\alpha_2^2 + f_3\varphi_3\alpha_3^2$$
$$- (f_1\alpha_1^2 + f_1\alpha_2^2 + f_3\alpha_3^2)(\varphi_1\alpha_1^2 + \varphi_2\alpha_2^2 + \varphi_3\alpha_3^2)$$

d'où facilement $h(\mu)$ et $h(\lambda + 2\mu)$, et le calcul de $G(0)$ conduit au résultat suivant (où $\nu$ désigne le rapport de Poisson)

$$\frac{G(0)}{4\mu^2(\lambda+\mu)r} = \frac{1-\nu}{2\mu g} + [f_2\varphi_2 + f_3\varphi_3 + \nu(f_2\varphi_3 + f_3\varphi_2)]\alpha_1^4 + \cdots$$
$$+ [2f_3\varphi_3 + f_1\varphi_2 + f_2\varphi_1 + \nu(f_1 + f_2)\varphi_3$$
$$+ \nu(\varphi_1 + \varphi_2)f_3]\alpha_1^2\alpha_2^2 \qquad (3.9)$$

les coefficients des termes non écrits en $\alpha_2^4$, $\alpha_3^4$, $\alpha_2^2\alpha_3^2$, $\alpha_3^2\alpha_1^2$ s'obtiennent par permutation circulaire à partir des coefficients de $\alpha_1^4$ et $\alpha_1^2\alpha_2^2$.

Nous allons faire la discussion dans le cas du critère de Mohr, critère indépendant de la contrainte principale intermédiaire $\sigma_2$, d'où $f_2 = 0$, et en admettant en outre que $\varphi_2 = 0$.

## 4. Cas du critère de Mohr

Les contraintes principales étant rangées dans l'ordre $\sigma_1 > \sigma_2 > \sigma_3$ et la fonction $f$ mise sous la forme $f \equiv \sigma_1 - H(\sigma_3)$, on a:

$$f_1 = 1, \qquad f_2 = 0, \qquad f_3 = -j$$

en posant $j = H'(\sigma_3)$. D'après les propriétés de la courbe enveloppe de MOHR : $0 \leq j \leq 1$, et l'on pose souvent

$$j = \frac{1 - \sin \varrho}{1 + \sin \varrho} = \tan^2 \left( \frac{\pi}{4} - \frac{\varrho}{2} \right).$$

En supposant qu'on ne se trouve pas sur une arête de la surface de charge ($\sigma_2$ différent de $\sigma_1$ et $\sigma_2$) on devrait avoir :

suivant la théorie de MOHR (déformation par glissement d'où $\varphi_2 = 0$ et invariance du volume) : $\varphi_1 = 1$, $\varphi_2 = 0$, $\varphi_3 = -1$.

suivant la théorie du potentiel plastique : $\varphi_1 = 1$, $\varphi_2 = 0$, $\varphi_3 = -j$.

Nous engloberons ces deux cas en prenant : $\varphi_1 = 1$, $\varphi_2 = 0$, $\varphi_3 = -k$. Nous supposerons $0 < k \leq 1$.

Avec ces valeurs principales des tenseurs $f$ et $\varphi$, nous obtenons

$$\frac{1}{r} = \frac{1}{g} + \lambda(1 - k)(1 - j) + 2\mu(1 + kj) = \frac{1}{g} + K(1 - k)(1 - j)$$
$$+ \frac{2\mu}{3}(2 + k + j + 2kj).$$

$K = \lambda + \frac{2\mu}{3}$ désignant le module de compression élastique. $K$ et $\mu$ étant positifs, $r$ est positif.

D'autre part :

$$f_{12}\varphi_{12} + f_{13}\varphi_{13} = \alpha_1^2 \alpha_2^2 (1 + j)(1 + k) + \alpha_2^2(\alpha_1^2 + jk\alpha_3^2) \geq 0$$

donc $rh(\mu) \geq 0$, d'où $G(\mu) \leq 0$. La fonction $G(S)$ a donc deux racines réelles séparées par $\mu$ (lorsque $\alpha_1 = \alpha_3 = 0$ l'une de ces racines est égale à $\mu$).

a) La position de la plus grande racine $S_1$ par rapport à $\lambda + 2\mu$ dépend du signe de :

$$(\lambda \Phi + 2\mu \varphi_{11})(\lambda F + 2\mu f_{11})$$
$$= \frac{4\mu^2}{(1 - 2\nu)^2} [(1 - \nu - \nu j)\alpha_1^2 + \nu(1 - j)\alpha_2^2 + (\nu - j + \nu j)\alpha_3^2]$$
$$\times [(1 - \nu - \nu k)\alpha_1^2 + \nu(1 - k)\alpha_2^2 + (\nu - k + \nu k)\alpha_3^2].$$

L'annulation de chacun des crochets fournit un cône de second degré, réel si $\nu/1 - \nu$ est inférieur à $j$ (ou $k$), et entourant l'axe $\sigma_3$. Pour les normales comprises à l'intérieur du cône réel s'il en existe un, ou entre les deux cônes réels s'il en existe deux, $h(\lambda + 2\mu)$ est positif, d'où $S_1 > \lambda + 2\mu$. On obtient des ondes de célérité supérieure à la plus grande des célérités élastiques, (circonstance exclue par la coïncidence des 2 cônes dans le cas du potentiel plastique).

b) La position de la plus petite racine $S_3$ par rapport à 0 dépend du signe de $G(0)$, soit d'après 3—9 de :

$$\frac{1-\nu}{2\mu g}\,(\alpha_1^2 + \alpha_2^2 + \alpha_3^2)^2 + (\alpha_3^2 - k\alpha_1^2)(\alpha_3^2 - j\alpha_1^2)$$
$$+ \alpha_2^2[2jk\alpha_1^2 + (1+jk)\alpha_2^2 + 2\alpha_3^2] - \nu(j+k)\alpha_2^2(\alpha_1^2 + \alpha_2^2 + \alpha_3^2).$$

L'annulation de cette expression détermine les directions pour lesquelles $S_3 = 0$ d'où $\Omega_3 = 0$ (surface de discontinuité stationnaire) c'est à dire les directions des normales aux surfaces caractéristiques du problème statique. Pour $g = +\infty$ (absence d'écrouissage) on obtient un cône du 4 ème dégré réel dont la section par le plan $\alpha_2 = 0$ est constituée par les droites $\alpha_3 = \pm \sqrt{j}\,\alpha_1$, $\alpha_3 = \pm \sqrt{k}\,\alpha_1$ dans lesquelles on reconnait les normales aux lignes caractéristiques du problème statique de déformation plane [2]. La section par le plan $\alpha_1 = 0$ est imaginaire, la section par le plan $\alpha_3 = 0$ formée par 4 droites, réelles si $\nu(j+k) > 2\sqrt{jk}$, imaginaires dans le cas contraire. Dans ce dernier cas l'aspect du cône est celui de 4 trompes assemblées en $X$ par leurs sommets.

Dans le cas de l'écrouissage on a encore un cône du 4ème dégré réel tant que

$$\mu g > 2(1-\nu)\,\frac{(1+k)(1+j)}{(k-j)^2}\,,$$

formé de trompes de plus en plus effilées à mesure que croît l'écrouissage et qui disparaissent lorsque $\mu g$ devient inférieur à la valeur précédente.

Pour les directions de normale intérieures au cône on a: $S_3 < 0$, donc *une des trois célérités des ondes plastiques disparait*. Cette circonstance ne se produit pas dans le cas du potentiel plastique, parce que dans ce cas le cône se réduit aux 2 droites réelles $\alpha_3 = \pm\sqrt{j}\,\alpha_1$, $\alpha_2 = 0$ lorsqu'il n'y a pas écrouissage, et disparait entièrement lorsqu'il y a écrouissage.

## 5. Discussion des résultats

Nous supposons ici que le milieu est initialement en équilibre limite et que, dans une couche mince dont la normale est prise comme axe $0x_1$, lui soit imposée une petite perturbation définie par :

$$v_{11} = V_1, \qquad v_{12} = V_2/2, \qquad v_{13} = V_3/2, \qquad v_{22} = v_{33} = v_{23} = 0,$$

$V$ désignant le vecteur propre correspondant à la valeur propre $S$ de la matrice $B$ relative à la normale $0_{x1}$, matrice définie par :

$$B_{ik} = \mu_{i1,\,1k}.$$

Alors d'après (2.1):

$$\dot\sigma_{i1} = \mu_{i1,hk} v_{hk} = \mu_{i1,1k} V_k = B_{1k} V_k = S V_i$$

ou:

$$\dot\sigma_{11} = S v_{11}, \qquad \dot\sigma_{21} = 2 S v_{21}, \qquad \dot\sigma_{31} = 2 S v_{31}.$$

a) Une valeur négative de $S$ implique que, pour maintenir l'équilibre de la couche, les forces extérieures qui lui sont appliquées diminuent, plus exactement que leurs *variations* aient un travail négatif ($d\sigma_{ij} d\varepsilon_{ij} < 0$, $d\varepsilon_{ij}$ désignant le tenseur de déformation). Si elles sont maintenues fixes, il y a *instabilité* entraînant une grande et brusque déformation de la couche: glissement ou rupture (et non propagation d'onde). Ceci explique 1°/l'existence de lignes de glissement même lorsqu'il y a un certain écrouissage 2°/leur disparition lorsque l'écrouissage est assez important (il n'y a plus de racine $S$ négative) 3°/ le fait maintes fois signalé qu'en déformation plane elles font avec la direction de $\sigma_3$ un angle mal défini, compris entre $(\pi/4) - (\varrho/2)$ et $\pi/4$ (pinceau d'instabilité).

b) Soit maintenant une valeur $S$ supérieure à $\lambda + 2\mu$. Pour simplifier considérons une perturbation longitudinale dans une couche perpendiculaire à la direction de $\sigma_3$. On aura: $\dot\sigma_3 = S v_3$ avec $S > \lambda + 2\mu$ si $j < \nu/(1-\nu) < k$. Si la déformation est ensuite renversée (toujours sans dilatation transversale) elle devient élastique et s'effectue avec le module $\lambda + 2\mu$ inférieur au module plastique. Au total au cours de l'aller-retour le travail des variations des forces extérieures est négatif ($d\sigma_{ij} d\varepsilon_{ij}^p < 0$, $d\varepsilon_{ij}^p$ désignant la déformation permanente produite par $d\sigma_{ij}$) mais ceci n'est pas en contradiction avec le principe de Carnot. En effet, celui-ci fait intervenir le travail des forces extérieures (et non de leurs variations) au cours d'un cycle fermé (ce qui n'est pas le cas, $\sigma_1$ et $\sigma_2$ ne reprenant pas leurs valeurs initiales).

Notons que la stabilité n'exige pas que $d\sigma_{ij} d^p\varepsilon_{ij}$ soit positif (postulat de Drucker) mais seulement que $d\sigma_{ij} d\varepsilon_{ij}^p$ le soit ($d\varepsilon_{ij}$ déformation totale)[1]. En définitive la possibilité d'ondes de célérité supérieure à $[(\lambda + 2\mu)/\varrho]^{1/2}$ ne nous semble pas devoir être rejetée.

## 6. Surfaces caractéristiques des équations de l'équilibre élastoplastique

La condition $G(0) = 0$, où $G(0)$ est donnée par (3.9), fournit les directions caractéristiques des équations de l'équilibre élastoplastique lorsqu'on admet les relations (3.3). On obtient pour les normales aux

---

[1] Cette dernière condition n'est pas générale, mais est applicable aux variations qui correspondent à un mouvement propre, c'est-à-dire à une racine $S$.

facettes caractéristiques un cône de 4ème degré, qu'on a étudié dans le cas du critère de MOHR.

Revenons sur le cas du potentiel plastique ($\varphi = f$) pour montrer que, dans ce cas, le cône peut au plus (et exceptionnellement) avoir deux génératrices réelles.

En introduisant les modules $K$ et $\mu$ tous deux positifs, on voit d'abord que $r$, donné par (3.4), est positif, puis en utilisant les expressions (3.7), (3.8) où $0x_1$ est la normale à la facette, on obtient:

$$\frac{G(0)}{\mu^2 r} = 2K[(2\mu g)^{-1} + f_{22}^2 + f_{33}^2 + 2f_{23}^2 + (f_{22} + f_{33})^2]$$
$$+ \frac{4\mu}{3}[(\mu g)^{-1} + (f_{22} - f_{33})^2 + 4f_{23}^2].$$

$G(0)$ n'est donc pas négatif et ne peut s'annuler que dans le cas suivant: $g = +\infty$ (absence d'écrouissage) et $f_{22} = f_{33} = f_{23} = 0$.

Ces conditions signifient que les vitesses de déformation dans le plan d'une facette caractéristique sont nulles, ce qui en général n'est possible que pour des états de contrainte particuliers. En effet choisissons dans le plan de la facette l'axe $x_2$ de telle manière que $f_{12} = 0$. Les relations $f_{21} = f_{23} = 0$ montrent que la direction $x_2$ est alors une direction principale. En désignant par $\sigma_2$ la contrainte principale correspondante, la condition $f_{22} = 0$ signifie que dans l'espace des contraintes principales le point $\sigma_1$, $\sigma_2$, $\sigma_3$ se trouve sur le contour apparent de la surface de charge parallèlement à l'axe $0\sigma_2$*. Elle n'est donc réalisée que pour des états de contrainte exceptionnels, sauf dans le cas du critère de MOHR.

Enfin la condition $f_{33} = 0$ précise l'orientation des facettes caractéristiques lorsque la condition précédente est réalisée. Ce sont les facettes passant par $\sigma_2$ et pour lesquelles la vitesse de dilatation plastique est nulle dans la direction perpendiculaire à $\sigma_2$. Il existe deux facettes répondant à la question si les valeurs principales $f_1$ et $f_3$ du tenseur $f$ sont de signes contraires.

## 7. Frontières de charge et de décharge

On ramène leur étude à celle des ondes plastiques par le procédé décrit en [1]. L'indice 1 désignant la région située en avant de la frontière dans le sens de da propagation, l'indice 2 la région située en arrière, $\psi$ la vitesse de charge, posons:

$$\psi_2 = \varkappa \psi_1.$$

---

* Elle montre aussi que la déformation plastique est plane. C'est pourquoi dans les problèmes de déformation plane relatifs à un milieu rigide parfaitement plastique on trouve des caractéristiques réelles.

Il suffit alors de remplacer $1/r$ par:

$$\frac{1}{r'} = \frac{1 - \varkappa}{g} + \lambda F \Phi + 2\mu f_{\sigma\tau}\varphi_{\sigma\tau}, \quad \varkappa \leq 0 \quad \text{pour une frontière de décharge,}$$

$$\frac{1}{r''} = \frac{1 - \varkappa^{-1}}{g} + \lambda F \Phi + 2\mu f_{\sigma\tau}\varphi_{\sigma\tau}, \quad \varkappa \geq 0 \quad \text{pour une frontière de charge.}$$

Supposant $\lambda F \Phi + 2\mu f_{\sigma\tau}\varphi_{\sigma\tau} > 0$, on a $0 \leq r' \leq r$, tandis que $r''$ appartient au complément de l'intervalle $]o, r[$ sur la droite réelle.

L'une des valeurs de $\varrho\Omega^2$ est toujours égale à $\mu$. Pour les deux autres, en considérant l'intersection de la parabole:

$$y = (\lambda + 2\mu - S)(\mu - S)$$

avec les droites:

$$y = r\,h(S), \qquad y = r'\,h(S), \qquad y = r''\,h(S),$$

on voit que:

pour une frontière de décharge les valeurs de $\varrho\Omega^2$ sont comprises dans les intervalles $]S_3, \mu[$ et $]S_1, \lambda + 2\mu[$,

pour une frontière de charge les valeurs de $\varrho\Omega^2$ appartiennent au complément des intervalles $[S_3, \mu]$ et $[S_1, \lambda + 2\mu]$.

## 8. Conclusions

L'abandon de l'hypothèse du potentiel plastique entraîne dans l'étude des ondes plastiques un certain nombre de conséquences remarquables: l'orthogonalité des vecteurs propres ne subsiste pas; le nombre des ondes peut devenir inférieur à 3 (ce cas paraissant lié à la formation de surfaces de glissement ou de rupture); la célérité des ondes plastiques les plus rapides peut dépasser celle des ondes élastiques les plus rapides; enfin il existe, tant que l'écrouissage reste assez faible, des surfaces caractéristiques réelles pour les équations de l'équilibre élastoplastique (ondes de célérité nulle) et leurs normales en un point formant un cône du 4ème degré.

En ce qui concerne les frontières de charge et de décharge les propriétés énoncées dans le cas du potentiel plastique restent valables pour les carrés de leurs célérités.

### Références

[1] Mandel, J.: Journal de Mécanique 1, 3—30 (1962).
[2] Geiringer, H.: Advances in Applied Mechanics, Vol. 3, New York: Academic Press 1953, p. 227—246.

# Subject Index